Housing and Sustainability - Achieving a Sustainable Future

Edited by Sérgio António Neves Lousada

Published in London, United Kingdom

Housing and Sustainability - Achieving a Sustainable Future
http://dx.doi.org/10.5772/intechopen.1003294
Edited by Sérgio António Neves Lousada

Contributors
Adiba Shafique, Anaid Yerena, Andrii Khorolskyi, Carla Chiarantoni, Cletus Alexander Umerie, Dafni Mora, Elyas Jahanshahi, Fazlur Rahman, Fredy Nicolás Molina Rodríguez, Giuliano Galluccio, Guillermo Yorel Noriega Aquise, Hatvany Gómez Concepción, Miguel Chen Austin, Mohd Arshad Ameen, Noraslinda Abdul Rahman, Olga Yuil Valdés, Puteri Nur Farah Naadia Mohd Fauzi, Rosemary C. Nnaemeka-Okeke, Rui Liu, Sayyed Javad Asadpoor, Sergio Russo Ermolli, Silvia Vilčekova, Simona Pugliese, Suneela Ahmed, Svitlana Delehan, Sérgio Lousada, Thasnee Solano, Yazmín Mack-Vergara, Yu Chen

First published in London, United Kingdom, 2025 by IntechOpen
IntechOpen is the global imprint of INTECHOPEN LIMITED, registered in England and Wales, registration number: 11086078, 167-169 Great Portland Street, London, W1W 5PF, United Kingdom

For EU product safety concerns: IN TECH d.o.o., Prolaz Marije Krucifikse Kozulić 3, 51000 Rijeka, Croatia, info@intechopen.com or visit our website at intechopen.com.

British Library Cataloguing-in-Publication Data
A catalogue record for this book is available from the British Library

Housing and Sustainability - Achieving a Sustainable Future
Edited by Sérgio António Neves Lousada
p. cm.

This title is part of the Civil Engineering Book Series, Volume 13
Topic: Construction Engineering
Series Editor: Assed Haddad
Topic Editor: Amin Akhnoukh

Print ISBN 978-0-85466-689-8
Online ISBN 978-0-85466-688-1
eBook (PDF) ISBN 978-0-85466-690-4
ISSN 3029-0287

If disposing of this product, please recycle the paper responsibly.

IntechOpen Book Series
Civil Engineering
Volume 13

Aims and Scope of the Series

Civil engineering is a traditional field of engineering from which most other branches of engineering have evolved. It comprises traditional sub-areas like transportation, structures, construction, geotechnics, water resources, and building materials. It also encompasses sustainability, risk, environment, and other concepts at its core. Historically, developments in civil engineering included traditional aspects of architecture and urban planning as well as practical applications from the construction industry. Most recently, many elements evolved from other fields of knowledge and topics like simulation, optimization, and decision science have been researched and applied to increase and evolve concepts and applications in this field. Civil engineering has evolved in the last years due to the demands of society in terms of the quality of its products, modern applications, official requirements, and cost and schedule restrictions. This series addresses real-life problems and applications of civil engineering and presents recent, cutting-edge research as well as traditional knowledge along with real-world examples of developments in the field.

Meet the Series Editor

Professor Assed N. Haddad is a Civil Engineer with a degree from the Federal University of Rio de Janeiro (UFRJ) earned in 1986, as well as a Juris Doctor degree from the Fluminense University Center earned in 1993, and a Master's degree in Civil Engineering from the Fluminense Federal University (UFF) obtained in 1992. He completed his Ph.D. in Production Engineering from COPPE / Federal University of Rio de Janeiro in 1996. Professor Haddad's academic pursuits have taken him to postdoctoral stays at the University of Florida, USA in 2006; at the Universitat Politècnica de Catalunya, Spain in 2010; and at the University of New South Wales Sydney, Australia in 2019. Currently, he serves as a Full Professor at the Federal University of Rio de Janeiro. He has held visiting professorships at various institutions including the University of Florida, Universitat Politècnica de Catalunya, Universitat Rovira i Virgili, and Western Sydney University. His research expertise encompasses Civil, Environmental, and Production Engineering, with a primary focus on the following topics: Construction Engineering and Management, Risk Management, and Life Cycle Assessment. He has been the recipient of research grants from the State of Rio de Janeiro, Brazil: CNE FAPERJ from 2019 to 2022 and from 2023 to 2025. Additionally, his research grants obtained from the Brazilian Government CNP since 2012 last to this date. Professor Haddad has been involved in several academic endeavors, being the Guest Editor of the International Journal of Construction Management; MDPI's Sustainability, Energies, and Infrastructures; Associate Editor at Frontiers in Built Environment / Sustainable Design and Construction; Guest Editor at Frontiers in Built Environment / Construction Management; and Academic Editor of the Journal of Engineering, Civil Engineering Section of Hindawi. He is currently a Professor of the Environmental Engineering Program at UFRJ and the Civil Engineering Program at UFF.

Meet the Volume Editor

Sérgio António Neves Lousada holds an International Ph.D. in Civil Engineering – Hydraulics from the University of Minho (UMinho) in Portugal and an International Ph.D. in Sustainable Territorial Development from the University of Extremadura (UEx) in Spain. He has published some articles and books and participated in events mainly in the areas of Hydraulics, Urban Planning and Land Management. Furthermore, he collaborates with the Research Group on Environment and Spatial Planning (MAOT) - University of Extremadura (UEx), Badajoz, Spain; VALORIZA - Research Center for the Enhancement of Endogenous Resources, Polytechnic Institute of Portalegre (IPP), Portugal; CITUR-Madeira - Centre for Tourism Research, Development and Innovation, Madeira, Portugal; RISCO - Civil Engineering Department of University of Aveiro, Aveiro, Portugal; OSEAN – *Outermost Regions Sustainable Ecosystem for Entrepreneurship and Innovation*, Madeira, Portugal. Moreover, he holds an International Interuniversity Master's Degree in Ports and Coasts Engineering from the University of Las Palmas de Gran Canaria (ULPGC) and the University of La Laguna (ULL), both in Spain.

Contents

Preface

Housing defines the areas where communities develop and futures are created, making it a fundamental component of both human progress and environmental stewardship. The need for careful and sustainable housing practices has never been greater in a time of swift urban change and growing environmental concerns.

This book, "Housing and Sustainability – Achieving a Sustainable Future," is the result of a convergence of collaborative expertise and visionary insights aimed at rethinking the function of our living spaces. These pages provide a thorough examination of cutting-edge housing strategies that incorporate energy efficiency, green building techniques, and intelligent urban planning, providing new insights and useful methods suited to the changing demands of contemporary society.

The trip starts with a profound understanding of the difficult balancing act between conservation and development. We understand that creating resilient communities that value social inclusion, environmental responsibility, and technological innovation is as important as building sustainable housing. This discussion highlights the transformative potential of combining sustainable design with community involvement by reflecting on the crucial interaction between human ingenuity and the natural world.

Each chapter in this book emphasizes the dynamic interaction between cutting-edge approaches and workable solutions in the housing sector. Whether discussing cutting-edge building methods or avant-garde urban renewal plans, the concepts presented here are intended to motivate academics, policymakers, urban planners, and concerned citizens.

With an open mind and a drive for advancement, we encourage readers to explore these pages as we jointly steer toward a time when sustainable housing practices form the cornerstone of long-term well-being and global resilience.

This edited book consists of twelve chapters focusing on housing and sustainability. A summary of each part is given below.

Chapter One, "A Smart Multi-Criteria Assessment of Housing Energy Efficiency Relevant to Occupants' Socio-Demographic Characteristics: A Concept Paper", focuses on AI-driven multi-criteria evaluations, passive tactics, and carbon mitigation while examining occupant demographics, design, environment, and climate in relation to energy efficiency.

Chapter Two, "Applying a Zero Carbon Footprint Strategy to Improve Environmental and Energy Security", examines the effectiveness of post-war buildings in Ukraine using life cycle assessment and environmental screening. Increasing certifications Adopting cutting-edge technologies, Leadership in Energy and Environmental

Design (LEED), and the Building Research Establishment Environmental Assessment Method (BREEAM) minimize emissions, maximize energy use, and harmonize laws with international norms.

Chapter Three, "Data-Driven Technologies for Eco-Efficient Dwelling: Light Steel Framing Volumetric Schemes", demonstrates how BIM and CAD/CAM can speed up cold-formed steel volumetric housing, minimize waste, and maximize resources. Irondom Liguria's environmentally friendly prototype in seismic zones serves as an example of integrated design for cost-effective, scalable solutions.

Chapter Four, "Housing Overcrowding: Relationships and Challenges in the Contemporary Era", investigates how policies are influencing new developments, population changes, and crowding. It emphasizes inclusive methods for promoting sustainable communities, lowering overcrowding, and improving well-being.

Chapter Five, "Integrating Nature-Based Solutions and Green Building Standards in Healthcare Facilities in Winnipeg, Canada," demonstrates how nature-based solutions (such as healing gardens and green roofs) and LEED and BREEAM promote sustainable healthcare. Collaboration for resilient environments that benefit patients, employees, and the surrounding ecosystems is guided by an integrated framework.

Chapter Six, "Introduction of Vernacular Imaginary in Reimagining Affordable Housing: A Sustainable Approach in Urban Regeneration in Johor, Malaysia", highlights how Johor, Malaysia's vernacular traditions in affordable housing promote social cohesion, environmental sustainability, and cultural sensitivity while striking a balance between contemporary demands and heritage preservation.

Chapter Seven, "Reducing Poverty by Protecting Properties through Insurance", examines disaster mitigation, housing protection, insurance for vulnerable populations, and poverty alleviation in line with the 2030 Agenda.

Chapter Eight, "Study on the Future Development Direction of Full-Cycle Real Estate Development Based on Carbon Neutral Background," investigates the impact of carbon neutrality on real estate by examining emissions, policies, and challenges. It emphasizes methods for ecological, intelligent, and sustainable development, emphasizing a full-cycle strategy to reduce environmental impact and achieve climate goals.

Chapter Nine, "Sustainable Low-Income Housing: Analyzing Green Building Initiatives within the LIHTC Program across U.S. States", analyzes Housing Finance Agency data and Qualified Allocation Plans (QAPs) from 2015 to 2023 to show how affordable housing green standards are changing and how the Low-Income Housing Tax Credit (LIHTC) promotes environmentally friendly, low-income building.

Chapter Ten, "Sustainable Makerspace Pedagogy in Architecture Education", shows how to integrate makerspace into third-year design studios at Karachi's Indus Valley School of Art and Architecture to promote sustainable, adaptable solutions. Through the use of inquiry by design, this practical method enhances architectural education and equips students to tackle modern, environmentally conscious problems.

Chapter Eleven, "Sustainable and Resilient Housing in Tropical Climates: Best Practices for Construction and Energy Security", examines building innovations for resilient housing, climate adaptation, and ecosystem services. A framework for architects, legislators, and stakeholders in tropical housing is provided by case studies that highlight energy security and construction strategies to slow down climate change.

Chapter Twelve, "Techniques and Procedures That Reduce Energy Consumption and Minimize the Negative Environmental Impact of a Sustainable House", emphasizes solar capture, waste management, and in-home wastewater treatment to generate electricity, biogas, and water, promoting sustainable urban living.

We hope the selected chapters will help readers understand the intricate tapestry of housing innovation and sustainability. This book aspires to serve as a unifying thread, interweaving the latest insights, diverse perspectives, and groundbreaking strategies. As we embark on this collective journey toward a sustainable future, may the pages that follow inspire a shared commitment to responsible development, eco-friendly practices, and the harmonious balance between human progress and environmental preservation.

Sérgio António Neves Lousada
UMa – Faculdade de Ciências Exatas e da Engenharia,
Campus Universitário da Penteada,
Funchal, Portugal

Chapter 1

A Smart Multi-Criteria Assessment of Housing Energy Efficiency Relevant to Occupants' Socio–Demographic Characteristics: A Concept Paper

Sayyed Javad Asadpoor and Elyas Jahanshahi

Abstract

Housing energy efficiency has a critical role in greenhouse gas and carbon emission mitigation. Previous studies have seldom considered the interrelationships among the different architectural aspects, i.e. human, building, environment, and climate in impacting housing energy efficiency. Given this gap, a literature review was conducted to explain the conceptual platform of the study. Then detailed cross-comparisons among five recent studies were made to develop the variable setting of the study. The study then addressed the implications of artificial intelligence in a multi-criteria assessment of housing energy efficiency. In conclusion, the overall improvement of housing energy consumption depends on a balanced interplay between the different architectural aspects, with an especial emphasis on the role of occupants' socio-demographic characteristics. To achieve this goal it is necessary to develop a smart multi-criteria energy efficiency assessment tool, which also assists in putting passive climatic principles in practice, and in reducing fossil fuel dependency.

Keywords: energy efficiency, housing, sustainability, occupants' socio-demographic characteristics, smart multi-criteria assessment, artificial intelligence

1. Introduction

Energy efficiency is a key concept in mitigating climate change and global warming. Among different industries, the housing sector has a significant share of energy consumption.

In some developed countries, e.g. USA, UK, and Australia, the government and housing authorities aim for zero-carbon housing. To achieve this goal, a variety of sustainability action plans, e.g. building codes, measures, regulations, technologies, and initiatives, have been developed (e.g. [1–6]). The implementation of such action plans has significantly impacted the housing market and the perceptions of buyers in these countries, and hence has significantly impacted the attention of different stakeholders, e.g. architects, designers, and builders towards the running costs and environmental impacts of housing energy consumption.

One critical challenge, however, in improving the EE is the lack of proper attention to the impacts of the interplay between human factors, building characteristics, environmental dimensions, and climatic information, which may even result in relatively higher energy consumption of a residential building, even though it may built based on sustainability codes and principles [7–10].

A recent study indicated that providing comprehensive sustainability information plays a crucial role for promoting sustainable housing [11]. The provision of comprehensive information improves public knowledge and awareness about the different aspects of sustainability, thereby has the potential to encourage people's positive perceptions, attitudes, and behaviours [10]. EE at the building scale may not result in EE during the operation phase because of human behaviours, which are influenced by socio-demographic characteristics and neighbourhood environmental conditions [8, 9]. Gower [8] indicated that EE at the building scale may not adequately address occupants' demands and desires at the unit scale, potentially may worsening their energy vulnerability. She believed that this happened because of their socio-demographic characteristics and the impact of some secondary environmental factors, and therefore, it is highly important to proactively alleviate energy vulnerability. Some studies [12, 13] have recently developed multi-criteria housing energy-efficiency (HEE) assessment platforms. The main gap in such research is the lack of attention to the human aspect of the equation and the substantial roles of people's perceptions and attitudes reflected in their socio-demographic characteristics.

The dispersion of housing sustainability information and the lack of proper attention to the interplay between the different architectural aspects, especially in relation to human factors, detract from the outcomes of the implemented EE action plans. This would in turn prevent occupants from meeting their needs rooted in their perceptions and attitudes, which would also negatively impact their energy consumption behaviours.

Affluent people are urban populations with more energy consumption and heat generation than vulnerable low-income people regardless of the differences in urban environmental conditions and housing quality of these two population groups [9]. Gower [8] indicated that although the currently implemented energy efficiency codes provide EE at building scale, the energy performance at the unit scale might be differ during the operation phase due to the differences in occupants' socio-demographic characteristics and urban environment dimensions (e.g. orientation, density, sun radiation, wind direction, vegetation). Such conditions eventually impact the opportunity for proper utilisation of passive climatic principles and the dependency on active energy solutions, which can compromise the health and well-being of vulnerable occupants [8].

The currently implemented measures have been more applicable and beneficial for affluent people, whereas vulnerable people who are compromised by the impacts of environmental challenges and experience higher discomfort and health-related problems from environmental pollution and UHI are at high-risk groups [8]. The conducted studies revealed that poor building condition is highly critical in increasing energy consumption and heat transfer of housing, and good design quality with the emphasis on the passive climatic principles is more impactful in dealing with these deficiencies than using energy-efficient technologies [7, 8]. Accordingly, good design quality is a highly important factor in dealing with energy vulnerability and in moving towards energy justice [8]. Good design quality depends not only on low-cost, simply available measures suitable for ordinary people at building scale but also on careful consideration of the congruity between occupants' needs, housing unit characteristics, and environmental dimensions [14].

Given this, achieving EE in housing is more impactful when it aligns with occupants' expectations and priorities, which is addressed as a serious scientific gap in this study. A potential solution to bridge this gap is to develop a concrete explanation of the interrelationships between the different architectural aspects of HEE, i.e. Human characteristics, building characteristics, environmental characteristics, and climate information.

In other words, given the necessity of achieving zero carbon housings studying the interplay between the different architectural aspects of housing EE would provide a good opportunity for the representation of occupant' concerns, worries, and needs; and the explanation of the differences between different population groups in relation to their different socio-demographic characteristics, neighbourhood environmental dimensions, and physical attributes of dwelling units. Such a study provides a suitable platform to assist the different people to meet their demands and desires based on their perceptions, attitudes, and affordability, and hence enhances their proactive participation and pro-environmental behaviours.

This paper includes two different stages, first identifying the research background and variable setting in relation to the different architectural aspects of housing EE, and then developing the conceptual and methodological platform for a smart multi-criteria energy efficiency assessment, as will be discussed in the next sections.

2. Literature review

An overview of the literature shows that the implemented energy efficiency (EE) measures have not been able to reduce the overall energy consumption (e.g. [15–17]). Partial improvements in EE without paying proper attention to all aspects of the concept often result in greater use of services and products or the emergence of new expectations and demands [18]. Comprehensive and integrated improvements of the entire system [19] are essential for the overall EE. To achieve this goal, some recent research works has addressed energy sufficiency, conservation, and saving through the enhancement of all energy-related aspects of the system [17, 20].

A balanced enhancement of the whole energy-related aspects of a system has the potential to provide a practical platform for the reduction of overall energy consumption of the system, and that is why some studies stress the necessity of developing a hybrid approach that is a combination of energy efficiency and energy, sufficiency, or a combination of bottom-up human-centric and top-down techno-centric approaches [16, 21–24]. This hybrid approach provides a suitable platform for a balanced interplay between the different energy-related aspects of the system, with especial emphasis on the human side of the equation, and hence results in people's positive perceptions, attitudes, and sense of control, promoting their active participation and pro-environmental behaviours [25, 26].

Given the share of housing industry as the main source of energy consumption in many countries, and the role of the different architectural aspects, e.g., people's characteristics, building characteristics, and environmental dimensions in the energy performance of the industry [7, 9, 27, 28], it is highly critical to develop such a blended approach for residential buildings.

In some countries, e.g., the UK and Australia, there is a vast amount of information in the area of housing sustainability, and hence there is a proper level of public and general knowledge and awareness. As a result, people have gained positive attitudes towards sustainable housing [29]. This would ensure the implementation of housing sustainability codes, initiatives, and regulations in the housing sector. For instance, the implementation of building sustainability codes makes it possible for all new houses to meet the requirements of sustainability in the UK and Australia [29, 30]. In Australia, a variety of governmental initiatives that aim to encourage the energy efficiency of existing houses, e.g., solar panels and insulation initiatives have been also developed [30]. A few websites, e.g. Your Home, Home Energy Use, and Energy Efficient Home Design have also been developed to provide energy efficiency information, recommendations, and tips [31–33]. These online information provision platforms aim at providing some tips and recommendations, which help occupants to improve their housing energy efficiency.

The snowballing effects of the abovementioned measures play a critical role in invariably improving the Australian people's knowledge and awareness about housing sustainability. Judge et al. [30] revealed that a strong and serious predictor of intention to purchase a sustainable dwelling in Australia is related to some other subjective norms; e.g. people's attempts to promote environmental values and to be 'green consumers'. A wide range of empirical evidence in Australia shows that people hold positive norms and attitudes towards sustainability of their dwelling units so that a dwelling unit with a higher level of energy efficiency has a higher market value (e.g. [6, 30, 34–36]).

Despite these efforts, there is an abundance of evidence that shows a steady increase in the overall energy consumption of the housing sector [29, 37, 38], rooted in the current direction of housing development and household behaviours and daily life activities [18, 28, 29, 37, 38]. Rashed et al. [11] indicated that the sustainability improvement of the building industry remains suboptimal, and stressed the lack of comprehensive, easy access, simple, and integrated sustainability information as the main reason of unsatisfactory progress of sustainable housing. This means that although the EE of housing is technologically and physically achieved at the building scale, there are serious insufficiencies in EE at the unit scale [8], because people are not able to find an energy-efficient house in relevance to their demands, needs, desires, and expectations; hence, the occupied house is not able to meet their needs. This would in turn put serious pressures, e.g., comfort, health, and well-being, on the households, resulting in inappropriate energy consumption habits and detracting from the housing energy performance [7].

Therefore, a good housing energy performance depends on the provision of accountability and ambition in energy conservation/saving of dwelling units via a balanced interplay between different sides, of housing energy consumption e.g., human, building, and environment. To achieve this goal, developing more sensitive and profound measures, i.e., online databases, rating tools, and evaluation instruments that makes it possible to identify the interplay between the different architectural aspects of HEE is highly essential. Such a decision-making tool gives the occupants the chance to achieve more congruent housing in relevance to their perceptions, affordability, demands, and desires, resulting in broad and proactive participation.

Given this, the overall increase in energy consumption is indeed a sign of serious insufficiency in the housing market, which is mainly related to the lack of comprehensive and integrated information platforms. The platform assists the public in assessing the level of congruity of energy-related features of residential units with their needs and desires, which enriches the leverage of energy codes, enhances the energy performance of units, and improves people's energy consumption habits and activities [14, 28]. Therefore, enhancing HEE would depend seriously on the provision of suitable assessment opportunities that make it possible to evaluate the interplay among the different architectural aspects related to the energy performance of housing.

Regarding the necessity of a balanced interplay between the different architectural aspects in achieving housing EE, the next step is to develop the variable setting of the study in relation the different architectural aspects of housing. For this purpose, the following paragraphs are dedicated to have a look at some empirical studies conducted on the interrelationships among the different architectural aspect in impacting HEE.

Looking at environmental equity and energy justice, environmental challenges impact at-risk vulnerable people more seriously [7]. Hence, it remains very important to pay proper attention to these population groups in the development of sustainability measures to provide environmental equity in the mitigation programmes [7]. Regarding the inequitable distribution of environmental challenges in different urban districts, the lack of attention to the different architectural aspects of housing EE has resulted in serious environmental inequity. For instance, among the different vulnerable groups, low-income people living in poor housing conditions are the most at-risk, and they require particular attention [8]. A number of recent studies have stressed the impacts of environmental challenges on low-income occupants living in poor and low-standard housing with high thermal transfer and energy waste. Such conditions put higher pressures on their life conditions and compromise their health, well-being, and comfort (e.g. [7–9]). In contrast, affluent people have the chance to simply consume higher amounts of energy regardless of their opportunity to live in better urban environmental conditions with good housing design quality [8].

Referring to energy justice, Gower [8] stated that low-income people experience higher pressure and discomfort; because the energy costs seriously restrict their ability to consume the proper amount of energy. She indicated that poor and low-standard physical conditions of dwelling units of these population groups not only result in high thermal transfer and energy waste but also amplify the negative impacts of environmental challenges. Antonopoulos et al. [9] stressed that affluent people consume higher amounts of energy, whereas vulnerable low-income occupants are not able to consume a proper amount of energy because of the related costs and expenditures. They indicated that vulnerable occupants confront a high level of energy waste because of the high thermal transfer in their housing units. This occurs because they live in relatively low-standard, high-density, small, and poorly planned and designed houses, while living in warmer urban districts with poor environmental conditions, which increase their energy needs and requirements [9].

Sari [10] addressed the effect of different climatic dimensions, e.g., wind speed, wind direction, temperature, ventilation, and humidity, on building energy consumption. She also stated that some human factors e.g. information, communication, awareness, and education, would assist in improving energy consumption behaviours, e.g., relaxing in shade rather than simply using air- conditioning. Asadpoor et al. [14] stressed that the occupants' perception of housing EE depend on their bioclimatic preferences and sense of control, which impact their perceptions and attitudes towards different energy-related housing attributes. Rashed et al. [11] indicated that the availability of reliable, simple, and informative information in relation to the different architectural aspects of housing sustainability for different groups of occupants has essential role in promoting housing sustainability. **Table 1** addresses the different architectural aspects of HEE, e.g., building characteristics, occupant characteristics, urban environmental dimensions, and climatic data, alongside the various variables related to each aspect mentioned by these recently conducted studies, including Rashed et al. [11], EPA [7], Sari [10], Gower [8], and Antonopoulos et al. [9]. The table encompasses 105 variables categorised into four aspects factors, i.e. human characteristics, building characteristics, environmental characteristics (site level and neighbourhood), and climatic data.

These research works together have stressed the provision of a multi-criteria information platform about HEE, especially in relevance to occupants' socio-demographic characteristics as a crucial requirement of housing EE. The provision of such a platform depends on a concrete explanation of the interrelationships among occupants' socio-demographic characteristics, building physical characteristics, energy consumption, urban environmental characteristics, and climatic data. Therefore, to improve housing EE, developing a more sensitive and innovative information platform relevant to the occupants' perceptions of EE is substantially necessary. Such an information provision would be beneficial not only for the engagement of occupants in adjusting, modifying, changing, and retrofitting their housing units in relation to their energy needs and demands but also for the enhancement of their EE consumption habits and behaviours by improving their sense of control, pro-active participation and pro-environmental behaviours [39–41]. Such a database would also provide a suitable platform for housing authorities to enhance housing EE in relevance to the different groups of occupants.

3. Conceptual platform of the study

The Paris Climatic Agreement stressed that long-term sustainable development depends deeply on the development of a blended hybrid approach, which is a combination of bottom-up flexibility and top-down fixed rules [21]. Moser and Uzzell [22] stated that a combination of up-down needs-based approach with top-down right-based approach is essential to properly respond to environmental challenges through pro-environmental behaviours, proactive participation, and people's involvement in decision-making. Biloria [16] also indicated that in achieving the expected EE goals, the integration of techno-centric and human-centric approaches is highly necessary. Orr [42] believed that paying attention to technological aspects is just the beginning, and real long-term sustainability depends on paying proper attention to all aspects of environmental sustainability. He recommended ecological (earth-centered) approach instead of the current technological approach. Given the concept of energy equity, Gower [8] also stressed that it is necessary to proactively alleviate energy vulnerability by considering the different aspects of energy consumption, e.g. human factors, environmental factors, and building characteristics. Such a comprehensive, hybrid, and proactive approach is necessary for a balanced interplay between the different aspects, especially in relation to human factors. This would, in turn, promote pro-environmental behaviours, and hence enhance the overall outcomes of the implemented measures by encouraging individual and collective environmentally friendly behaviours [41, 43].

Such a multi-criteria approach in the area of HEE makes it possible to combine efficiency and sufficiency (energy conservation/saving/consumption reduction) concepts, and hence to reduce the overall energy use in housing industry [15, 16]. Energy conservation/saving depends mainly on the reduction of the overall energy demands

Factors	Variables	Rashed et el. [11]	EPA [7]	Sari [10]	Gower [8]	Antonopoulos et al. [9]
Human Characteristics	Information	+	O	+	O	O
	Awareness	+	O	+	O	O
	Education	+	O	+	O	+
	Interest	+	O	O	O	O
	Comfort (Visual, Thermal, Acoustic)	+	O	O	O	+
	Willingness	+	O	O	O	O
	Knowledge	+	O	O	O	O
	Economy	+	O	O	O	O
	Trust	+	O	O	O	O
	Communication	+	O	O	O	O
	Opinion	+	O	O	O	O
	Perception	+	O	O	O	O
	Priorities	+	O	O	O	O
	Expectations	+	O	O	O	O
	Preferences	+	O	O	O	O
	Needs	+	O	O	O	O
	Cost	+	O	O	O	O
	Budget	+	O	O	O	O
	Uptake	+	O	O	O	O
	Benefits	+	O	O	O	O
	Concerns	+	O	O	O	O
	Motivations	+	O	O	O	O
	Curiousity	+	O	O	O	O
	Time	+	O	O	O	O
	Channel	+	O	O	O	O
	Format (Simple & Informative)	+	O	O	O	O
	Well-being	+	O	O	O	O
	Health	+	+	O	O	O
	Satisfaction	+	O	O	O	O
	Income	O	+	O	+	+
	Age	O	+	O	O	+
	Gender	O	+	O	O	O
	Race	O	+	O	O	+
	Job	O	+	O	O	O
	Households size	O	+	O	O	+

Factors	Variables	Rashed et el. [11]	EPA [7]	Sari [10]	Gower [8]	Antonopoulos et al. [9]
	Homelessness	O	+	O	O	O
	Occupancy Status	O	O	O	+	+
	Residents' Experiences	O	O	O	+	O
	Residential scale	O	O	O	+	O
	Length of Residence	O	O	O	+	O
	Ethnicity	O	O	O	O	+
	Ownership/Housing tenure	O	O	O	O	+
	Life stage	O	O	O	O	+
Building Characteristics	Building type	+	O	O	O	O
	Heat insulation	+	O	O	O	+
	Housing Features	+	O	O	O	+
	Dwelling Type	+	O	O	+	O
	Poor building condition	O	+	O	O	O
	Low standard building	O	+	O	O	O
	Dimension & spacing buildings	O	+	O	O	O
	Height	O	+	O	O	O
	Façade	O	+	O	O	O
	Roof	O	+	+	O	O
	Greenery	O	+	O	O	O
	Building Height	O	+	O	O	O
	Solar Chimney	O	+	O	O	O
	Window size	O	O	+	O	O
	Shading devices	O	O	+	O	O
	Greenery	O	O	+	O	O
	Thermal mass	O	O	+	O	O
	Orientation	O	O	+	O	O
	Louvre	O	O	+	O	O
	Green roof	O	O	+	O	O
	Cool roof	O	O	+	O	O
	White roof	O	O	+	O	O
	Orientation	O	O	O	+	O
	Size	O	O	O	+	O
	Design quality	O	O	O	+	O

Factors	Variables	Rashed et el. [11]	EPA [7]	Sari [10]	Gower [8]	Antonopoulos et al. [9]
	Building scale	O	O	O	+	O
	Thermal envelope	O	O	O	O	+
	Housing size (Floor Area)	O	O	O	O	+
	Density	O	O	O	O	+
	Age	O	O	O	O	+
	Structural materials	O	O	O	O	+
Environmental Characteristics (Site level factors, Neighbourhood factors)	Location	+	O	O	O	O
	Transpiring & Evaporating water	O	+	O	O	O
	Shade	O	+	O	O	O
	Thermal Mass	O	+	O	O	O
	Trees & Vegetation	O	+	O	O	O
	Water Bodies	O	+	O	O	O
	Urban Materials	O	+	O	O	O
	Hard & Dry Surfaces	O	+	O	O	O
	Impervious & Covered Surfaces	O	+	O	O	O
	Impermeable Surfaces	O	+	O	O	O
	Narrow Streets	O	+	O	O	O
	Geographic Features	O	+	O	O	O
	Passive Climate Control	O	O	O	O	+
	Neighbourhood density (Population & Building)	O	+	O	O	+
	Land use	O	O	O	O	+
	Landscape characteristics	O	O	O	O	+
	Physical geography	O	O	O	O	+
Climatic data	Humidity	+	O	O	O	O
	Indoor Air Quality	+	O	O	O	O
	Ventilation	+	O	+	+	O
	Temperature	O	O	+	O	O
	Wind speed	O	O	+	O	O
	Wind direction	O	O	+	+	O
	Relative humidity	O	O	+	O	O
	Thermal performance	O	O	+	O	O

Factors	Variables	Rashed et el. [11]	EPA [7]	Sari [10]	Gower [8]	Antonopoulos et al. [9]
	Sunlight (Radiant Heat/Natural Light)	O	O	O	+	O
	Damp & Mould	O	O	O	+	O
	Wind flow	O	+	O	O	O
	Weather Conditions	O	+	O	O	O
	Urban micro climate	O	O	O	O	+
	Ambient average urban temperature	O	O	O	O	+

+ Mentioned 0 Not Mentioned.

Table 1.
Different Architectural aspects of HEE.

[15], which is strongly related to pro-environmental behaviours rooted in the enhancement of positive perceptions and attitudes among different groups of people [22, 24]. **Table 2** represents an overall picture of the hybrid multi-criteria approach addressed by various scholars and researchers, e.g. Orr, [42] Moser and Uzzell [22], Centre for Climate Change and Energy Solutions (C2ES) [21], Biloria, [16], Antonopoulos et al. [9], EPA [7], Sari [10], Gower [8].

According to **Table 2**, all these researchers, scholars, and experts have addressed the interrelationships between different aspects, although their conceptual platforms, assumptions, and contributions are different. They emphasised balanced interrelationships between various aspects as the key-determinant of environmental equity, energy justice, pro-environmental behaviours, proactive participation, assistance to vulnerable urban populations, and proper responses to a variety of ecological deficiencies. A reliable informational platform about the interrelationships between various aspects is, therefore, essential to enhance public awareness, and to help them meet their needs while considering ecological concerns and requirements.

These recently conducted studies provide a suitable platform for developing a hybrid multi-criteria approach to HEE (**Figure 1**). Given the figure, such an approach depends initially on a sensitive and careful explanation of the interplay between the different architectural aspects, i.e., human characteristics, building characteristics, environmental characteristics, and climatic information, in impacting housing energy performance. The explanation of the interplay between the different aspects assists in the reduction of the overall energy consumption by improving the congruity of housing energy-related attributes with occupants' socio-demographic characteristics. Such a reliable information platform is beneficial not only for putting passive climatic principles in practice more effectively and reducing fossil fuel dependency but also for encouraging pro-environmental behaviours, which lead to an overall energy consumption reduction in housing sector.

4. Methodological platform of the study

This research was developed on the premise that identification of the interplay between the different architectural aspects is highly critical for housing EE. To meet

Study	Year	Hybrid approaches to EE	Conceptual platform
Orr	2002	Ecological (earth-centred) approach with technological approach	Radical reconceptualization of Sustainability: An intense consciousness of land The careful meshing of human purposes with the larger patterns and flows of the natural world. An integrative principle requiring a radical reconceptualization.
Moser & Uzzell	2003	Bottom-up right-based approach with up-down needs- based approach	Pro-environmental behaviours: Proactive participation, Self-determination, transparency, accountability, informed community, involvement of people in decision-making
C2ES	2015	Bottom-up flexibility with top-down fixed rules	Long-term sustainable development: Climate change mitigation, Accountability and ambitious through a broad participation.
Antonopoulos, et al.	2019	The interplay between the different aspects in impacting EE	Environmental inequity: To assist vulnerable groups by considering their limitations and affordability and the impacts of climatic conditions on their health and wellbeing.
Biloria	2020	Human-centric approach with techno-centric approach	Emphatic city: A combination of efficiency and sufficiency concepts. Energy saving and conservation
EPA	2021	The interplay between the different aspects in impacting EE	Environmental equity: the impact of buildings and urban environmental characteristics on changing neighbourhods climatic conditions and its impacts on vulnerable people
Sari	2021	The impact of building, urban planning and people information, awareness, education and communication on energy consumption and UHI mitigation.	Changing land cover, poorly designed building, thermal discomfort, and lack of awareness, the increase in energy consumption and UHI.
Gower	2021	To proactively alleviate the energy vulnerability by taking into account the different aspects impacting energy consumption in the predesign and design stages.	Energy justice: Building physical condition, Design quality, Passive climatic control, building regulations to protect vulnerable people, energy efficiency at building scale vs. residential scale, the different energy performance of different apartments in relation to the units physical dimensions and occupants' characteristics.

Table 2.
Hybrid multicriteria approach to HEE reflected in several recent studies.

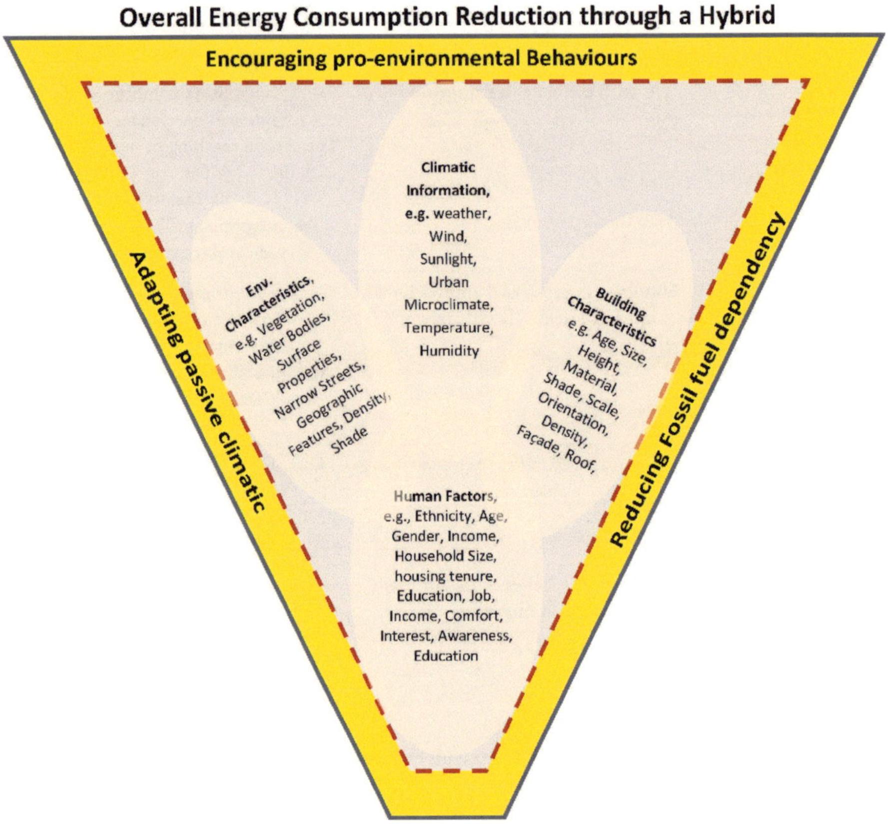

Figure 1.
Interplay between the different architectural aspects incorporated in the hybrid multi-criteria approach to HEE.

this aim, this research develops a multi-method approach through a combination of qualitative and quantitative methodologies by adopting different data collection methods and analysis techniques, which are discussed in this section [44].

Given the complexity of the interrelationships among the different architectural aspects in impacting housing EE, it is also necessary to address the implication of AI in explaining these interrelationships, with especial emphasis on the role occupants' socio-demographic characteristics. There is an abundance of evidence that shows the impact of physical characteristics on housing EE, and people's perceptions (e.g. [38, 45–47]). Environmental physical characteristics convey a wide range of information to people that can have significant impacts on their perceptions, affecting their daily life activities and choice behaviours [22]. Therefore, it is initially necessary to objectively explain people's perceptions of housing physical characteristics.

One of the first attempts to develop a concrete approach to environmental perceptions is environmental image (EI) [48]. The concept of EI has been developed by various researchers and scholars [22] and has been recently employed in many empirical studies to objectively address human environmental perceptions (e.g. [38, 49–51]). Such studies, despite differences in their research areas, theoretical platform, and assumptions, have also provided a suitable platform to objectively explain the interrelationships between different aspects e.g. human, environment, building, and climate, in responding to the different environmental challenges.

These studies employed photo-based methodologies to explain people's environmental perceptions [38, 49], some of which developed AI platforms using deep learning [50, 51]. Their efforts are based on the identification of people's perceptions of environmental attributes through a semantic representation of the attributes presented in different photos using deep learning methodologies. The semantic representation is derived from the content of the environmental attributes in the photos, as determined from people's expressions of the meanings and messages conveyed by these attributes. Identifying people's expressions of the meanings and messages conveyed by the attributes, alongside semiotic interpretations of the attributes meanings results in identification of a number of semantic categories of the attributes [50]. Such a semantic categorisation of the environmental attributes provides a proper platform for determining the multifunctional and multidimensional content of environments and the similarities and differences among them. The interpretations of the interrelationships among different semantic categories, and similarities in their functions and meanings lead to identification of a number of high-level attribute-based factors, explaining people's environmental perceptions. Analyses at this level provide a proper platform for a semantic categorisation of the different environmental photos, and then make it possible to develop an AI platform for the prediction of meaning and functionality of other photos from unknown environments using deep learning methodology [50, 51].

Sallesses et al. [52] also employed the concept of environmental image to develop a GIS-based platform for mapping people's equity and safety perceptions using geo-tagged urban photos. This is a step forward to develop an AI platform for mapping people's EI in different urban areas and neigbourhoods. Most recently, Takagi and Nishimato [53] also developed an AI platform to predict people's thoughts through people's expressions about various photos. Looking at such research works is a promising indication of the potential implications of AI in predicting people's environmental perceptions.

In response to the gap addressed by this study, the next step is to transfer the concept of EI into the area of housing EE. Back to the abovementioned studies, EI is a suitable conceptual platform for explaining residents' perceptions of housing EE [46, 54]. Accordingly, energy efficient housing image (EEHI) would have the potential to concretely explain occupants' perceptions of housing EE, and to identify the determinants of EEH perceptions [54]. Therefore, EEHI provides a suitable platform for determining the of people's perceptions of the interrelationships between building characteristics, environmental characteristics, and climatic data in impacting HEE.

EEHI depends on a clear representation of HEE perceptions derived from occupants' assessment of the role of different energy-related attributes/factors on housing EE and a specific explanation of the rationales of their assessments [46]. The assessment can be made with a photo-based methodology, which would end up in developing a deep learning methodology for the prediction of occupants' perceptions of HEE for unknown housing photos. The employment of geo-tagged photos would make it also be possible to map occupants' perceptions of HEE in the different urban areas and neighbourhoods in a GIS-based database.

This study emphasises the external aspects of residential buildings, i.e., architectural composition principles and envelope physical features, to determine occupants' HEE perceptions. External aspects are the most salient aspects of buildings, which have significant impacts on people's perceptions, have critical impacts on the energy performance of a dwelling unit [46], and as mentioned above, many studies have recently addressed the proper potential of photos from sceneries, open spaces, exterior environments, and building outside views in developing an AI platform.

The third step is to identify the interrelationships among the different architectural aspects of housing EE, e.g., housing characteristics, occupants' characteristics, environmental characteristics, and climatic information. Hsu [55] developed an AI platform to predict the level of housing energy efficiency from outside by employing Google Street view photos, aerial photos, and satellite-based measurements of building heat loss [55]. The study of Gower [8] and Antonopoulos, et al. [9] also employed primary and secondary real data to address the interrelationships between the different architectural aspects in impacting housing EE.

In the study conducted by Antonopoulos et al. [9], variables were selected from the literature. Data were collected from various reliable sources, e.g. electric and gas bills at the block group and individual household levels, as well as national census data. The researchers then applied several inferential analysis techniques, i.e. pearson's correlation, linear regression, spatial autocorrelation, and spatial regression model [9]. Gower [8] developed a case study method and conducted content analysis by observing design and planning documents alongside the actual spatial design impacts. Her study developed a coding technique for the qualitative rating of the variables, which classified them into three scales, i.e. optimal, moderate, and poor.

Some efforts have also been made to develop an AI platform for multi-criteria HEE evaluations based on real and stated housing data [13, 14]. Accordingly, two types of data were collected, including secondary data, e.g. online databases and websites, national census, as well as design and planning documents, and primary data, e.g. field surveys, observations, and experimental measurements, stimulation software, and satellite-based measurements [8–13]. The analyses of the collected data have provided a suitable platform for addressing the interrelationships among the different architectural aspects in impacting housing EE. The development of a smart multi-criteria database would be a product of the analyses, which has the potential to address the interplay among the different architectural aspects of housing EE.

Putting these three steps together provides a reliable platform for developing an AI-based multi-criteria decision making in the area of HEE. The platform would also provide a balanced interplay between different aspects of housing EE, i.e., building characteristics, human characteristics, environmental characteristics, and climatic information from a building's outside view. The AI platform is a GIS-based database with a set of various primary and secondary data relevant to building, environment, climate, and occupants, with geo-tagged photos of housing outside views, which provides comprehensive and integrated housing energy information for different occupants in relation to their socio-demographic characteristics.

This AI multi-criteria platform is beneficial for the assessment of the different architectural aspects of housing EE in relevance to occupants' perceptions. It provides specific and comprehensive EE recommendations, tips, and suggestions tailored to different occupants based on their socio-demographic characteristics. The platform assists in modifying, changing, adjusting, and retrofitting dwelling units based on occupants' socio-demographic characteristics, improving sense of control, proactive participations, and pro-environmental behaviours, and finally enhancing energy equity and justice. Furthermore, this research is beneficial for the different stakeholders in the housing industry, e.g. government, planners, architects, and designers, as it helps enhance the overall HEE by enabling them to enhance the energy-related attributes based on occupants' perceptions while considering the influence of their socio-demographic characteristics in the multi-aspects equation of housing EE.

5. Conclusion

Regarding the main objective of this study which is to explain the implications of AI in a multi-criteria energy efficiency assessment of housing based on occupants' socio-demographic characteristics, this paper initially made a literature review to explain the research background. The conducted review stressed the importance of a balanced interplay between the different architectural aspects in providing housing EE. Then the research made a cross comparison among five recently conducted studies, i.e. Rashed et al. [11], EPA [7], Sari [10], and Gower [8] to develop the variable setting of the study in relation to the different architectural aspects which has significant impacts on housing EE. The cross comparison resulted in identification of 105 variables in 4 areas including environment, building, human, and climatic data (**Table 1**).

The paper then developed the conceptual platform of the research by looking at some scholars' and experts' empirical and conceptual research works, i.e. (C2ES) [21], Moser and Uzzell [22], Biloria, [16], Antonopoulos et al. [9], EPA [7], Sari [10], and Gower [8] (**Table 2**). Despite differences in the research areas, theoretical platforms, and assumptions, these research works all together stressed the critical roles of the interrelationships among the different aspects in properly responding to environmental challenges. They indeed stressed a hybrid approach which is a combination of techno-centric and human-centric approaches as the main solution for the environmental challenges. Transferring this approach to housing, the overall energy consumption reduction depends on a balanced interplay among the different architectural aspects, by special emphasis on the occupants' socio-demographic characteristics (**Figure 1**), which enhances sense of control and pro-environmental behaviours. This approach also assists in reducing fossil fuel dependency, and in putting passive climatic principles in practice more effectively.

The last section of the paper was dedicated to the explanation of methodological platform of the study, which was made in three steps. The first step explained how to represent people's environmental perceptions through the concept of EI based on previous studies (e.g. [22, 38, 49–53]). This concept was then transferred to EEHI, by looking at some other studies (e.g. [46, 54]). The third step considered some other research works (e.g. [8–13, 55]) and explained the methods and techniques that can be adapted in the identification of the interrelationships among different architectural aspects in impacting HEE. Putting all three parts together provides a proper platform to develop a smart multi-criteria EE assessment platform for housing based on occupants' perceptions and attitudes expressed through their socio-demographic characteristics. In the future, conducting the operation phase of the research with a proper sampling design will make it possible to develop a smart multi-criteria EE evaluation platform for housing based on occupants' socio-demographic characteristics.

Acknowledgements

The authors would like to express their highest appreciation and gratitude to Islamic Azad University – Mashhad Branch, Iran for all the supports provided for this research.

Author details

Sayyed Javad Asadpoor* and Elyas Jahanshahi
Department of Architecture, Mashhad Branch, Islamic Azad University, Mashhad, Iran

*Address all correspondence to: sja.1075@gmail.com

References

[1] Department for Communities and Local Government (DCLG). Building a Greener Future: Policy Statement. London, UK: Department for Communities and Local Government (DCLG); 2007

[2] Department for Communities and Local Government (DCLG). Code for Sustainable Homes, Technical Guide. London, UK: DCLG; 2010

[3] Department of Energy and Climate Change (DECC). The Green Deal: A Summary of the Government's Proposals. London, UK: Department of Energy and Climate Change (DECC); 2010. Available from: www.decc.gov.uk/ [Accessed: May 18, 2011]

[4] Australian Building Codes Board (ABCB). National Construction Code: Free for all from 2015. Plumbing Connection. Canberra, Australia: Australian Building Codes Board (ABCB); 2014

[5] Australian Building Codes Board (ABCB). BCA History. Canberra, Australia: Australian Building Codes Board (ABCB); 2020. Available from: https://ncc.abcb.gov.au/ncc-online/About/BCA-History

[6] Department of Environment, Water, Heritage and the Arts (DEBWHA). Energy Efficiency Rating and, House Price in the Act. Canberra, Australia: Commonwealth of Australia; 2008

[7] United States Environmental Protection Agency (EPA). Learn About Heat Island. 2021. Available from: https://www.epa.gov/heatislands/learn-about-heat-islands.

[8] Gower A. Energy justice in apartment buildings and the spatial scale of energy sustainable design regulations in Australia and the UK. Frontiers in Sustainable Cities. 2021;**3**:27

[9] Antonopoulos C, Trusty A, Shandas V. The role of building characteristics, demographics, and urban heat islands in shaping residential energy use. City and Environment Interactions. 2019;**V3**:100021

[10] Sari DP. A review of how building mitigates the urban heat island in Indonesia and tropical cities. Earth. 2021;**2**(3):653-666

[11] Rashed Y, Omrani S, Nilsson D, Drogemuller R. Sustainability information provision (SIP) framework: A review of the promotion of sustainability in the residential sector. Building and Environment. 2023;**229**:1-13. DOI: 10.1016/j.buildenv.2022.109930

[12] Bortolini R, Rodrigues R, Alavi H, Vecchia LFD, Forcada N. Digital twins' applications for building energy efficiency: A review. Energies. 2022;**15**(19):7002. DOI:10.3390/en15197002

[13] Mazur Ł, Bac A, Vaverková MD, Winkler J, Nowysz A, Koda E. Evaluation of the quality of the housing environment using multi-criteria analysis that includes energy efficiency: A review. Energies. 2022;**15**:7750. DOI: 10.3390/en15207750

[14] Asadpoor J, Goh YW, Thorpe D. A human-centric participatory approach to energy-efficient housing based on occupants' collaborative image. Open House International. 2021;**46**(4):615-635. DOI: 10.1108/OHI-11-2020-0163

[15] Harris J, Diamond R, Iyer M, Payne C, Blumstein C,

Siderius HP. Towards a sustainable energy balance: Progressive efficiency and the return of energy conservation. Energy Efficiency. 2008;**1**(3):175-188. Available from: https://pubarchive.lbl.gov/islandora/object/ir%3A150324/datastream/PDF/view

[16] Biloria N. From smart to empathic city. Frontiers of Architectural Research. 2020;**10**(1):3-16. DOI: 10.1016/j.foar.2020.10.001

[17] Thomas S, Thema J, Brischke L, Leuser L, Kopatz M, Spitzner M. Energy sufficiency policy for residential electricity use and per-capita dwelling size. Energy Efficiency. 2019;**12**:1123-1149. DOI: 10.1007/s12053-018-9727-4

[18] Decker. Bedazzled by Energy Efficiency: To Focus on Energy Efficiency is to Make Present Ways of Life Non-Negotiable. Low-Tech Magazine. 2018. Available from: https://solar.lowtechmagazine.com/2018/01/bedazzled-by-energy-efficiency/

[19] Sorrell S. Reducing energy demand: A review of issues, challenges and approaches. Renewable and Sustainable Energy Reviews. 2015;**47**:74-82. Available from: http://www.sciencedirect.com/science/article/pii/S1364032115001471

[20] Bertoldi P. Are current policies promoting a change in behaviour, conservation and sufficiency? An analysis of existing policies and recommendations for new and effective policies. In: ECEEE Summer Study. Toulon: ECEEE; 2017. pp. 201-211. Available from: https://ec.europa.eu/jrc/en/publication/are-current-policies-promoting-change-behaviour-conservation-and-sufficiency-analysis-existing

[21] Centre for Climate Change and Energy Solutions (C2ES). Outcomes of the U.N. Climate change conference in Paris. In: 21st Session of the Conference of the Parties to the United Nations Framework Convention on Climate Change (COP 21). Arlington, USA: C2ES; 2015. Available from: http://www.c2es.org/international/negotiations/cop21-paris/summary>

[22] Moser G, Uzzell D. Environmental psychology. In: Millon T, Lerner MJ, editors. Comprehensive Handbook of Psychology. New York: John Wiley and Sons, Inc.; 2003

[23] Ramkissoon H, Smith LDG, Weiler B. Testing the dimensionality of place attachment and its relationships with place satisfaction and pro-environmental behaviours: A structural equation modelling approach. Tourism Management. 2013;**36**:552-566

[24] Ramkissoon H, Smith LDG, Weiler B. Relationships between place attachment, place satisfaction and pro-environmental behaviour in an Australian national park. Journal of Sustainable Tourism. 2013;**21**(3):434-457

[25] Ajzen J. The theory of planned behaviour: Reactions and reflections. Psychology and Health. 2011;**26**(9):1113-1127. DOI: 10.1080/08870446.2011.613995

[26] Goh YW, Jeevanandam L, Jeevanandam N. Missing pieces in sustained sustainability – Where do the behavioural gaps lie? Singapore Psychologist. 2021;**Q2**(7):54-58

[27] Ahvenniemi H, Häkkinen T. Households' potential to decrease their environmental impacts: A cost-efficiency analysis of carbon saving measures. International Journal of Energy Sector

Management. 2020;**14**(1):193-212. DOI: 10.1108/IJESM-02-2019-0009

[28] Asadpoor J, Thorpe D. Pro-environmental behaviours in improving energy performance of the Australian housing. International Journal of Environment and Sustainable Development. 2017;**V16**(4):315-335. DOI: 10.1504/IJESD.2017.087247

[29] Kellet J, Jowsey E, J. The contribution of housing to carbon emissions and the potential for reduction: An Australia-UK comparison. In: 18th Annual Pacific-Rim Real Estate Society Conference. Adelaide, Australia: Pacific Rim Real Estate Society; 2012

[30] Judge M, Warren-Myer G, Paladino A. Using the theory of planned behaviour to predict intentions to purchase sustainable housing. Journal of Cleaner Production. 2019;**215**:259-267. DOI: 10.1016/j.jclepro.2019.01.029

[31] The Government of South Australia: Department of Energy and Mining (SA. GOV). Home Energy Use. 2019. Available from: https://www.sa.gov.au/topics/energy-and-environment/using-saving-energy/home-energy-use

[32] The Government of South Australia: Department of Energy and Mining (SA. GOV). Energy Efficient Home Design. 2017. Available from: https://www.sa.gov.au/topics/energy-and-environment/energy-efficient-home-design

[33] Australian Government [online]. Your Home. 2019. Available from: https://www.yourhome.gov.au/

[34] Bond S. Barriers and drivers to green buildings in Australia and New Zealand. Journal of Property Investment and Finance. 2011;**29**(4/5):494-509. DOI: 10.1108/14635781111115036

[35] Hurst N. Energy efficiency rating systems for housing: An Australian perspective. International Journal of Housing Markets and Analysis. 2012; **5**(4):361-376. DOI: 10.1108/17538271211268501

[36] The CRC for Low Carbon Living (CRCLCL). Consumers want energy-efficiency facts on homes – new research shows. 2016. Available from: http://www.lowcarbonlivingcrc.com.au/news/news-archive/2016/08/media-release-consumers-want-energy-efficiency-facts-homes---new-research

[37] Department of Environment, Water, Heritage and the Arts (DEBWHA). Energy Use in the Australian, Residential Sector, 1990–2020. Canberra, Australia: Commonwealth of Australia; 2008

[38] Beyond Zero Emissions (BZE). Zero Carbon Australia: Buildings Plan, BZE. Melbourne: Melbourne Energy Institute, The University of Melbourne; 2013. Available from: https://bze.org.au/buildings

[39] Counted V, Cowden RG, Ramkissoon H. Pro-environmental behavior, place attachment, and human flourishing: Implications for post-pandemic research, theory, practice, and policy. In: Place and Post-Pandemic Flourishing. Cham: Springer; 2021. pp. 93-108

[40] Ramkissoon H. Perceived social impacts of tourism and quality-of-life: A new conceptual model. Journal of Sustainable Tourism. 2020;**31**(2):1-17

[41] Ramkissoon H. COVID-19 place confinement, pro-social, pro-environmental behaviors, and residents' wellbeing: A new conceptual framework. Frontiers in Psychology. 2020;**11**:2248

[42] Orr DW. Four challenges of sustainability. Conservation Biology. 2002;**16**(6):1457-1460

[43] Ramkissoon H, Mavondo F, Uysal M. Social involvement and park citizenship as moderators for quality-of-life in a national park. Journal of Sustainable Tourism. 2018;**26**(3): 341-361

[44] Creswell JW. Research Design. Qualitative, Quantitative, and Mixed Methods Approaches. Thousand Oaks, CA: SAGE Publications; 2003

[45] Chandel SS, Sharma V, Marwah BM. Review of energy efficient features in vernacular architecture for improving indoor thermal comfort conditions. Renewable and Sustainable Energy Reviews. 2016;**65**:459-477

[46] Granadeiro V, Duarte JP, Correia JR, Leal VM. Building envelope shape design in early stages of the design process: Integrating architectural design systems and energy simulation. Automation in Construction. 2013;**32**:196-209. Available from: http://www.sciencedirect.com/science/article/pii/S0926580512002415

[47] da Luz Reis AT, Dias Lay MC. Internal and external aesthetics of housing estates. Environment and Behavior. 2010;**42**(2):271-294. DOI: 10.1177/0013916509334134

[48] Lynch K. The Image of the City. Vol. 11. Harvard-MIT Joint Center for Urban Studies Series. Cambridge, Massachusetts: MIT Press; 1960

[49] Chen W. Decode the City: A Methodological Study Responding to the New Trend of City "re-image": Montreal as a Case. 2009. Available from: https://books.google.com.au/books?id=Txu7ZwEACAAJ

[50] Patterson G, Xu C, Su H, Hays J. The sun attribute database: Beyond categories for deeper scene understanding. International Journal of Computer Vision. 2014;**108**(1–2):59-81. Available from: http://link.springer.com/article/10.1007/s11263-013-0695-z

[51] Zhou B, Liu L, Oliva A, Torralba A. Recognizing city identity via attribute analysis of geo-tagged images. In: European Conference on Computer Vision. Zurich, Switzerland: Springer International Publishing; 2014. pp. 519-534

[52] Salesses P, Schechtner K, Hidalgo CA. The collaborative image of the city: Mapping the inequality of urban perception. PLoS One. 2013;**8**(7): e68400. DOI: 10.1371/journal.pone.0119352

[53] Takagi Y, Nishimoto S. High-resolution image reconstruction with latent diffusion models from human brain activity. In: IEEE/CVF Conference on Computer Vision and Pattern Recognition. 2023. pp. 14453-144463. DOI: 10.1109/CVPR52729.2023.01389

[54] Asadpoor J, Thorpe D. An overview of factors affecting housing energy efficiency image. In: 3rd International Conference on Sustainable Ecological Engineering Design for Society (SEEDs). UK: Leeds Sustainability Institute, Leeds Beckett University: LSI Publishing; 2017. pp. 338-351. ISBN: 978-0-9955690-2-7

[55] Hsu J. AI Can Tell Which Buildings Are Energy Efficient from the Outside. London, UK: New Scientists, Technology; 2022. Available from: https://www.newscientist.com/article/2341034-ai-can-tell-which-buildings-are-energy-efficient-from-the-outside/

Chapter 2

Applying a Zero Carbon Footprint Strategy to Improve Environmental and Energy Security

Sérgio Lousada, Svitlana Delehan, Silvia Vilčekova and Andrii Khorolskyi

Abstract

This article presents an empirical study on enhancing the energy efficiency of buildings in Ukraine, particularly in the context of post-war reconstruction. The research focuses on implementing measures based on environmental screening and life cycle assessment (LCA). Using LCA software, the study compares various methods for measuring environmental impacts, confirming significant potential for energy savings and emission reductions. The comprehensive LCA considers all stages of a building's life cycle, including material extraction, production, construction, operation, maintenance, and final disposal or recycling. This analysis provides an objective assessment of the environmental impacts of construction activities. The findings highlight that transitioning from an energy certificate C to B is a highly effective strategy for improving energy efficiency. Recommendations include adopting advanced technologies and materials to optimise energy consumption and reduce emissions, supporting sustainable development and natural resource conservation. Additionally, the study emphasises the importance of incorporating leading building certification technologies such as Building Research Establishment Environmental Assessment Method (BREEAM) and Leadership in Energy and Environmental Design (LEED) to ensure high standards in sustainable building practices. The analysis also identifies shortcomings in Ukrainian environmental legislation, underscoring the necessity of rebuilding according to global and European construction standards. These efforts are crucial for improving environmental conditions amidst the substantial rebuilding demand following the conflict.

Keywords: carbon footprint, construction, CO_2 emissions, energy consumption, environmental screening, life cycle assessment, sustainable development

1. Introduction

1.1 State of the art of the problem

The problem of ensuring the country's energy and environmental security is a basic one. Humanity is trying to reduce the consumption of energy resources that are harmful to the environment. In addition, it is impossible to talk about a developed society without addressing the issue of improving the energy efficiency of buildings. The sooner the transition from energy sources that are inefficient in terms of generation and environmental impact takes place, the sooner the social situation will improve. These problems are common. The common problems are shared by established global practices that include certification of construction [1], decarbonisation of the economy [2, 3], diversification of activities in regions where minerals were extracted [4, 5], risk assessment and development of post-mining strategies for Eastern European countries [6, 7], and use of environmentally friendly materials [8–10]. However, there are specific problems that, in light of recent global events, are becoming urgent for many countries. These problems include military operations and the destruction of industrial energy generation capabilities. If at the beginning of the twenty-first century, the problems of military operations were local and covered the Third World, since 2014, and especially since 2022, these problems have become relevant for the countries of Europe and the Middle East. Given the geopolitical instability, it can be argued that the relevance of the study will remain. Our study has two dimensions. The first dimension covers the problem of ensuring the country's energy stability, which requires reducing energy consumption by improving the energy efficiency of buildings. The second dimension covers the tasks associated with reducing the carbon footprint, thereby improving environmental safety.

Ukraine was chosen as the object of study. This is due to a number of factors:

First, active military operations are taking place on the territory of the country, which is destroying the energy infrastructure; in our opinion, the trend of increasing the share of "green" and "clean" energy sources will continue; in addition, there will be a transition from powerful thermal power plants, nuclear power plants to smaller, local, within a separate settlement, industrial facility [11, 12];

Second, based on previous studies [13–15], it has been established that it is currently possible to replace materials with more environmentally preferable ones that have a smaller carbon footprint; work [16] notes that both the conditions of Slovakia and Ukraine, the methodology for assessing the carbon footprint can be applied at the construction stage, and the available software [17, 18] will allow the development of safe building projects.

Our results can be scaled up and applied in the European Union. It is worth noting that not only in Ukraine are buildings inadequate in terms of humidity and air quality—a similar situation exists in other countries, including Portugal [13].

Since Ukraine entered the world arena as an independent state, the prevailing challenge for its economic system has been the urgent task of overcoming the high degree of energy dependence, which undoubtedly affects the overall economic and energy sustainability of the country. According to the Energy Strategy of Ukraine until 2035 entitled "Security, Energy Efficiency, Competitiveness" and the Zero Carbon Footprint strategy, the international level is witnessing a transformation of approaches to the formation of energy and construction policies. This transformation involves a shift from the outdated model of the energy sector and construction industry to a new approach that includes creating a competitive environment, an even distribution of opportunities for the development of different areas, and minimising the dominance of any single energy production or fuel source. This new approach places great emphasis on improving energy efficiency and the use of renewable energy sources, as well as on measures to counteract and adapt to climate change. These global energy development priorities are forcing Ukraine to face new challenges, while opening up new opportunities for finding and implementing innovative solutions in the field of energy production, transformation, supply, and consumption. This, in turn, necessitates the formation of a new energy policy adapted to modern realities.

Given the difficult geopolitical situation in recent years, in particular Russia's aggression against Ukraine since 2014 and the invasion in 2022, as well as damage to critical infrastructure, there is an urgent need for a comprehensive analysis and research of issues related not only to energy security but also to achieving Ukraine's energy independence, ensuring its energy sovereignty and sustainable development, as well as ensuring guaranteed human and civil rights and freedoms in the face of the above challenges. In the current context of the urgency of achieving a zero carbon footprint, the construction sector is a key player with a negative impact on the environment at all stages of its life cycle. Recent scientific studies indicate that the construction industry plays an important role in energy consumption and greenhouse gas (GHG) emissions worldwide [14–21]. This requires special attention and highlights the relevance of research on decarbonisation processes in this sector [22–24].

In 2021, the Government of Ukraine began to fulfil its obligations under the Memorandum of Understanding on the Settlement of Problematic Issues in the Renewable Energy Sector of Ukraine, concluded in June 2020 following mediation at the Energy Community Dispute Resolution Centre between the Government of Ukraine and the National Energy and Utilities Regulatory Commission (NEURC), on the one hand, and the Ukrainian Wind Energy Association and the European-Ukrainian Energy Agency, on the other. The gradual repayment of debts to renewable energy producers that had accumulated in previous years has begun. At the same time, the NGO Antitrust League [25] attempted to recognise the feed-in tariff as illegal state support or as having been adopted in an unconstitutional manner [26].

The President of Ukraine joined the Global Wind Energy Manifesto at COP26 and committed to completely eliminate domestic coal consumption by 2035 [27] and to begin the gradual decommissioning of thermal generation starting in 2022, while the Government of Ukraine continued to focus all efforts on maintaining the outdated nuclear energy infrastructure and adopted the State Programme for the Development of the Nuclear Power Industry until 2026 [28]. This resulted in the Verkhovna Rada of Ukraine supporting and the President of Ukraine signing the Law of Ukraine "On Amendments to Certain Laws of Ukraine on the Development of Energy Storage Systems," which made it possible to make generation from renewable energy sources more stable through the construction of energy storage systems [29]. At the same time, renewable energy producers were forced to resist the initiative of some MPs to introduce an excise tax on electricity from renewable sources [30].

The large-scale war launched by Russia on the territory of Ukraine in February 2022 left the renewable energy sector in a state of anticipation and uncertainty. This situation was exacerbated not only by active hostilities, damage and occupation of energy facilities but also by the artificial creation of additional problems and challenges in the market by certain government agencies.

1.2 Analysis of the Ukrainian renewable energy sector on the eve of active hostilities

In 2019, Ukraine was recognised as one of the top 10 countries globally for renewable energy development, and in 2020, it ranked among the top five European nations in solar energy advancement. Furthermore, according to the 2019 Climate Scope rating by Bloomberg New Energy Finance (Bloomberg NEF), Ukraine improved its position to 8th place (up from 63rd) out of 104 countries regarding investment attractiveness for the development of low-carbon energy sources and the construction of a green economy. In 2021, Ukraine was ranked 48th in overall investment potential among 136 countries in the Bloomberg NEF ranking [31].

Beginning in 2019, investments in new renewable energy projects in Ukraine consistently surpassed those in fossil fuel projects. Over the past decade, international and Ukrainian investors in renewable energy have secured over USD 12 billion in foreign direct investment. By the end of 2021, foreign investors accounted for over 35% of the installed capacity in Ukraine's renewable energy sector, highlighting the sector's competitiveness and openness. Major international lenders and investors in Ukraine's renewable energy sector include the European Bank for Reconstruction and Development, the Black Sea Trade and Development Bank, the American International Development Finance Corporation (DFC), the Federal Bank of Bavaria Bayern LB, the Investment Fund for Developing Countries (IFU), the Nordic Environment Finance Corporation (NEFCO), among others. As reported by the National Energy and Utilities Regulatory Commission (NEURC), by 31 December 2021, the installed capacity of the renewable energy sector in Ukraine reached 9655.9 MW, or 8450.8 MW excluding solar installations for private households (SPH) [32].

In 2021, only relatively active development was observed in the segment of home solar power plants (SPPs). During this period, their capacity increased by 426.1 MW, which is 36.4% of the total increase in renewable energy capacity commissioned in 2021. Thus, the total capacity of all residential solar systems reached 1205.1 MW by the end of 2021. In contrast, industrial solar power showed less impressive results. In 2021, the capacity of industrial solar generating plants increased by only 305.5 MW (26.1% of the total increase in renewable energy capacity commissioned in 2021). This is 3.6 times less than in 2020, when 1123.6 MW of new capacity was commissioned. Thus, the total capacity of the industrial solar energy sector as of the end of 2021 was 7586.3 MW (including home solar power plants) [33].

The wind energy sector remains the second largest renewable energy sector in Ukraine, after solar energy, in terms of total installed capacity. It should be noted that Ukraine's wind energy sector has made the largest contribution to the country's green energy development. Wind power plants accounted for 30.6% or 358.8 MW of the new renewable energy capacity commissioned in 2021. This is 2.5 times more than in 2020, when 144.2 MW of new wind power capacity was commissioned. Thus, the total installed capacity of the wind energy sector as of the end of 2021 was 1672.9 MW. Prior to the outbreak of the large-scale war, Ukraine had 34 wind farms or 699 wind turbines with an average capacity of 3.5 MW.

In 2021, renewable energy sources (RES) contributed 8.1%, equivalent to 12.8 TWh, of the total electricity production. Of this amount, 56% was generated from solar energy, 33% from wind power, nearly 8% from biomass and biogas, and 3% from small hydropower systems.

Overall, RES facilities generated 12,804 million kWh of clean electricity in 2021, marking an increase of 1941.9 million kWh or 17.8% compared to 2020 [34]. Specifically:

- Wind power plants produced 3866 million kWh, representing an increase of 614.4 million kWh over 2020, and accounted for 2.97% of total electricity generation.

- Solar power plants generated 7670 million kWh, a 4.8% increase from the previous year, with an additional 1065.4 million kWh.

- Small hydropower plants increased their output by 56.1 million kWh, reaching 276 million kWh, which constitutes 0.17% of the overall electricity balance.

- Biopower plants produced 992 million kWh, an increase of 206 million kWh compared to the previous year.

It is notable that 2021 was a significant year for the national RES sector, as on 11 May 2021, daily electricity generation from RES surpassed that of thermal power plants for the first time, with 79 million kWh produced by RES compared to 77 million kWh from thermal sources [35].

1.3 Study of the problem of energy poverty in Ukraine in the context of the threat to the sustainability and stability of Europe

Ukraine has suffered severe human and material losses due to the war, which has resulted in the destruction of civilian, industrial and military infrastructure. According to official figures, almost 1 million residential buildings, tens of thousands of non-residential buildings, and thousands of kilometres of roads, railways and bridges were destroyed or damaged. The construction industry suffered heavy losses and partially lost its raw material base and production. Ukraine was one of the top 10 countries in Europe in terms of installed power generation capacity and is the third largest gas producer, with the largest underground gas storage facilities in Europe. Extensive and reliable gas, oil and petroleum products transportation and electricity transmission systems connect Ukraine's neighbouring EU countries and Moldova.

In Ukraine, about 70% of electricity is produced by nuclear, hydro and renewable generation [36].

Russia's war against Ukraine has led to a global energy crisis, as well as to complex and unpredictable impacts on climate issues at the international level. One of the international consequences of this war was a conscious understanding of the need to develop energy independence for Ukraine and the EU. Ukraine, due to its large territory compared to other European countries, has significant potential for wind, solar and biomass energy production, and Ukraine's connection to ENTSO-E opens up opportunities for exporting electricity from renewable sources, which will generate revenue and help reduce carbon emissions in the European Union.

Russia's war against Ukraine has caused major changes in the global energy system. This situation has caused significant difficulties, such as high energy prices, and has led to policy changes in the European continent, with a focus on combating energy poverty. In the face of global challenges such as climate change, these countries are aiming to achieve a just energy transition and ensure that every citizen has access to basic energy services. These goals include minimising the level of energy poverty by 2030 [37].

Energy poverty is defined as the lack of access to modern energy services [38]. It is important to note that energy poverty should not be evaluated based on a country's raw material potential or the availability of energy resources. Rather, it is fundamentally linked to the level of access to contemporary energy services. In essence, a modern and developed society should aim to utilise safe and efficient energy supply systems. The concept of "energy poverty" first emerged in the early 1990s [39]. Initially, it primarily referred to the insufficient capacity for heating and cooling within homes. However, perspectives on this issue have evolved over time. Currently, energy poverty is recognised as a complex systemic inequality, characterised by barriers to accessing affordable and modern energy services. This issue is challenging to quantify due to its dynamic nature, which varies over time and across different regions. Moreover, energy poverty now encompasses social and cultural dimensions, as it directly impacts the quality of life by affecting the fulfilment of basic human needs.

Addressing energy poverty is crucial for enhancing the quality of life and is a cornerstone for economic development. Ensuring access to affordable, reliable, sustainable, and modern energy for all is one of the key goals of sustainable development [40]. This issue is being addressed not only by national governments but also by supranational institutions and organisations [41–43].

In 2022, the European Economic and Social Committee (EESC) presented an opinion on tackling energy poverty and ensuring the sustainability of the European Union from a socio-economic perspective [44]. In this document, the EESC calls on EU countries to support a common approach to combating energy poverty, considering this phenomenon as a result of geopolitical, economic, environmental and social factors. The EESC also recommends the development of an EU strategy to ensure consumers' right to energy and set targets for overcoming energy poverty in the long term.

One of the tools for implementing the European Union's New Industrial Strategy is the introduction of a number of so-called "green alliances," which include the European Raw Materials Alliance, the European Battery Alliance, the Clean Hydrogen Alliance and the Circular Economy Alliance. The main goal of these green alliances is to create an innovative, competitive and sustainable value chain in a specific area. In other words, these alliances are aimed at developing a specific sector of the economy—from research and raw material needs to production and consumption.

These alliances create the conditions for the implementation and financing of large projects that can have additional positive effects for both businesses and governments, as they use the knowledge of small and medium-sized enterprises, large companies, research institutions, and regional organisations to overcome barriers to innovation and ensure the integrity of public policy. In this process, Ukraine should actively participate and be a key player to create the necessary conditions for the development of a single energy-independent European space [45].

2. Methodology

2.1 Analyse the legislative framework and institutional foundations for the application of life cycle assessment (LCA) in Ukraine and the EU

The initial step involved a comprehensive review of the existing legislative and institutional frameworks governing the application of LCA methodologies in Ukraine and the EU. This analysis included examining relevant laws, regulations, and policies that impact the implementation of LCA in construction projects. The goal was to identify any legal or institutional barriers and to understand the regulatory environment that influences the adoption of energy efficiency measures and sustainable practices in building modernisation.

2.2 Investigate the construction object for opportunities to improve its energy efficiency and increase compliance with modern environmental standards

Detailed inspections and evaluations were conducted on the selected construction object to identify areas where energy efficiency improvements could be made.

This involved assessing the current state of the building's energy systems, materials, and overall design. The investigation focused on pinpointing specific components or practices that could be upgraded to meet modern environmental standards, thereby reducing energy consumption and environmental impact.

2.3 Make a calculation for a real object using modelled energy costs, by individual parameters, using the One-Click Life Cycle Assessment software application

Utilising the One-Click LCA software, energy cost calculations were performed for the real object under study. This step involved creating detailed models that simulate the building's energy performance under various scenarios. The software allowed for the input of specific parameters, such as building materials, energy sources, and operational practices, to accurately estimate the potential energy savings and CO_2 emissions reductions that could be achieved through proposed energy efficiency measures.

2.4 Draw conclusions and propose further possible steps to improve the energy efficiency of the studied facility by individual parameters

Based on the data obtained from the One-Click LCA simulations, conclusions were drawn regarding the effectiveness of different energy efficiency measures. The analysis provided insights into which specific interventions would yield the most significant improvements in energy performance and environmental impact. The research concluded with a set of actionable recommendations for further enhancing the energy efficiency of the studied facility, tailored to the unique characteristics and needs of the building.

By following this detailed methodology, the study ensured a systematic and thorough approach to evaluating and improving the energy efficiency of the construction object, providing a robust framework for future research and practical applications in sustainable building practices.

3. Overview of the legal framework

3.1 Green certification of Ukraine's civil and industrial infrastructure based on the zero carbon footprint principle as a way to ensure Ukraine's energy security in the post-war period

The environmental criteria used in certification and labelling systems are essential standards for assessing the life cycle and environmental benefits of construction projects. Systems such as BREEAM and LEED offer transparent and reliable standards that ensure compliance with the zero carbon footprint principle [1], which is vital for Ukraine's post-war reconstruction efforts.

Environmental certificates issued under ISO 14024 confirm the compliance of products or services with environmental criteria, supporting green procurement [46]. The EU Directive 2014/24 mandates the use of public procurement criteria to increase the market presence of environmentally certified products [47]. However, Ukraine is still developing its legislative requirements for the environmental safety of construction materials [48]. The introduction of comprehensive bylaws and regulations is necessary to ensure the safety of these materials throughout their life cycle.

Ukraine's implementation of the EU Energy Efficiency Directive and the Energy Performance of Buildings Directive aims to bring the country closer to creating net-zero energy buildings [49]. The Law on Energy Efficiency in Buildings, introduced in 2018, aligns with Directive 2010/31/EU and mandates energy certification for new construction projects, promoting significant utility sector savings [50].

Adopting the ISO 15392 standard, which balances economic, environmental, and social aspects of construction, is crucial for sustainable development [10]. Moreover, the use of LCA software like One-Click Life Cycle Assessment ensures compliance with key standards (ISO 14040, ISO 14044, EN 15978) and provides a quantitative assessment of environmental impacts [51].

In summary, Ukraine's legal framework is evolving to support energy-efficient and environmentally safe construction [47], but further development and implementation of detailed standards and regulations are essential to meet international benchmarks [20]. The adoption of advanced building certification systems and the identification of legislative shortcomings are crucial steps towards sustainable rebuilding efforts that align with global and European standards [52].

3.2 Features of LEED and BREEAM certification application

The LEED (Leadership in Energy and Environmental Design) system, introduced in the United States in 1998, and the BREEAM (Building Research Establishment Environmental Assessment Method) system, established in the United Kingdom in 1990, are two of the most widely recognised global standards for evaluating the sustainability of buildings [53]. These assessment systems use formal criteria, categorised into several key areas, to evaluate buildings, with the highest level of certification awarded to those demonstrating optimal environmental performance.

LEED emphasises energy and water management, encouraging efficiency improvements and the use of eco-friendly technologies [54]. In contrast, BREEAM focuses on ecological aspects and environmental issues, assessing factors like energy efficiency, water use, material selection, and more. Both systems aim to enhance building sustainability and energy efficiency, boosting their competitiveness in the real estate market and aligning with increasing environmental demands and construction industry goals.

While LEED and BREEAM share common goals, they exhibit significant differences in their approaches and emphases. LEED's rigorous standards focus on energy and water management, while BREEAM's assessments delve into broader ecological considerations [54]. LEED-certified buildings undergo scrutiny from certification auditors during development, ensuring compliance from the outset. Conversely, BREEAM employs qualified professionals to conduct comprehensive assessments tailored to each building's location and characteristics.

Practicality and adaptability also distinguish LEED and BREEAM. BREEAM certification accommodates various building types, including offices, malls, and residential complexes, demonstrating its flexibility across diverse construction sectors [54]. This adaptability reflects the system's ability to meet varied industry needs and requirements.

In Ukraine, the implementation of energy efficiency measures, including the adoption of international environmental certifications like BREEAM, signals a proactive approach to sustainability [55]. Financial incentives, such as tax exemptions for companies engaged in green construction, further promote energy-efficient initiatives, positioning Ukraine as a burgeoning hub for sustainable building practices.

In summary, while LEED and BREEAM share common sustainability goals, their nuanced differences in approach and implementation cater to diverse building contexts and industry needs, driving advancements in global environmental standards and practices.

4. Results of research on the development of a building modernisation project

4.1 Description of the programme, modernisation objectives, and research object

The Higher Education in Ukraine project is aimed at reconstructing and rehabilitating research, teaching and support facilities owned by various state higher education institutions in Ukraine. The main goal of the project is to improve the energy efficiency of these facilities and provide other types of investment for the development of education that go beyond energy efficiency.

In December 2017, the National Environmental and Energy Agency (NEFCO) signed a EUR 30 million loan agreement with the Ministry of Finance of Ukraine. The agreement provides financing for the Higher Education in Ukraine project. Ukraine, represented by the Ministry of Finance and the Ministry of Education and Science, is the recipient of the loan. The Ministry of Education and Science (MES) is the main responsible body for coordinating the project implementation. Universities and other public higher education institutions that intend to renovate their educational, research and ancillary facilities to improve energy efficiency and receive other types of investment for the development of education are the final beneficiaries of the project [56].

The National Environmental and Energy Agency (NEFCO) is financing the implementation of energy efficiency measures at Uzhhorod National University. The University receives a loan from the Government of Ukraine through the Ministry of Finance of Ukraine and the Ministry of Education and Science of Ukraine for two purposes:

1. Improving the energy efficiency of classrooms and dormitories;

2. Conducting research on the presence of hazardous materials in buildings, with a particular focus on asbestos.

The expected key outputs of the project include:

1. Quantification of energy-saving potential as a result of cost-effective energy efficiency measures;

2. An improved policy and institutional framework for managing the reduction and disposal of hazardous materials such as asbestos, lead and/or mercury.

It was planned to carry out major renovation works in the following classrooms to modernise them:

- Dormitory No. 4, 12 Universytetska Street;

- Academic buildings A, B, C, 14 Universytetska Street, A-B;

- Academic building D, 14 D Universytetska Street.

The project involved the modernisation and implementation of energy efficiency measures across three buildings at the University, specifically the educational and laboratory buildings A, B, C, and D (referred to as BAM), as well as student Dormitory No. 4. The total budget for the project was UAH 153,749,396, with Uzhhorod National University contributing UAH 25,624,899 as part of the funding.

The project includes the modernisation of three buildings/structures. As part of the main component of the project, an environmental and social assessment was carried out and an Environmental and Social Management Plan (ESMP) was developed, which includes the identification of measures required to implement modern methods of reducing the volume and disposal of hazardous materials during the energy efficiency works. Attention is focused on the main hazardous materials, such as asbestos-containing materials (ACMs), which can be loosely or firmly bound, lead-based paints (LBPs) and compact fluorescent lamps containing mercury (CFLs).

Due to Russia's full-scale invasion, the European partners were forced to suspend the project. Three educational institutions in the western region—Uzhhorod National University, Ivan Franko National University of Lviv and Vasyl Stefanyk Precarpathian National University—received partial funding of 20% of the contract amount.

4.2 Results of environmental screening of buildings to be modernised

The information obtained in the course of the environmental screening will allow us to offer recommendations for reducing carbon dioxide emissions and modernising buildings.

As part of this project, monitoring visits were carried out by environmental experts from the State Higher Educational Institution "Uzhhorod National University" and environmental experts from NEFCO (Northern Environmental Financial Corporation). The purpose of these visits was to check the environmental condition of the buildings and assess the basic conditions of waste management in accordance with international standards.

Experts from Uzhhorod National University, acting as the responsible environmental experts for this project, used their scientific and practical knowledge to conduct research and assess the environmental condition of the sites to be monitored. NEFCO's experts, in turn, contributed to the process by adhering to international standards and requirements in the field of ecology.

All the monitoring visits and environmental assessments carried out provided an objective picture of the project's environmental impact and allowed us to develop appropriate recommendations for improving environmental performance in accordance with the defined standards.

4.2.1 Environmental screening of asbestos-containing materials (ACM)

Environmental experts conducted a visual inspection of critical areas, such as attics/technical floors, individual heating points, basements and technical rooms.

During the visual inspection, the experts examined the condition of each site in detail, paying attention to various aspects such as the presence of possible sources of contamination, compliance with environmental standards, the degree of wear and tear on equipment and the general appearance. The results of this review have been systematised and presented in **Table 1**, which provides a detailed analysis of the environmental condition of the critical sites.

Categories of products containing asbestos						
Category product 1	**Category product 2**	**Category product 3**	**Category product 4**	**Category product 5**	**Category product 6**	**Category product 7**
Asbestos cement roof or other products. For example, sewerage (dismantling without destruction)	Crushed asbestos cement (and contamination dust)	Insulation or insulation coating of the main pipes heating	Insulation or insulation coating of secondary heating pipes	Flanges/ compensators expansion joints/fire doors	Insulation coating or asbestos containing pipe textiles heating (cannot be dismantled and destroyed without destruction)	Insulation or insulation coating of external pipes heating
Firmly bound	Tightly bound and pieces loosely bound dust	Poorly bound	Poorly bound	Firmly bound	Firmly bound and poorly bound	Firmly bound and poorly bound
			Like a category 3 product, but in a smaller package quantity			

Table 1.
Categories of asbestos-containing products identified during monitoring inspections (Source: Authors).

The assessment was carried out by an asbestos specialist through visual inspection of suspected hazardous materials and sampling. Four samples (material samples) were taken to test the materials for asbestos content. All samples were analysed in the laboratory of a NEFCO partner in Vienna.

Product categories have been introduced to enable further assessment and identification of the asbestos-containing material (ACM) removal methods used.

Poorly bonded asbestos, often called "dust asbestos" or asbestos material, differs from other types of asbestos in its structure and asbestos content, which is higher than 25%. This material is composed of soft and loose asbestos fibres or asbestos dust, which can be exposed to the environment.

Because loosely bound asbestos has a high asbestos content, it is the subject of special attention due to its potentially negative impact on human health and the environment. The use and handling of this material requires specific personal protective measures to ensure the safety of workers and others who may come into contact with weakly bound asbestos.

Particular attention should be paid to health and safety requirements and the use of appropriate protective equipment such as respirators and protective clothing. In addition, measures must be taken to prevent the release of asbestos particles into the air and their entry into water or soil, which may occur during the handling or disposal of this material.

Protecting the environment from loosely bound asbestos involves proper disposal and consideration of its potential impact on the ecosystem. This may include appropriate waste disposal and treatment methods to reduce the risks of negative impacts on human health and the environment.

Based on an inspection of the buildings and sites, we identified possible sources of contamination, compliance with environmental standards, the degree of deterioration of equipment and general appearance.

4.2.2 Environmental screening of mercury-containing lamps (mercury CFLs)

Within the European Union, the Directive on the Restriction of the Use of Certain Hazardous Substances in Electrical and Electronic Equipment (Hazardous Substances Directive) prohibits the use of mercury in specific electrical and electronic products and restricts the mercury content in other products to less than 1000 ppm. In the EU, the maximum allowable mercury content in compact fluorescent lamps (CFLs) is 3.5 mg per lamp, a limit that was further reduced to 2.5 mg (<30 W) after 2012. Consequently, all modern high-energy-efficiency lamps contain minimal amounts of mercury. **Table 2** outlines the recommended number of lamps that should be replaced.

A survey of the existing lighting systems for all these buildings was carried out during the site visits. To obtain a general picture of the specific number of lamps to be replaced with LEDs, the consultant used the information on each individual facility provided by CES in their 2019 energy audits.

According to the Sanitary and Epidemiological Norms and Regulations 2.2.7.029-99, mercury is classified as class I hazardous waste and requires special handling by licenced companies. Prior to selecting such a licenced company and signing a contract with it for the disposal/burial of such mercury-containing lamps, a specially equipped temporary storage area must be arranged. In Zakarpattia, there are three licenced companies engaged in the collection and storage of fluorescent lamps, including one company licenced to carry out the entire range of works to reduce their quantity: collection, transportation, storage, delivery, disposal and burial.

Object	Type	Number of units
Dormitory #4	T8 fluorescent lamps	440
Educational buildings A, B, C	T8 fluorescent lamps	2725
Educational building D	T8 fluorescent lamps	1817

Table 2.
Exact number of mercury-containing lamps to be replaced and further recycled/recycled (Source: Authors).

4.2.3 Environmental screening of lead-based paints (PBS)

Lead-based paints present a significant risk of lead poisoning during the production, handling, and disposal of components coated with such paints. The primary hazards arise from the inhalation of aerosols and abrasive dust generated during these processes. In Germany, the use of lead-based paints has been restricted since 1993. Historically, due to its excellent anti-corrosive properties, lead-based paint was widely applied to steel structures in industrial buildings, particularly on bridges. Despite its primary use in industrial contexts, lead-based paint has also been used on walls and facades due to its durability.

In relation to the energy efficiency programme, no areas of potential contamination were identified within the buildings. The collected screening data provides a basis for formulating recommendations to reduce carbon dioxide emissions.

4.3 Calculation of emission reductions CO_2

Table 3 outlines the estimated annual reduction in CO_2 emissions resulting from proposed energy efficiency measures for various buildings. The calculations were performed by CES Clean Energy Solutions and were included in the Environmental Assessment reports submitted in 2019. For Dormitory #4, the implementation of energy efficiency measures is projected to reduce CO_2 emissions by 387.48 tonnes per year. This reduction is significant, reflecting the impact of targeted energy-saving interventions in residential buildings. For educational buildings A, B, and C, these buildings are estimated to achieve a combined annual CO_2 emissions reduction of 493.23 tonnes. This higher reduction rate indicates the substantial potential for energy savings in educational institutions, which often have extensive and continuous energy demands. For educational building D, the estimated annual CO_2 emissions reduction is 320.5 tonnes. This value underscores the effectiveness of the proposed energy efficiency measures tailored to the unique energy consumption patterns of this facility. The total estimated annual CO_2 emissions reduction for all the buildings combined is

Object	Savings rate CO_2—(тCO_2/y)
Dormitory #4	387.48
Educational buildings A, B, C	493.23
Educational building D	320.5
Total:	1201.21

Table 3.
Amount of annual CO_2 emissions reduction after implementation of the recommended energy efficiency measures (Source: Authors).

Fuel	Emission factor CO_2—kg CO_2/kWh
Centralised heating	0.260
Natural gas	0.220
Electricity	0.800

Table 4.
CO_2 emission factors (Source: Authors).

1201.21 tonnes. This aggregate figure highlights the cumulative impact of the energy efficiency measures across multiple facilities, contributing significantly to the overall reduction of greenhouse gas emissions. These reductions demonstrate the potential benefits of adopting energy efficiency measures in both residential and educational buildings. By implementing these measures, not only do we reduce environmental impact, but we also achieve cost savings and improve energy security.

The following CO_2 emission factors were used for the calculations, which are presented in **Table 4**. These factors represent the amount of CO_2 emitted per unit of energy consumed from various sources. The CO_2 emission factor for centralised heating is 0.260 kg CO_2 per kWh. This value reflects the emissions associated with the generation and distribution of heat in centralised systems, which often rely on fossil fuels. The CO_2 emission factor for natural gas is 0.220 kg CO_2 per kWh. Natural gas is commonly used for heating and electricity generation due to its relatively lower emissions compared to coal and oil. The CO_2 emission factor for electricity is 0.800 kg CO_2 per kWh. This higher value indicates the significant emissions associated with electricity generation, particularly in regions where fossil fuels are the primary energy source. Understanding these emission factors is critical for accurately assessing the environmental impact of energy consumption. By using these factors, we can quantify the potential CO_2 savings from energy efficiency measures and prioritise actions that yield the greatest environmental benefits.

A detailed analysis of the estimated CO_2 emission reductions provides important insight into the environmental impact of dismantling and retrofitting. Using data from tables such as the one above, we can develop and implement effective waste management and greenhouse gas reduction strategies. This approach not only helps to meet regulatory requirements but also advances sustainability goals by contributing to broader efforts to combat climate change and improve energy efficiency.

5. Modelling the energy efficiency of a building using One-Click LCA

The idea of utilising the results from a completed building modernisation project to model improvements in energy efficiency is both innovative and practical. By leveraging the One-Click LCA application, we can simulate and analyse various parameters that influence energy consumption and environmental impact. This approach allows us to gain valuable insights into the effectiveness of different energy efficiency measures and their potential benefits. Through detailed modelling and assessment, we can identify the most impactful strategies for enhancing energy performance, ultimately contributing to more sustainable and environmentally friendly building practices.

This work is particularly important because it addresses both theoretical and practical aspects of energy efficiency improvements. The study provides a clear connection between previous research on the environmental impact of building modernisation and the comprehensive methodology used in this analysis. By explaining the different steps taken in the research, including data collection, simulation, and analysis using the One-Click LCA application, we highlight the robustness and reliability of our findings.

The calculations of the environmental impact are based on previous studies, ensuring a solid foundation for our work. However, we also aim to clarify the link between our specific research and the broader methodological framework. Our conclusions demonstrate the tangible benefits of implementing energy efficiency measures, including significant reductions in CO_2 emissions. This research not only validates the proposed strategies but also offers concrete recommendations for future projects, emphasising the importance of sustainable building practices in the context of global environmental challenges.

Our goal was to analyse the carbon footprint of buildings constructed from different building materials. The carbon footprint indicator was not chosen randomly, as the share of the carbon footprint in the structure of the ecological footprint varies between 33% and 65%. In the case of Ukraine, the share of the carbon footprint is 62.7%. In the European Union, it is 59%. The concept of an ecological footprint was developed by William Rees and Mathis Wackernagel in the 1990s of the twentieth century. The simplest expression of the ecological footprint is the amount of goods and resources required to produce the goods and services that people need. The carbon footprint, mentioned earlier, is part of the ecological footprint and is quantified using the CO. The advantage of this indicator is that it can be calculated for different countries, groups of people, regions, etc. The annual ecological footprint is calculated by the International Organisation Global Footprint Network [57]. The methodology and calculation results for specific countries are presented in [58]. The criteria and basic methodology for calculating the ecological footprint for different sectors of the economy are given in [59]. The ecological footprint is divided into consumption categories (food, housing, goods, services, etc.). It can also focus on types of employment (land, forests, marine areas). However, life cycle analysis is used to analyse the production of goods and services (in this case, construction) [60]. Software is used to assess life cycles. Such software can convert energy consumption into a normalised land area, the so-called global hectares (a unit of measurement for carbon footprint) [59, 60].

The main purpose of modelling the building energy efficiency outcome using One-Click LCA is to determine the state of energy use by the facility and develop a consistent plan of energy modernisation measures and their feasibility study. It also analyses opportunities to reduce heating costs.

During the building survey, information on energy consumption should be systematically collected and summarised. This should also include reviewing existing technical and operational documentation, performing an instrumental inspection in accordance with the approved measurement programme, determining the energy efficiency class of the building, the level of the energy metering system, and building an energy balance for heat use. Based on this information, an energy profile of the facility should be formed (**Table 5**), and a number of energy efficiency measures should be proposed to reduce energy consumption in this building.

In the course of modelling the energy efficiency outcome of the building, the environmental benefits of each package of measures were assessed. The results are presented in **Table 6**.

The actual calculated energy consumption of a building is the energy consumption that takes into account the actual operating conditions, specific parameters

Energy budget of the building				
The article of the budget	**Estimated to be executed ES**	**Measured (actual values)**	**Prior to the execution of the EA, the base values are**	**After performing the EA**
	[kWh/year]	**[kWh/year]**	**[kWh/year]**	**[kWh/year]**
Heat energy				
Heating	523,329	—	919,711	37,902
Ventilation				
Hot water supply	—	—	42,130	32,130
Electric energy				
Lighting	86,518	—	162,602	162,602
Other	18,352	—	15,304	16,927
Total	607,199	—	1,029,747	527,150

Table 5.
Energy profile of the facility (Source: Authors).

Environmental benefits from the implementation of measures		
Actions	**Energy savings, kWh/year**	**Reduction of CO_2 emissions, tonnes per year**
Insulation of the attic floor	120,645	26.5
Replacement of windows	105,543	23.2
Door replacement	8861	1.95
Installation of an individual heating point	134,469	29.6
Installation of thermostatic regulators on radiators of the heating system	19,485	5.07
Installation of automatic balancing valves	15,920	3.5
Installation of a ventilation system	12,385	21.6
Total:	503,220	111.5

Table 6.
Modelling of the building energy efficiency outcome and environmental benefits from the implementation of measures (Source: Authors).

of the microclimate and internal thermal comfort in the building (baseline energy consumption adjusted to the actual operating conditions).

Figure 1 shows the results of the assessment of heat loss through the building envelope before the modernisation measures.

Energy consumption after implementation of measures is the estimated energy consumption in buildings after implementation of energy modernisation measures (**Figure 2**), which were proposed as a result of the survey. This takes into account compliance with the regulatory parameters of microclimate, internal thermal comfort and regulatory conditions of the building.

The modelling of the building's energy efficiency using One-Click LCA revealed opportunities for significant energy consumption reduction and limitation of air

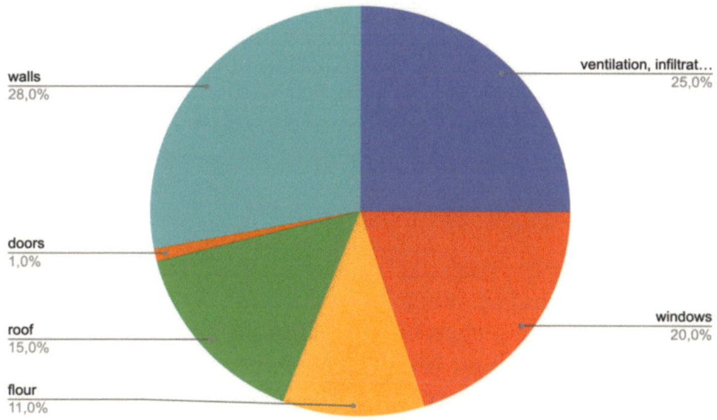

Figure 1.
Estimated distribution of heat losses through the building envelope (Source: Authors).

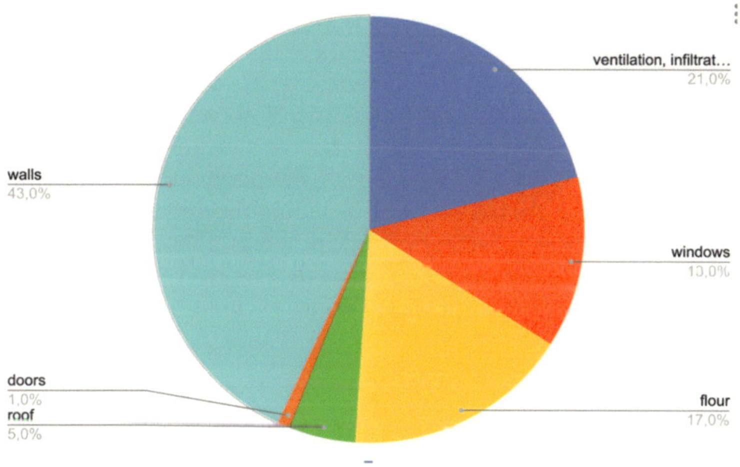

Figure 2.
Estimated distribution of heat loss through the building envelope after energy efficiency measures (Source: Authors).

emissions. The assessment of the modelling results allows us to identify specific measures aimed at improving the sustainability and environmental performance of the building. It was found that by implementing specific energy efficiency measures, such as improving insulation, installing high-efficiency heating and air conditioning systems, using renewable energy sources, and others, it is possible to significantly reduce energy consumption and, consequently, reduce emissions. These measures not only help to improve the energy efficiency of a building but also make it more resilient to climate change and other negative environmental impacts. In addition, they will make a significant contribution to reducing greenhouse gas emissions, contributing to a more sustainable development and conservation of natural resources.

As a result of the proposed measures, it is possible to increase the level of the energy certificate from class C to class B. This not only indicates an increase in the energy efficiency of the building but can also affect its value, attracting the attention of investors, tenants and other stakeholders. This approach demonstrates the importance and benefits of sustainable construction, which aims to reduce the carbon footprint and optimise energy efficiency.

6. Discussion and conclusions

Despite significant strides towards sustainability, Ukraine faces several barriers in adopting international sustainability standards such as LEED and BREEAM. These standards are globally recognised for assessing and certifying the environmental performance of buildings, yet their implementation in Ukraine is challenged by a combination of legislative, institutional, economic, and geopolitical factors.

1. Legislative and institutional barriers

   While Ukraine's legislative framework is gradually aligning with European Union standards, significant gaps remain. For instance, according to Directive 2010/31/EU, all new buildings in EU member states are required to meet near-zero energy consumption (NZEB) standards. However, Ukraine has yet to fully define and implement this criterion. The lack of clear legislative directives and standards hinders the widespread adoption of energy-efficient building practices. Moreover, institutional barriers, including the ongoing military conflict and economic challenges, further impede progress. These conditions limit the capacity to enact necessary reforms and to allocate the resources needed for the transition to sustainable building practices.

2. Economic and resource constraints

   The adoption of international sustainability standards requires significant financial investment and access to advanced technologies and materials. Ukraine's current economic situation, exacerbated by the ongoing conflict, limits both public and private sector investments in sustainability. The economic instability also affects the availability of skilled labour and the ability to import or develop the necessary materials and technologies to meet LEED or BREEAM criteria. This results in a slower pace of modernisation and energy efficiency improvements across the building sector.

3. Geopolitical challenges

   The geopolitical context, particularly the ongoing war with Russia, has had a profound impact on Ukraine's ability to implement modern energy-efficient technologies. The destruction of infrastructure due to military actions necessitates not only restoration but also modernisation to align with current energy efficiency standards. However, the need for immediate reconstruction often takes precedence over the long-term goal of achieving sustainability, thus delaying the adoption of international standards.

Recommendations for overcoming barriers:

1. Strengthening legislative frameworks: Ukraine needs to accelerate the alignment of its legislative framework with European standards. This includes defining NZEB criteria and ensuring that building codes and regulations reflect the latest sustainability standards. Clear and enforceable laws are crucial for encouraging the construction industry to adopt energy-efficient practices.

2. Institutional support and capacity building: Developing institutional capacity is essential for overcoming the barriers to implementing sustainability standards. This includes training professionals in the construction industry, increasing awareness of LEED and BREEAM standards, and providing technical support to developers and builders. Strengthening institutions will also involve creating incentives for the adoption of sustainable building practices, such as tax breaks or subsidies for green building projects.

3. Securing financial investments: Attracting both domestic and international investments is key to overcoming economic constraints. This can be achieved by creating a favourable investment climate through policy reforms and by securing financial support from international organisations and partners. Additionally, the promotion of public-private partnerships can mobilise the necessary resources for the large-scale adoption of sustainability standards.

4. Leveraging the reconstruction process: The reconstruction of infrastructure presents a unique opportunity to integrate sustainability into the rebuilding process. Ukraine should focus on using modern, energy-efficient technologies and materials in its reconstruction efforts. International aid and collaboration can play a critical role in this, providing both the financial resources and technical expertise required.

5. Promoting the use of "One-Click LCA": The "One-Click LCA" tool offers a streamlined method for assessing the energy efficiency of buildings, which is particularly relevant in the context of limited resources and urgent reconstruction needs. By adopting such innovative tools, Ukraine can accelerate the evaluation and implementation of energy efficiency measures, ensuring that new developments meet international sustainability standards.

This study highlights the importance of addressing the barriers to the adoption of international sustainability standards in Ukraine. By strengthening the legislative framework, building institutional capacity, securing investments, and leveraging reconstruction efforts, Ukraine can enhance the energy efficiency of its buildings and contribute to global sustainability goals. The findings of this research provide valuable insights for policymakers and practitioners, emphasising the need for a coordinated and comprehensive approach to overcoming the challenges of adopting LEED and BREEAM standards. Through such efforts, Ukraine can improve its energy independence, reduce its environmental impact, and set an example for other countries facing similar challenges.

7. Limitations of the study and future research directions

A limitation of this study is the relatively small number of buildings that can be assessed at present. This is due to the fact that the number of construction projects has decreased as a result of military operations and unsatisfactory socio-economic conditions in Ukraine. However, the practice of life cycle assessment for buildings constructed in Slovakia shows the prospects of this methodology. It is also worth noting that in the case of Ukraine, the regulatory framework is incomplete, and there are discrepancies between national building and certification standards. In the future, statistically significant measurements of indoor environmental quality parameters will be carried out, carbon footprint will be determined, and life cycle cost analysis will be carried out. The main goal of the study is to offer recommendations on how to achieve carbon neutrality throughout the life cycle of a building.

Acknowledgements

This publication has been possible thanks to funding granted by the "Consejería de Economía, Ciencia y Agenda Digital" (Ministry of Economy, Science and Digital Agenda) of Extremadura govern, and by the European Regional Development Fund of the European Union through the reference grants GR21135, Research Group on Environment and Spatial Planning.

Conflict of interest

The authors declare no conflict of interest.

Author details

Sérgio Lousada[1,2,3,4,5]*, Svitlana Delehan[6], Silvia Vilčekova[7] and Andrii Khorolskyi[8]

1 Department of Civil Engineering and Geology (DECG), Faculty of Exact Sciences and Engineering (FCEE), University of Madeira (UMa), Funchal, Portugal

2 VALORIZA—Research Centre for Endogenous Resource Valorization, Portalegre, Portugal

3 RISCO—Department of Civil Engineering, University of Aveiro, Aveiro, Portugal

4 CITUR—Centre for Tourism Research, Development and Innovation, Madeira, Portugal

5 Research Group on Environment and Spatial Planning (MAOT), University of Extremadura, Badajoz, Spain

6 Centre for Interdisciplinary Scientific Research, Uzhhorod National University, Ukraine

7 Institute for Sustainable and Circular Construction, Faculty of Civil Engineering, The Technical University of Košice, Košice, Slovakia

8 Laboratory of Mining Problems, Branch for Physics of Mining Processes of the National Academy of Sciences of Ukraine, Dnipro, Ukraine

*Address all correspondence to: slousada@staff.uma.pt

References

[1] Godinho J, Hoefnagels R, Braz CG, Sousa AM, Granjo JFO. An economic and greenhouse gas footprint assessment of international maritime transportation of hydrogen using liquid organic hydrogen carriers. Energy. 2023;**278**:13. DOI: 10.1016/j.energy.2023.127673

[2] Stojilovska A, Dokupilová D, Gouveia JP, Bajomi AZ, Tirado-Herrero S, Feldmár N, et al. As essential as bread: Fuelwood use as a cultural practice to cope with energy poverty in Europe. Energy Research and Social Science. 2023;**97**:5. DOI: 10.1016/j.erss.2023.102987

[3] Loures L, Castanho RA, Gómez JMN, Lousada SA, Fernández-Pozo L, Cabezas J, et al. Impactos Socioculturais da Cooperação Transfronteiriça (CT) no Espaço Europeu. Fronteiras. 2019;**8**(3):292-312. DOI: 10.21664/2238-8869.2019v8i3.p292-312

[4] Bazaluk O, Ashcheulova O, Mamaikin O, Khorolskyi A, Lozynskyi V, Saik P. Innovative activities in the sphere of mining process management. Frontiers in Environmental Science. 2022;**10**:878977. DOI: 10.3389/fenvs.2022.878977

[5] Barbosa R, Vicente R, Santos R. Climate change and thermal comfort in Southern Europe housing: A case study from Lisbon. Building and Environment. 2015;**92**:440-451. DOI: 10.1016/j.buildenv.2015.05.019

[6] Khorolskyi A, Hrinov V, Mamaikin O, Fomychova L. Research into optimization model for balancing the technological flows at mining enterprises. E3S Web of Conferences. 2020;**201**:01030. DOI: 10.1051/e3sconf/202020101030

[7] Lousada S, Cabezas J, Castanho RA, Gómez JMN. Land-use changes in insular urban territories: A retrospective analysis from 1990 to 2018. The case of Madeira Island—Ribeira Brava. Sustainability. 2022;**14**(24):16839. DOI: 10.3390/su142416839

[8] Lousada S, Silva PSD, Castanho RA, Naranjo-Gómez JM. Modelação de sistemas de abastecimento de água. O caso de Ilha da Madeira. Bitácora Urbano Territorial. 2019;**29**(2):89-98. DOI: 10.15446/bitacora.v29n2.69381

[9] Tsai SB, Xue Y, Zhang J, Chen Q, Liu Y, Zhou J, et al. Models for forecasting growth trends in renewable energy. Renewable and Sustainable Energy Reviews. 2017;**77**:1169-1178. DOI: 10.1016/j.rser.2016.06.001

[10] Xiao Y, Ma D, Zhang F, Zhao N, Wang L, Guo Z, et al. Spatiotemporal differentiation of carbon emission efficiency and influencing factors: From the perspective of 136 countries. Science of the Total Environment. 2023;**879**:163032. DOI: 10.1016/j.scitotenv.2023.163032

[11] Stankevich A. Explaining the consumer decision-making process: Critical literature review. Journal of International Business Research and Marketing. 2017;**2**(6):7-14. DOI: 10.18775/jibrm.1849-8558.2015.26.3001

[12] Petlovanyi M, Sai K, Malashkevych D, Popovych V, Khorolskyi A. Influence of waste rock dump placement on the geomechanical state of underground mine workings. IOP Conference Series: Earth and Environmental Science. 2023;**1156**(1):012007. DOI: 10.1088/1755-1315/1156/1/012007

[13] Delehan S, Melehanych H, Khorolskyi A. The Traditions and Technologies of Ecological Construction in Portugal. In: B: The 4th International Conference on Advances in Environmental Engineering. Basel, Switzerland: Licensee MDPI; 2023. p. 23. DOI: 10.3390/engproc2023057023

[14] Van Den Bergh JCJM, Grazi F. Reply to the first systematic response by the global footprint network to criticism: A real debate finally? Ecological Indicators. 2015;**58**:458-463. DOI: 10.1016/j.ecolind.2015.05.007

[15] Brienen RJW, Phillips OL, Feldpausch TR, Gloor E, Baker TR, Lloyd J, et al. Long-term decline of the Amazon carbon sink. Nature. 2015;**519**(7543):344-348. DOI: 10.1038/nature14283

[16] Kaposztasova D, Lavkova Cakyova K, Vertal M, Vranayova Z, Burdova EK. Active green constructions and their impact on gray infrastructure. Buildings. 2024;**14**(2):306. DOI: 10.3390/buildings14020306

[17] Cherubini F, Guest G, Strømman AH. Application of probability distributions to the modeling of biogenic CO_2 fluxes in life cycle assessment. GCB Bioenergy. 2012;**4**(6):784-798. DOI: 10.1111/j.1757-1707.2011.01156.x

[18] Vilcekova S, Burdova EK, Budajova J, Mesaros P. Building materials circularity of green roofs. In: Proceedings of CEE 2023; Lecture Notes in Civil Engineerin. Vol. 438. 2024. pp. 465-475. DOI: 10.1007/978-3-031-44955-0_47

[19] Fei R, Guo Y, Li J, Hu B, Yang L. An improved BPNN method based on probability density for indoor location. IEICE Transactions on Information and Systems. 2023;**E106.D**(5):773-785. DOI: 10.1587/transinf.2022DLP0073

[20] Wang J, Xu Z, Zheng X, Liu Z. A fuzzy logic path planning algorithm based on geometric landmarks and kinetic constraints. Information Technology & Control. 2022;**51**(3):499-514. DOI: 10.5755/j01.itc.51.3.30016

[21] Li J, Tang H, Li X, Dou H, Li R. LEF-YOLO: A lightweight method for intelligent detection of four extreme wildfires based on the YOLO framework. International Journal of Wildland Fire. 2024;**33**(1):044. DOI: 10.1071/WF23044

[22] Kochmar I, Karabyn V, Stepova K, Stadnik V, Sozanskyi M. Thermal impact on heavy metal bioavailability in burnt rocks of waste heap of chervonohradska coal-preparation plant (Lviv Region, Ukraine). Geomatics and Environmental Engineering. 2024;**18**(1):117-133. DOI: 10.7494/geom.2024.18.1.117

[23] Bondarenko VI, Kovalevska IA, Podkopaiev SV, Sheka IV, Tsivka YS. Substantiating arched support made of composite materials (carbon fiber-reinforced plastic) for mine workings in coal mines. IOP Conference Series: Earth and Environmental Science. 2022;**1049**(1):012026. DOI: 10.1088/1755-1315/1049/1/012026

[24] Zahrytsenko A, Podvigina O, Dereviahina N. Scientific and methodological foundations to develop numerical hydrodynamical models of mine fields in Donbas. E3S Web of Conferences. 2018;**60**:00034. DOI: 10.1051/e3sconf/20186000034

[25] Sodoma R, Sadura O, Kudak K, Kotys N, Panukhnyk O, Blishchuk K. Territorial communities of Ukraine in the context of decentralization. Management Theory and Studies for Rural Business and Infrastructure Development. 2023;**45**(3):306-313. DOI: 10.15544/mts.2023.31

[26] Shuryhin V, Karabyn V, Kuzyk A. Prediction of benzene migration parameters resulting from continuous flow in a Mountain River. Ecological Engineering & Environmental Technology. 2023;**24**(8):73-81. DOI: 10.12912/27197050/171529

[27] Mamaikin O, Sotskov V, Demchenko Y, Prykhorchuk O. Productive flows control in coal mines under the condition of diversification of production. E3S Web of Conferences. 2018;**60**:00008. DOI: 10.1051/e3sconf/20186000008

[28] Rudakov D, Inkin O. Evaluation of heat supply with maintaining a safe mine water level during operation of open geothermal systems in post-coalmining areas. Mining Of Mineral Deposits. 2022;**16**(1):24-31. DOI: 10.33271/mining16.01.024

[29] Plakitkina LS, Plakitkin YA, Dyachenko KI. World trends of coal industry development. Mining Industry Journal. 2019;**143**(1/2019):24-29. DOI: 10.30686/1609-9192-2019-1-143-24-29

[30] Mukhtarov S, Mikayilov JI. Could financial development eliminate energy poverty through renewable energy in Poland? Energy Policy. 2023;**182**:113747. DOI: 10.1016/j.enpol.2023.113747

[31] Azarova E, Jun H. Investigating determinants of international clean energy investments in emerging markets. Sustainability. 2021;**13**(21):11843. DOI: 10.3390/su132111843

[32] Aguirre M, Ibikunle G. Determinants of renewable energy growth: A global sample analysis. Energy Policy. 2014;**69**:374-384. DOI: 10.1016/j.enpol.2014.02.036

[33] Ragosa G, Warren P. Unpacking the determinants of cross-border private investment in renewable energy in developing countries. Journal of Cleaner Production. 2019;**235**:854-865. DOI: 10.1016/j.jclepro.2019.06.166

[34] Huenteler J, Niebuhr C, Schmidt TS. The effect of local and global learning on the cost of renewable energy in developing countries. Journal of Cleaner Production. 2016;**128**:6-21. DOI: 10.1016/j.jclepro.2014.06.056

[35] Suzuki M, Kanie N, Iguchi M. New approaches for transitions to low fossil carbon societies: Promoting opportunities for effective development, diffusion and implementation of technologies, policies and strategies. Journal of Cleaner Production. 2016;**128**:1-5. DOI: 10.1016/j.jclepro.2016.04.087

[36] Cabalu H, Koshy P, Corong E, Rodriguez UPE, Endriga BA. Modelling the impact of energy policies on the Philippine economy: Carbon tax, energy efficiency, and changes in the energy mix. Economic Analysis and Policy. 2015;**48**:222-237. DOI: 10.1016/j.eap.2015.11.014

[37] Shimbar A, Ebrahimi SB. Political risk and valuation of renewable energy investments in developing countries. Renewable Energy. 2020;**145**:1325-1333. DOI: 10.1016/j.renene.2019.06.055

[38] Shimbar A, Ebrahimi SB. The application of DNPV to unlock foreign direct investment in waste-to-energy in developing countries. Energy. 2017;**132**:186-193. DOI: 10.1016/j.energy.2017.05.098

[39] Morello TF. Carbon neutral merchant pig iron in Brazil: Alternatives that allow decoupling from deforestation. Energy for Sustainable Development. 2015;**27**:93-104. DOI: 10.1016/j.esd.2015.04.008

[40] Wang L, Skreiberg Ø, Becidan M, Li H. Sintering of rye straw ash and effect of additives. Energy Procedia. 2014;**61**:2008-2011. DOI: 10.1016/j.egypro.2014.12.063

[41] Wall R, Grafakos S, Gianoli A, Stavropoulos S. Which policy instruments attract foreign direct investments in renewable energy? Climate Policy. 2019;**19**(1):59-72. DOI: 10.1080/14693062.2018.1467826

[42] Tolliver C, Keeley AR, Managi S. Policy targets behind green bonds for renewable energy: Do climate commitments matter? Technological Forecasting and Social Change. 2020;**157**:120051. DOI: 10.1016/j.techfore.2020.120051

[43] Corrocher N, Cappa E. The role of public interventions in inducing private climate finance: An empirical analysis of the solar energy sector. Energy Policy. 2020;**147**:111787. DOI: 10.1016/j.enpol.2020.111787

[44] Opara V, Buzina I, Vynohradenko S. Environmental-economic efficiency of land use improvement reasoning. Geology Geography Ecology. 2017;**46**:21. DOI: 10.26565/2410-7360-2017-46-21

[45] Bera S, Rahut D. Technology development pathways: Enigmas of appropriate technology choice. Discover Sustainability. 2024;**5**(1):45. DOI: 10.1007/s43621-024-00222-5

[46] Jacques De Sousa L, Poças Martins J, Sanhudo L. The state of the sustainability factor for construction procurement: The case of Portugal. In: Lecture Notes in Civil Engineering. Vol. 444. 2024. pp. 576-586. DOI: 10.1007/978-3-031-48461-2_49

[47] Jolliet O, Frischknecht R, Bare J, Boulay AM, Bulle C, et al. Global guidance on environmental life cycle impact assessment indicators: Findings of the scoping phase. International Journal of Life Cycle Assessment. 2014;**19**(4):962-967. DOI: 10.1007/s11367-014-0703-8

[48] Da Costa JP, Silva AL, Barcelò D, Rocha-Santos T, Duarte A. Threats to sustainability in face of post-pandemic scenarios and the war in Ukraine. Science of the Total Environment. 2023;**892**:164509. DOI: 10.1016/j.scitotenv.2023.164509

[49] Zhabotynska S, Velivchenko V. New media and strategic narratives: The Dutch referendum on Ukraine—EU Association Agreement in Ukrainian and Russian Internet blogs. European Security. 2019;**28**(3):360-381. DOI: 10.1080/09662839.2019.1648253

[50] Widheden J, Ringström E. Life cycle assessment. In: Handbook for Cleaning/Decontamination of Surfaces. Amsterdam, Netherlands: Elsevier; Available from: https://linkinghub.elsevier.com/retrieve/pii/B9780444516640500218. [Accessed: April 14, 2024]

[51] Han Y, Zhang F, Huang L, Peng K, Wang X. Does industrial upgrading promote eco-efficiency?—A panel space estimation based on Chinese evidence. Energy Policy. 2021;**154**:112286. DOI: 10.1016/j.enpol.2021.112286

[52] Duleba A. Differentiated European integration of Ukraine in comparative perspective. East European Politics and Societies. 2022;**36**(2):359-377. DOI: 10.1177/08883254211005179

[53] Kirsch W. The distribution of power within the EU: Perspectives on a Ukrainian accession and a Turkish accession. International Economics and

Economic Policy. 2022;**19**(2):401-409. DOI: 10.1007/s10368-022-00541-w

[54] Anneveldt KJ, Nijholt IM, Schutte JM, Hehenkamp WJK, Veersema S, Huirne JAF, et al. Waste analysis and energy use estimation during MR-HIFU treatment: First steps towards calculating total environmental impact. Insights Into Imaging. 2024;**15**(1):83. DOI: 10.1186/s13244-024-01655-2

[55] Holubieva V, Akulenko V. Correlation of the International and National Institutional and Legal Regulations of applying some aspects of technical barriers in practices of Ukraine and the European Union. European Studies. 2020;**7**(1):325-359. DOI: 10.2478/eustu-2022-0060

[56] Barloese M, Petersen CL. Sustainable health care: A real-world appraisal of a modern imaging department. Clinical Imaging. 2024;**105**:110025. DOI: 10.1016/j.clinimag.2023.110025

[57] Global Footprint Network [Internet]. Data and Methodology. Available from: https://www.footprintnetwork.org/resources/data/ [Accessed: April 14, 2024]

[58] Lin D, Hanscom L, Murthy A, Galli A, Evans M, Neill E, et al. Ecological footprint accounting for countries: Updates and results of the National Footprint Accounts, 2012-2018. Resources. 2018;**7**(3):58. DOI: 10.3390/resources7030058

[59] Borucke M, Moore D, Cranston G, Gracey K, Iha K, Larson J, et al. Accounting for demand and supply of the biosphere's regenerative capacity: The National Footprint Accounts' underlying methodology and framework. Ecological Indicators. 2013;**24**:518-533. DOI: 10.1016/j.ecolind.2012.08.005

[60] Galli A, Wiedmann T, Ercin E, Knoblauch D, Ewing B, Giljum S. Integrating ecological, carbon and water footprint into a "footprint family" of indicators: Definition and role in tracking human pressure on the planet. Ecological Indicators. 2012;**16**:100-112. DOI: 10.1016/j.ecolind.2011.06.017

Chapter 3

Data-Driven Technologies for Eco-Efficient Dwelling: Light Steel Framing Volumetric Schemes

Sergio Russo Ermolli, Giuliano Galluccio and Simona Pugliese

Abstract

Data management during the entire building process using computational methodologies (CAD/CAM, building information modeling) represents the main issue to optimize resources, reducing wastes and delivery time. Considering the increasing demand of affordable housing, the scope of the chapter is to define a framework of possibilities for the use of volumetric system in cold formed steel to provide dwelling schemes, innovating the production, transport, and assembly phase through a close integration of the design and construction phases. A case study, still in progress, is presented. Developed with the Italian steel company Irondom Liguria, the volumetric scheme in cold formed steel technology is able to offer an experimental two-floor eco-efficient solution from small- to large-scale applications also in seismic areas.

Keywords: building information modeling, cold formed steel structures, off-site technologies, volumetric system, modular housing

1. Introduction

Construction with three-dimensional modules represented, among the various industrialized building systems, the one with the highest incidence of finishing and installation in factory and the least use of labor on site. In fact, the module represents the final result of the industrial prefabrication process: a box-shaped macro-component to put on-site completely finished and aggregated by juxtaposition and/or overlapping.

Since the beginnings of the Modern Movement, the idea of making dwellings through on-site assembly of as few factory-produced parts as possible has been a particularly evocative issue, capable of providing numerous insights into the creativity of designers. Taking this concept to extremes, the three-dimensional integrally prefabricated module to be juxtaposed and stacked on site to form multi-story buildings can be identified as the symbolic component of this trend [1]. A trend caused primarily by the desire to extend all the advantages of industrialized production to the residential construction sector, in order to obtain a high number of dwellings at a low price and in the shortest possible time [2]. This production trend has experienced considerable development since the mid-1960s, especially in the multifamily social housing sector. After a period of considerable euphoria in the late 1960s and mid-1970s toward such systems and, more generally, toward all prefabricated housing with large elements, the use of three-dimensional modules faded away [3]. The failure of the 'volumetric' 3D modular building could be understood in the wake of the broader and more generalized crisis of industrialization in the residential secotr since the early 1980s. This crisis might be related to the transformation of market conditions that effectively interrupted industrialization processes based boht on concentration and continuity of demand and on industrial "gigantism". To a greater extent than other systems, constructions made with three-dimensional modules have also shown obvious limitations and risks both morphologically and technologically, due to the repetitiveness of patterns, low flexibility of use, high formal predetermination of elements, and low level of performance offered [4].

Today, thanks to the potential offered by the so-called Digital Revolution, things seem to change.

2. Volumetric experiences in Industry 2.0

The experimental experiences carried out in the 1960s on the use of three-dimensional prefabricated modules to create multi-story residences—such as Moshe Safdie's Habitat '67 macrostructure or the cluster buildings of the Japanese Metabolists [5]—were expressions of creative individuality and unique prototypes, rather than concrete solutions to the problem of housing construction (**Figure 1**). Beginning in the 1960s, three-dimensional construction came out of the experimental phase, succeeding in taking advantage of the favorable conditions of that period: growing demand for low-priced housing, difficulty in finding site labor, and the availability, on the other hand, of significant amounts of affordable energy that made the industrial product competitive. Therefore, there has been a considerable development of three-dimensional element technology, mainly made of reinforced concrete and, to a lesser extent, steel, or wood, depending on the type of use and intervention [6].

Numerous manufacturers in different countries, while operating in very different socio-economic contexts, used three-dimensional reinforced concrete systems for new

Figure 1.
Left: Kisho Kurokawa, Nakagin Capsule Tower, Tokyo (1970. Demolished in 2022). Source: Yoshikazu Tsuno/AFP/ Getty Images. Right: Moshe Safdie, Habitat 67, Montréal (1967). Source: Getty Images/UIG.

multifamily social housing developments: the Sigma system in France, the Trucson in Great Britain, the Variel in Switzerland, the Ausa in Finland, and, finally, the Zanussi-Farsura in Italy (**Figure 2**). The modules of the various systems—whose weights ranged from 10 to 20 tons—were made up of five or more floors connected orthogonally to each other and were built with constant thicknesses in voluminous, hydraulically driven rotating formwork. In some cases, they were completed in the industry with all finishes (floors, wall coverings, fixtures, sanitary appliances, etc.). The modules were dry-assembled together by simple juxtaposition and overlapping, using centering pins or through wet casting. The modules were capable of bearing significant loads and transferring them to support points in the foundation or to another module.

Generally, the height of the buildings obtained by this type of assembly—unless using auxiliary support structures—did not exceed five to six stories, in order to limit the self-weight of the single elements and the load bearing on the units placed at the base of the building [7].

In addition to reinforced concrete, the production of three-dimensional elements involved the use of "lighter" materials, for permanent or temporary interventions,

Figure 2.
Example of Zanussi-Farsura volumetric system, Italy (1980–1982). Source: Studio Macola.

because of the requirements for transportability, mobility, and reuse arising from the weight and physical characteristics of the elements. Within the production of three-dimensional industrialized systems, the steel module has certainly represented a niche market, when compared with the large use of reinforced concrete modules. Compared to the latter, steel modules were lighter, easier to transport and move, and did not require demanding and expensive equipment during production. However, the heterogeneity and large number of parts constituting the module involved a complex series of operations to assemble them. The steel modules were usually made up of braced-element frames to which external and internal panels of different materials were connected, with the function of additional stiffening or simple envelope. The load-bearing structure was made by assembling, for the most stressed parts, hot-rolled elements in many cases with a closed section and lightweight cold-formed profiles as secondary elements [8].

The way the steel modules were assembled referred to two different types: the first involved the assembly of cantilevered modules from a main structure in which the vertical connections and service columns were located. This use—from a rather limited geographic and temporal point of view, since it was mainly developed in Japan in the 1970s—allowed for buildings with a high number of floors, the structure of the modules being decisive for overall stability, not so much the structure of the modules as that of the main construction that supported them. The second typology used the structural features of the module to make metal "wagons," capable of supporting both its own weight and limited loads, and which were used for low-density school or housing settlements, usually two-story, or, juxtaposed and stacked, for multi-story buildings.

Although it was possible in theory to implement a wide repertoire of typological and distributive solutions, in general, buildings formed by three-dimensional elements often presented repetitive and rigid design solutions, unable to correspond adequately to the specificity of the contexts in which they were located. Despite these limitations, the advantages of three-dimensional element technology were manifold. First of all, the transfer in factory of most of the operations for the realization of the building allowed greater control over the entire production process and, in particular, over some parts of the building—such as the MEP system—that required the intervention of numerous specialized operators. In addition, the speed of assembly made possible to cut construction time in half compared to traditional construction, resulting in significant economic savings. In fact, the substantial reduction in construction time allowed for the amortization of costs arising from production installations, transportation, and handling of elements. After a period of considerable euphoria—between the mid-1960s and mid-1970s—to such systems, the use of three-dimensional modules in the residential sector essentially disappeared.

The failure of "three-dimensional," which can be framed within the broader and more generalized crisis experienced by large-element prefabricated housing systems since the late 1970s, has numerous causes. These include the absence of continuous and stable housing demand, the rigidity of design schemes, the growing dissatisfaction with formal and performance outcomes, the limits in financing procedures, etc. Also, causes affecting the end of three-dimensional construction were the high costs of transportation, brought about by road transport regulations that set limits on the width of dimensions (in Italy, max 2.55 m), which can be exceeded only by "exceptional transport" escorted by traffic police. For the latter reason, the radius of action of the industries was around 100–200 km, beyond which the incidence of transport cost on the total of the operation increased significantly [9]. To overcome this problem, attempts had also been made to experiment with some technologies that tended to increase the size of the elements so as to reduce their number by making the maximum use of the capacities of the lifting equipment available at the construction site. But, regardless of some episodes of an emergency nature—as, for example, in the South of Italy in 1981/1985—the crisis of this prefabrication model appeared irreversible.

3. Volumetric construction in Industry 3.0

In the 1990s, industrial production of three-dimensional elements turns out to be mainly directed to the goods storage and transportation sector, shipbuilding, and various forms of housing assistance. In the latter sector, the one-piece corrugated steel container, entirely manufactured and assembled in the factory, represents the most widely used element in all those operations in which the properties of lightness, ease of transport and assembly, mobility, and reuse are fundamental.

The late 1990s saw a renewed interest in three-dimensional steel elements. In some countries, especially Japan, Scandinavia, and UK, the steel module continued to be a limited but interesting field of experimentation for both designers and industries. Starting precisely from the developments that have occurred in the housing container sector, research has been directed toward overcoming preconceived formal schemes and obvious performance limitations, in order to achieve unprecedented modes of expression and offer significant improvements in quality through the use of the increased potential of design and production processes.

From the technological point of view, the main innovation was the use for all structural parts of the module of cold formed steel (CFS) profiles, replacing traditional hot-rolled elements. These profiles are obtained by continuous profiling of galvanized steel coils of thickness from 1 to 3.2 mm and, assembled together, form different structural solutions [10].

The technical possibilities of forming CFS structural elements in relation to the optimization of stress responses and the use of high-strength steels make it possible to obtain modules that are lightweight, easy to handle, easy to transport, and easy to put in place. It is thus possible to have modular units with non-standardized dimensions, which can vary on demand and are mainly influenced by the maximum allowable transport dimensions without recourse to special permits.

Modules are made in the factory by assembling horizontal and vertical sandwich panels generally consisting of a grid of CFS braced profiles, with C or Z sections, forming the structural part of the system. The module can be subjected to more or less complete cladding, inspected, tested, covered with a reinforced plastic protective sheet, and finally transported to the construction site. The modules, containing one or more rooms, are lifted with small cranes, placed side by side and stacked, by juxtaposition and overlapping, on the top of each other to form buildings. Taking advantage of the strength by the form of metal "wagons" and the dimensional accuracy obtainable from manufacturing processes, the modules can form buildings—or their parts—up to four to five stories high [11].

In fact, the modules can be used without further structural modifications for buildings up to three to four stories; however, reinforcing and stiffening elements are required for taller realizations. Steel plates between 2 and 5 mm thick are placed in the foundation on which the first-level modules are placed. Using an alignment and gauging system, the modules of the upper floors are subsequently connected to those of the level below using special pins-guides and steel spacer plates. The presence of pre-drilled profiles at the corners of the module allows connection even horizontally. The MEP systems are already incorporated directly on site inside the sandwich panels that form the module; only the necessary connections are made on-site between the different modules. Construction therefore proceeds with speed and guarantees substantial time reductions, calculated at around 60%, compared to more conventional procedures [12].

The building is therefore completed using dry systems of cladding, according to the logic structure/envelope. The S/E construction uses dry connected layers: the outcome of a balanced overlap of prefabricated elements for which an exact response to physical stresses, calibrated layer by layer, as well as a planned maintenance according to the life cycle of the components, is foreseen. The S/E design and construction methodology is recognized as being capable of fostering a high degree of innovation in any intervention situation, as it is finalized to respond, with precision (and therefore with all the differentiations that from time to time ask the conditions of the context, the demands of the client, or the wishes of the designers) to specific design programs. A mode does not imply rigid sequencing or fixed hierarchies, rather a kind of network within which multiple paths can be traced with multiple combinations. Indeed, the S/E methodology can be seen as an approach of sequential logics typical of building industrialization, according to the idea of producing a building on site by assembling parts in rapid and economical ways.

It is important to emphasize the possibility of completing different building types using dry assembly practices that allow "dosing" the technological and environmental performance of the envelope according to design requirements. Drywall envelope and cladding operations generally involve the use of increasingly specialized and optimized catalog products offered by companies in the industry, but do not exclude the use of "customized" solutions. With a view to the broader customization of solutions, the experiences conducted highlight the multiple possibilities of formal and technological definition of envelope packages: a specific characterization obtained by applying the new interface tools between architectural design and industrial production capable of incorporating, not only virtually, products no longer rigidly bound by standardized shapes, geometries, and performance [13].

It is also possible to complete the construction using elements made with traditional techniques, or by executing on site even significant parts of the building such as, for example, the staircase-elevator block. Thus, there is an integration of construction elements with a high degree of precision and high-quality standard—the three-dimensional modules—with products and systems that are versatile and adaptable to different contexts, even with traditional techniques—the elements for completion. Such integration is perceived, mainly by users, as "added value," in that it is able to reconcile the demand for high-performance housing with the offer of a "home-product" from a formal point of view as far as possible from the container esthetic.

The numerous realizations carried out in those years highlight how the use of three-dimensional modules can also be developed in "semi-volumetric" form, that is, through combination with other construction procedures. This mode involves the use of three-dimensional modules only for those parts of the building—such as, for example, bathrooms and kitchens pods—that require more construction time and the intervention of highly skilled labor. In the latter case, buildings can reach even greater heights, as the overall stability can rely on the structural cooperation of different load-bearing systems.

The possibility of three-dimensional element systems to be integrated with different technologies intersected with an aspect that referred to the significant innovations that had occurred in manufacturing processes [14]. The use of modules, although strongly characterized from a formal and functional point of view, did not condition the design outcome in a decisive way, thanks to the possibilities offered in the industries by versatile production systems, capable of varying certain product characteristics on demand, in times and costs comparable to those of mass production. Indeed, using such systems makes it possible to quickly change the type of processing of the CFS elements that make up the airframe, offering diversified configurations of the final product. Although industries generally tended to supply prefabricated modules of standard dimensions (3.2–3.6 m wide and 8–12 m long), in reality, they were able to produce numerous types of modules with considerable margins of morphological, dimensional, and technological variability, depending, for example, on their different position in the building (at the base, top, or sides) or on the number and interior distribution [15].

The innovations that appeared in those years in construction techniques and production processes thus outlined a scenario that ended up significantly affecting design activity and its relationship to the industrialized product. In contrast to a period in which the design was also provided in the catalog, companies required the participation of the architect already in the initial phase of conception of the three-dimensional module. Through continuous dialog and discussion with the component industry, the designer would define the sizing of the module, the type of interior finishes, the interface with the exterior cladding, and the performance specifications. The outcome of this process, which can be described as "industrialized craftsmanship," was a building product whose formal, dimensional, and technological characteristics were "unique" while using automated production processes [16, 17].

It is no coincidence that the most interesting examples of multi-story constructions with steel modules appear in the United Kingdom: a country that historically has often played a proactive role with respect to technological innovations in the construction sector [18] and that in recent years, on the wave of the euphoria of the New Millennium and the revival of economic policy, has rewarded the architects most able to reconcile the conflicting relationship between tradition and innovation (**Figure 3**). The need to refer to foreign cases highlights the fact that in Italy, the exploitation of the potential offered by three-dimensional steel modules has not reached an adequate maturity and subsequent application, even if only experimental. Although there is a relevant presence of the Italian steel industry in Europe, this difficulty appears to be determined, not only by a limited knowledge on the part of construction contractors of the many performance and operational advantages ensured by steel but also by the permanence of a design culture closely linked to traditional construction technologies.

The considerable interest that has arisen in recent years around modular housing developments has contributed to the development of research and experimentation processes aimed at investigating the actual innovation margins of three-dimensional element technology, both referring to CFS modules and those made of wood and reinforced concrete. The latter are addressed, in particular, by Project Meteor, a British research project coordinated by Ove Arup, which uses the potential of computerized production systems to verify, through the creation of 1:1 scale prototype, the possibilities of modifying the morphological and dimensional characteristics of the modules, with a view to large-scale production [19].

The multi-story housing projects of those years highlighted how the evolution of the three-dimensional module could be read in terms of concomitant interferences between decision-making processes and innovations in production technologies, between experimental objectives and concrete market needs. It is precisely the high level of integration of these aspects that enabled a correct interpretation of three-dimensional prefabricated element technology and ensured a series of architectures of remarkable quality.

Figure 3.
Clockwise: Cartwright Pickard, Murray Grove, London, 1999 (source: Cartwright Pickard); AHMM, Raines Court, Stoke Newington, London, 2003 (source: AHMM); Feilden Clegg Bradley Studios, Beaufort Court, London, 2003 (source: FCB Studios); ShedKM Architects, Moho, Manchester, 2005 (source: ShedKM).

Among the various reasons that have contributed to the emergence of the use of CFS modules since the early years of the new millennium, the rapid spread of information technologies related to the manufacturing processes of thin steel elements should certainly be recognized. The use of technologies capable of responding to differentiated and time-varying demands is necessary as housing demand becomes increasingly interconnected with social, economic, and environmental aspects. The social complexity influencing the new housing issue is related to increasing phenomena, including an aging population, different forms of immigration, and diversification of family types. These categories require a housing supply with distinct types according to needs, but compatible with their economic resources. The structure of households is, in fact, progressively changing [20]: families are increasing, but their members are decreasing, resulting in the need for residential units that are functionally and dimensionally adequate, providing for the possibility, through appropriate additions, of being able to change housing over time as needed. Data from the last decade also show a particular social condition referred to as the "gray area," comprising individuals and households who are in an intermediate situation, that is, although they have income, they have difficulty finding a compliant housing solution in the private market. The overall performance required for housing from a dimensional and functional point of view intersects significantly with the requirements of overall environmental quality in terms of LCA, energy consumption, eco-efficiency, and the use of low-impact techniques and materials [21, 22]. The innovations happened in the digital environment in recent years define a scenario capable of proposing data-driven housing solutions that are more in line with these demands [23].

4. Design, simulation, fabrication: An Industry 4.0 perspective for volumetric construction

The new opportunities offered by digital innovation include a wide range of computational technologies for information management and architectural modeling through data-driven processes [24], such as building information modeling (BIM) [25]. BIM-based methodologies are incrementally, but deeply, reorienting the design practice toward the streamline of the decision-making processes and the information flows [26]. Moreover, the integration of BIM with file-to-factory technologies allows to manufacture highly specialized components, with a theoretical infinite degree of difference among the elements, at no added cost compared to traditional fabrication processes, with the "promise" to overcome the traditional dichotomy between craftsmanship and industry [27–29].

This new condition put designers and companies to cooperate, so to adjust the production line addressing with the customers' requirements and preferences. Also, the predictive potential of BIM and of the use of digital tools defines a design and manufacturing methodology that leverages simulation to generate iterative feedback loops and, so, to inform the design revision cycle [30]. Within this information flow, the building components to be manufactured and assembled are considered as data, and the revisions can be assisted by computer systems deputed to analyze and interpret the results, systematizing these data and providing inputs that can be read and implemented using the very same tools. In this way, file-to-factory workflows can be a solution to the standardization/customization dichotomy that has always limited the widespread of off-site prefabricated solutions, despite their benefits in terms of cost- and time-effectiveness, on-site impacts reduction and quality assurance [31].

More in detail, the benefits of a digital approach to volumetric construction include the following:

- Customization: production according to the demand, that can be exploited in the form of interoperable BIM Library, from which the designers themselves can select components, modify them within defined constrains, place them directly into their models and modify or substitute them;

- Simulation: performance control throughout the building or component life cycle, that allows to select and test technical solutions, products and materials based on the lowest environmental impact and verifying their effectiveness through environmental simulation;

- Automation: for the forming of the elements through CAD/CAM workflows and the integration of operations with CNC tools to increase precision, reduce waste and speed up manufacturing;

- Optimization: for the improvement of the assembly phase and the minimization of errors, extrapolating the drawings from the BIM model or directly linking the design models with a robotic assembly line;

- Management: as the ability to anticipate processes implicit in BIM, especially in the context of time control (BIM 4D) and quantity take-off (BIM 5D), it allows for a more effective control of material purchases and waste reduction.

Within this logic, the volumetric module represents no more a fixed constraint, but a parametric "genotype," that is, an elementary module from which the different design, technological, and aggregation possibilities can be explored in relation to quantitative and qualitative parameters, for example, number and typologies of users, height, area, architectural layout, interior and exterior materials, etc. [32].

An application of this concept has been tested in 2018 within the Eccelsa Research Project [33, 34], carried out by the Department of Structures for Architecture and Engineering and the Department of Architecture, University of Naples Federico II (IT) in partnership with Lameriedil, a specialized company for the production of lightweight modular housing.

Starting from a basic "cell," consisting of a volumetric steel frame module, it is therefore possible to implement numerous types of housing, different in terms of module aggregation, enveloping systems and finishes. In this process, the "generative" possibilities of computational technologies support the design phase, thanks to their ability to simultaneously manage multiple design parameters, with the aim of developing a range of possible design solutions (design optioneering), according to a logic of multidimensional analysis (**Figure 4**).

This concept has been further developed within the ReLUIS Research Project (2020–2021). The activity was carried out by the Department of Structures for Architecture and Engineering and the Department of Architecture, University of Naples

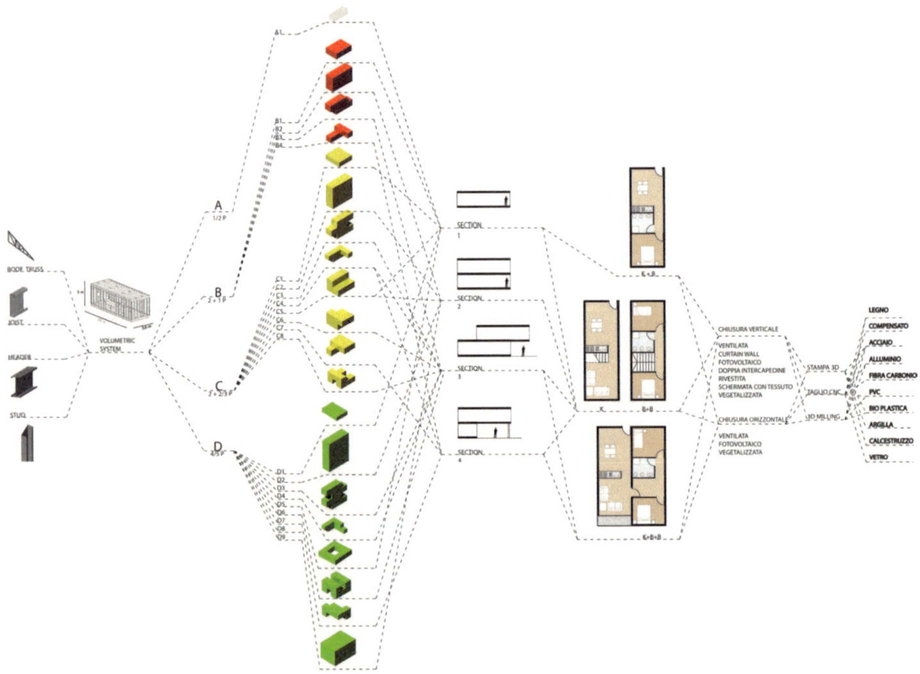

Figure 4.
Range of possible design solution starting from a single volumetric "genotype" developed within the Eccelsa Research Project. Source: Research Studies, Eccelsa/DiARC.

Federico II (IT) in partnership with the Department of Civil Protection in Italy and the ReLUIS consortium. The focus was to test a BIM-based methodology for the development of a post-disaster, seismic-resilient lightweight modular housing system, that could be adapted to different environmental, structural, or functional requirements depending on the site conditions [35]. In particular, the following example demonstrates the application of BIM technologies for the design of a volumetric steel frame module. In the first step, once the dimension of the "generator" volumetric module is established, for example, based on transport constraints (generally of the ordinary type), the basic structural elements are modeled, that is, the horizontal guides, the vertical rods, the edge beams, and the internal beams. The modeling of the components is carried out to create parametric elements, without having to draw the elements or even the structure itself from the beginning. In fact, by considering the elements as belonging to the same "family" (set of elements that share intrinsic characteristics), the different parametric types are created, whose dimensions are variable coefficients, which can be modified in real time (**Figure 5**).

The individual "instances" placed in the model therefore always refer to the type to which they belong and respond to the changes that, upstream, occur in the main characteristics of the type itself. In the case of more complex elements, such as trusses, the modeling first includes their components, usually C-, U-, or omega-shaped profiles, which are then introduced into the modeling environment of the beams themselves, thus constituting what is called a "nested family." The same principle used for structural elements also applies to enclosure elements: thanks to parametric modeling, it is in fact possible to instantly control the consequences that certain technical and design choices have on the thermo-hygrometric performance of the envelopes, allowing the performance of the building to be constantly monitored. In this way, the operation of optioneering, that is, evaluation between several design hypotheses, is also facilitated, based on the best possible choice among a range of variables. Finally, the placing of parametric objects such as doors and windows is facilitated by the parametric properties of the structural and closing elements modeled in this way: the load-bearing structure is able, on the basis of precise "genotypic" indications, provided to the relative family, to automatically modify its layout to accommodate, for example, a window, automatically generating, once the opening has been drawn, the perimeter

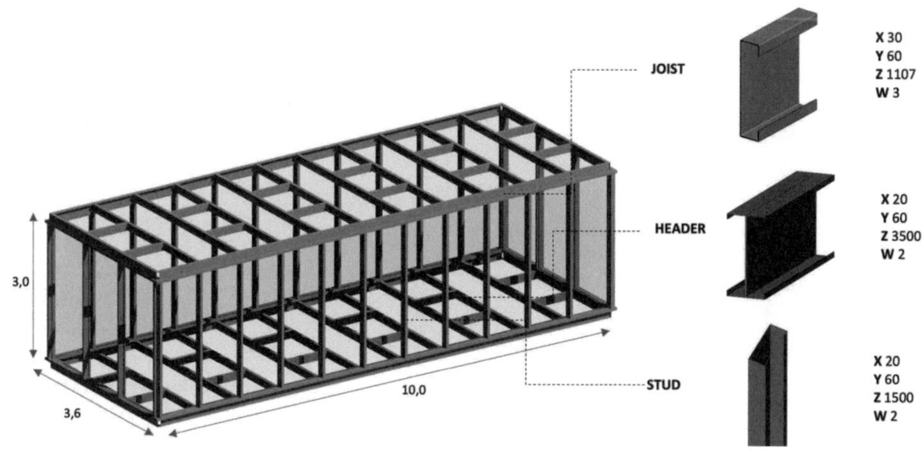

Figure 5.
BIM Modeling of a steel framing volumetric module, developed within the ReLUIS Research Project. Source: Research Studies, ReLUIS/DiARC.

profiles that contain it. The relationships that are created between the objects guarantee mutual "responsiveness" to the changes made (**Figure 6**).

Using a BIM modeling software, it is possible to retrieve, starting from a single file, all the drawings, models, and documentation useful for implementing the design

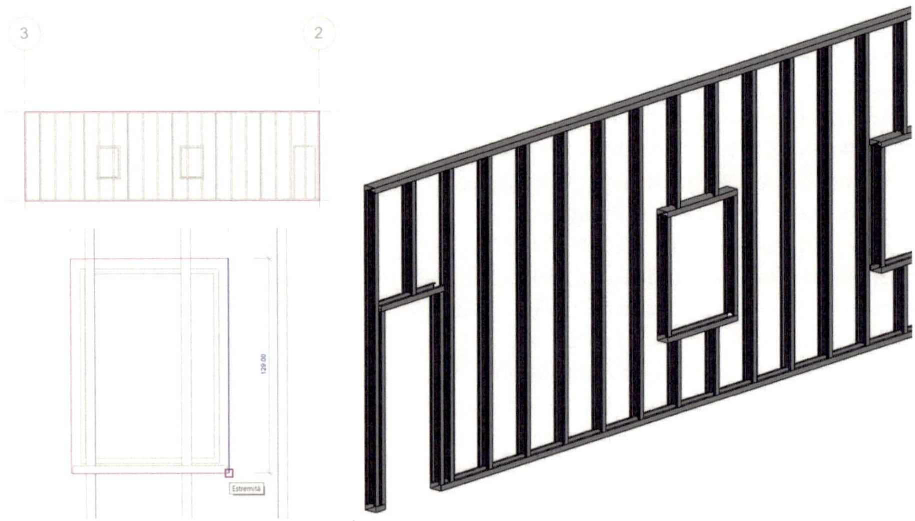

Figure 6.
Automatic adjustment of the steel frame grid according to the position of windows and doors element in the panel, developed within the ReLUIS Research Project. Source: Research Studies, ReLUIS/DiARC.

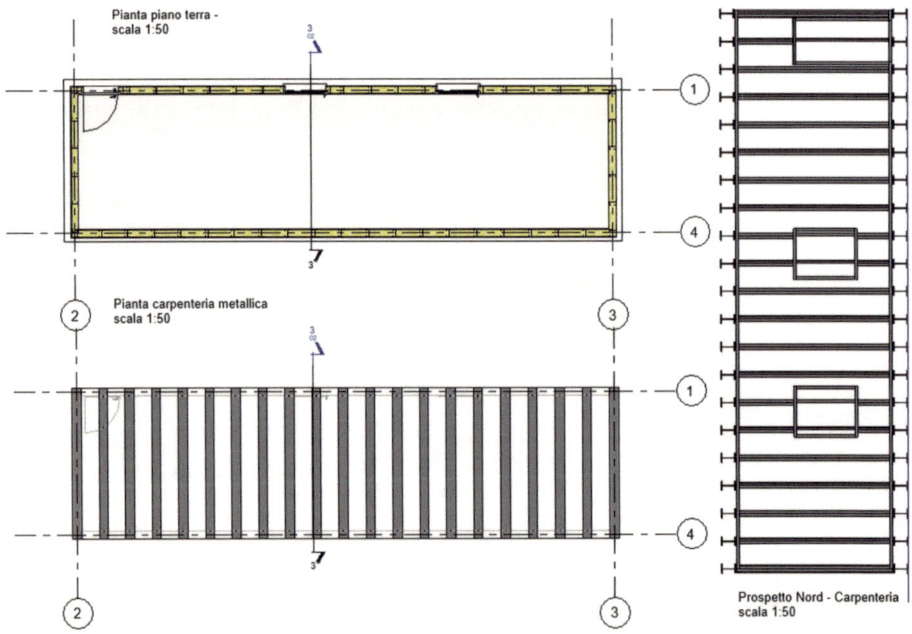

Figure 7.
Project sheet with automatically retrieved floor plan, section and front view drawings, developed within the ReLUIS Research Project. Source: Research Studies, ReLUIS/DiARC.

of the building. Within this parametric environment, the architectural volumes are modeled by importing "traditional" two-dimensional drawings, to which the structural components are connected, via "link," again modeled with the same software application, but at a different time. The project is thus detailed, gradually, from the first conceptual phases to the definition of the finishes, materials, furnishings, as well as the stratigraphy of the opaque and transparent, vertical, and horizontal envelope. Plans and sections are extrapolated from the three-dimensional model, and thanks to the property defined as "bidirectional associativity," all the changes made within a view (which can be defined as a real shot of the same subject) are automatically reflected in the others. This eliminates the need to update, at each revision, all the drawings already produced, thus making all the drawings congruent with each other and allowing considerable time savings, as well as greater accuracy of the graphic drawings (**Figure 7**).

Using the "Schedule" tool, it is possible to count the materials in the project, which are updated in real time with the progress of the project. In addition, since schedules are subject to the same principle of bidirectional associativity, that is, since these functions are also assimilated to common project "views," it is possible to make design changes (for example, by selecting a different type of profile or frame), which are instantly reflected in all drawings and in the model. However, the traditional two-dimensional drawing tools are also associated with the three-dimensional modeling functions, so that it is possible to add detailed drawings to scale to the model, such as for the definition of working and construction drawings (**Figure 8**).

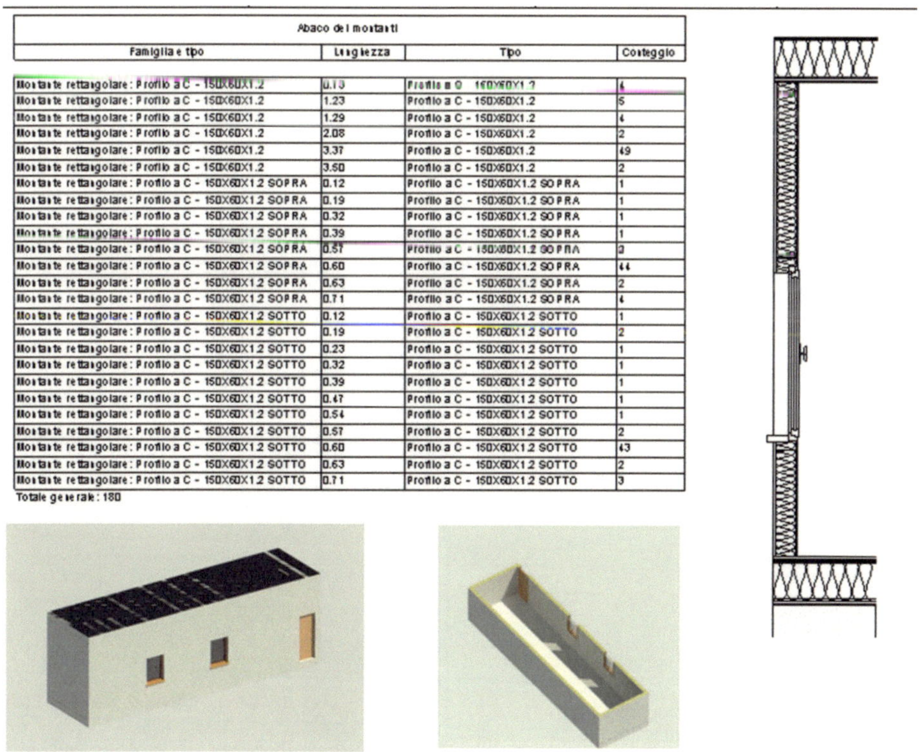

Abaco dei montanti

Famiglia e tipo	Larghezza	Tipo	Conteggio
Montante rettangolare: Profilo a C - 150X60X1.2	0.13	Profilo a C - 150X60X1.2	4
Montante rettangolare: Profilo a C - 150X60X1.2	1.23	Profilo a C - 150X60X1.2	5
Montante rettangolare: Profilo a C - 150X60X1.2	1.29	Profilo a C - 150X60X1.2	4
Montante rettangolare: Profilo a C - 150X60X1.2	2.08	Profilo a C - 150X60X1.2	2
Montante rettangolare: Profilo a C - 150X60X1.2	3.37	Profilo a C - 150X60X1.2	49
Montante rettangolare: Profilo a C - 150X60X1.2	3.50	Profilo a C - 150X60X1.2	2
Montante rettangolare: Profilo a C - 150X60X1.2 SOPRA	0.12	Profilo a C - 150X60X1.2 SOPRA	1
Montante rettangolare: Profilo a C - 150X60X1.2 SOPRA	0.19	Profilo a C - 150X60X1.2 SOPRA	1
Montante rettangolare: Profilo a C - 150X60X1.2 SOPRA	0.32	Profilo a C - 150X60X1.2 SOPRA	1
Montante rettangolare: Profilo a C - 150X60X1.2 SOPRA	0.39	Profilo a C - 150X60X1.2 SOPRA	1
Montante rettangolare: Profilo a C - 150X60X1.2 SOPRA	0.57	Profilo a C - 150X60X1.2 SOPRA	2
Montante rettangolare: Profilo a C - 150X60X1.2 SOPRA	0.60	Profilo a C - 150X60X1.2 SOPRA	44
Montante rettangolare: Profilo a C - 150X60X1.2 SOPRA	0.63	Profilo a C - 150X60X1.2 SOPRA	2
Montante rettangolare: Profilo a C - 150X60X1.2 SOPRA	0.71	Profilo a C - 150X60X1.2 SOPRA	4
Montante rettangolare: Profilo a C - 150X60X1.2 SOTTO	0.12	Profilo a C - 150X60X1.2 SOTTO	1
Montante rettangolare: Profilo a C - 150X60X1.2 SOTTO	0.19	Profilo a C - 150X60X1.2 SOTTO	2
Montante rettangolare: Profilo a C - 150X60X1.2 SOTTO	0.23	Profilo a C - 150X60X1.2 SOTTO	1
Montante rettangolare: Profilo a C - 150X60X1.2 SOTTO	0.32	Profilo a C - 150X60X1.2 SOTTO	1
Montante rettangolare: Profilo a C - 150X60X1.2 SOTTO	0.39	Profilo a C - 150X60X1.2 SOTTO	1
Montante rettangolare: Profilo a C - 150X60X1.2 SOTTO	0.47	Profilo a C - 150X60X1.2 SOTTO	1
Montante rettangolare: Profilo a C - 150X60X1.2 SOTTO	0.54	Profilo a C - 150X60X1.2 SOTTO	1
Montante rettangolare: Profilo a C - 150X60X1.2 SOTTO	0.57	Profilo a C - 150X60X1.2 SOTTO	2
Montante rettangolare: Profilo a C - 150X60X1.2 SOTTO	0.60	Profilo a C - 150X60X1.2 SOTTO	43
Montante rettangolare: Profilo a C - 150X60X1.2 SOTTO	0.63	Profilo a C - 150X60X1.2 SOTTO	2
Montante rettangolare: Profilo a C - 150X60X1.2 SOTTO	0.71	Profilo a C - 150X60X1.2 SOTTO	3

Totale generale: 180

Figure 8.
Example of a project sheet with automatically retrieved quantity take-off, detailed section and views of the steel frame module, developed within the ReLUIS Research Project. Source: Research Studies, ReLUIS/DiARC.

With the possibility of having a federated BIM model, that is, multidisciplinary, it is possible to modify the elements by providing, for example, slots in the structural profiles for the passage of technological systems. These slots, having to be positioned at precise points of the profiles, for structural reasons, are generated within the families of the modeled elements. These families can be exported in interoperable exchange formats (.dxf, for example) and processed inside CNC machines for automated manufacturing.

The BIM model therefore becomes, even at the scale of a small single-family residential building, a real digital archive, containing all the project information and above all interoperable with all the software for the simulation of the construction phases of the construction site, for clash detection, for cost estimation and automatic processing of bills of quantities, as well as for energy analysis and facility management [36].

5. iHomes industrial experimentation

Teaching and research activities developed within the Department of Architecture at the University of Naples Federico II (IT) include studies and experimentation focused on identifying the innovation drivers, the opportunities, the constraints and the strategies aimed at enhancing the CFS modular construction in the Italian building sector. In particular, the course Technological and Environmental Design Studio is finalized to develop eco-efficient off-site building design with cold formed steel structures and dry-construction envelopes (**Figure 9**). Since 2023, the Department of Architecture has been carrying out an activity of scientific support to the company

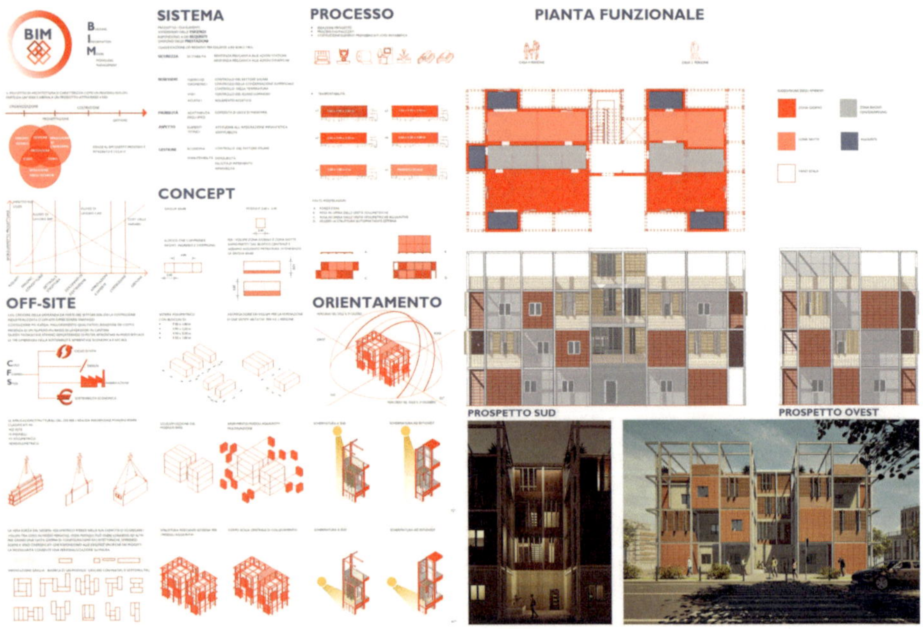

Figure 9.
Design solution of a housing scheme in volumetric light steel framing. Source: Teaching Studies, prof. S. Russo Ermolli, with S. Pugliese, G. De Luca. Students: S. Borrelli, F. Ceriello, G. Cinque (2024).

Irondom Liguria s.r.l. for the development of eco-efficient and seismic-resistant multi-story housing solutions made with light steel framing dry volumetric technologies, aimed at the technological optimization of production and construction processes, in a life-cycle logic, through digital methodologies. This activity stems from the Italian company's need to propose on the market a housing typology that can meet the variability of demand, completely off-site, customizable according to the specific requirements, capable of eliminating any risk of error during the construction phase and optimizing transport and assembly times. The company's previous volumetric experiences are aimed at the construction of single-story structures for school, residential, and commercial use: on the contrary, the type of demand that Irondom Liguria intends to respond to is that related to multi-story housing solutions, capable of advanced structural performance, in view of the presence in Italy of vast areas of high seismic risk.

Within this framework, the study has therefore identified the main aggregative logics starting from a basic module, capable of providing answers to multiple housing needs, also using additional modules (according to a plug-in logic) or of independent integrative structures connected to the facades. Technological experimentation was developed through a real BIM-oriented workflow, through the modeling of a volumetric module in CFS, with dimensions 4.70 × 12.77 × 3.40 m. This module is configured as an "open" model, that is, whose peculiar characteristic lies in being customizable, as well as for housing interventions at different scales. Therefore, a specific workflow has been defined for the design and realization of the building components, considering the constraints of the volumetric system as input and exploiting the generative possibilities of parametric flows for the simultaneous management of different hypotheses and product varieties.

The experimentation (still in progress) consisted of the following phases:

1. Elaboration of the design process at its different scales, of eco-efficiency characteristics of the housing system and of earthquake-resistant structural performance;

2. Development of a functional-spatial solution on two floors;

3. Modeling of structure and envelope elements;

4. Performance simulation and feedback analysis;

5. Drafting of working drawings (assembly kits) and preparation of files for the CAD/CAM production line;

6. Development of a catalog of customizable solutions, depending on geographic and climatic contextualization.

Starting from the basic volumetric module, defined in both its structure and envelope elements, the study is focused on the aggregative characterization of the project, which was done through a selection from different options (**Figure 10**).

A specific aggregative configuration has been chosen, related to the distributional aspects of the interior spaces, opaque and transparent envelope, furnishings and finishes, which were controlled by drawing up automated component abacuses and tables with two-dimensional construction details. The B option (**Figures 11** and **12**)

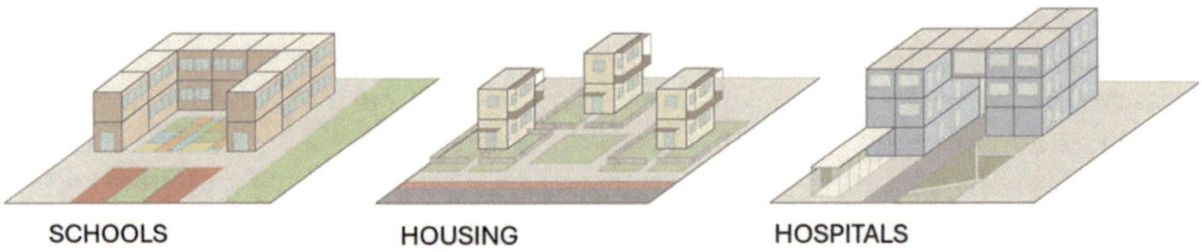

Figure 10.
Some possible aggregation of the modules. Source: Research Studies, S. Russo Ermolli, S. Pugliese with L. Fiorino, G. Maisto (2024).

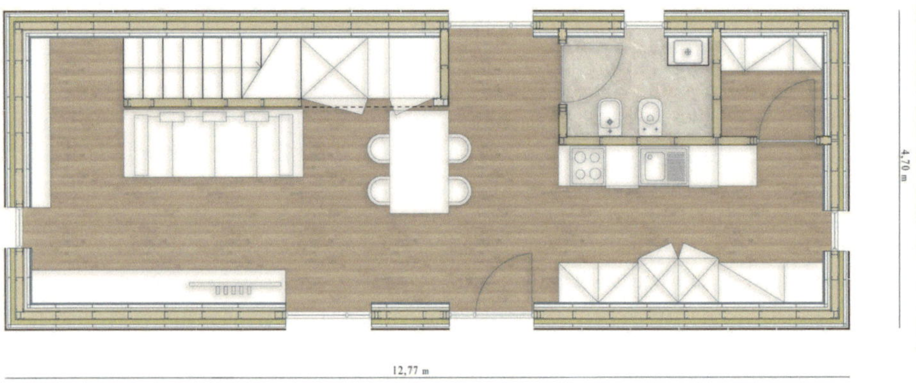

Figure 11.
Ground floor plan, day area. Source: Research Studies, S. Russo Ermolli, S. Pugliese with L. Fiorino, G. Maisto (2024).

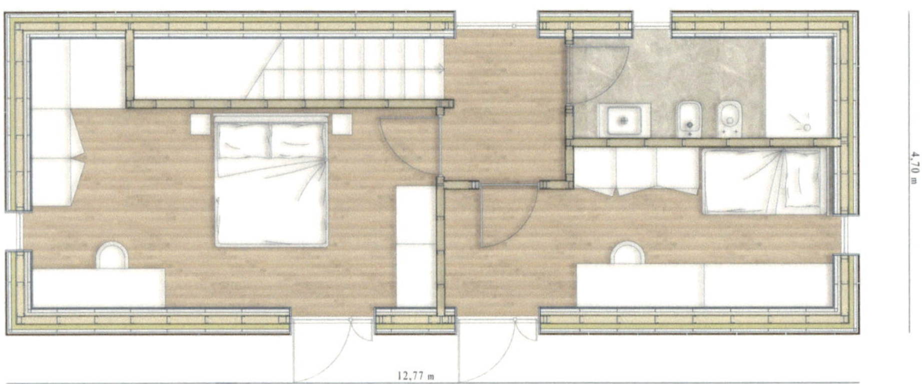

Figure 12.
First floor plan, night area. Source: Research Studies, S. Russo Ermolli, S. Pugliese with L. Fiorino, G. Maisto (2024).

has been selected for the experimentation, consisting of two volumetric modules arranged on two levels: the ground floor for the day area and the upper floor coinciding with the night area (**Figure 13**).

The building has an area of about 60 m^2, occupying a total volume of about 400 m^3. The large windows on the main front, protected by movable wooden brise-soleil elements, ensure the optimization of natural contributions and the control of solar radiation during the summer. The opposite front, on the other hand, has smaller

Figure 13.
Longitudinal section of the module. Source: Research Studies, S. Russo Ermolli, S. Pugliese with L. Fiorino, G. Maisto (2024).

windows useful for cross-ventilation. On the upper floor, a door gives access to a balcony on a self-loading steel structure that, on the ground floor, defines a paved space for outdoor activities. The roof, on which it is possible to place about 45 m^2 of photovoltaic panels and solar collectors, is presented, in this solution, with lattice trusses that give an inclination to the roof. The roof is completed by sandwich panels with rock wool insulation and covered externally with a galvanized steel sheet. The presence of a solar chimney in the roof will allow for increased air circulation to be used for cooling, taking advantage of natural ventilation mechanisms to dispose of warm air inside the rooms.

An important contribution in the control of the indoor microclimate and energy consumption is made by a heat recovery ventilation system (MVHR), which allows fresh air previously filtered from the outside to be taken from the building and, in winter, to be preheated before being fed into the interior rooms, avoiding the risk, on particularly cold days, of creating draughts in the rooms. The building will also be able to be integrated with a system for the recovery and reuse of rainwater and gray water, as well as an "intelligent" system capable of managing microclimatic and lighting conditions, as well as monitoring energy consumption and all possible entering via web connection (**Figure 14**).

The vertical and horizontal envelope of the building is designed according to the structure/envelope logic, that is, using dry construction procedures that allow to "dose" the technological and environmental performance according to the design requirements. In the S/E logic, it is possible to take advantage of a particularly wide range of elements for the envelope: in this specific case, the solutions developed

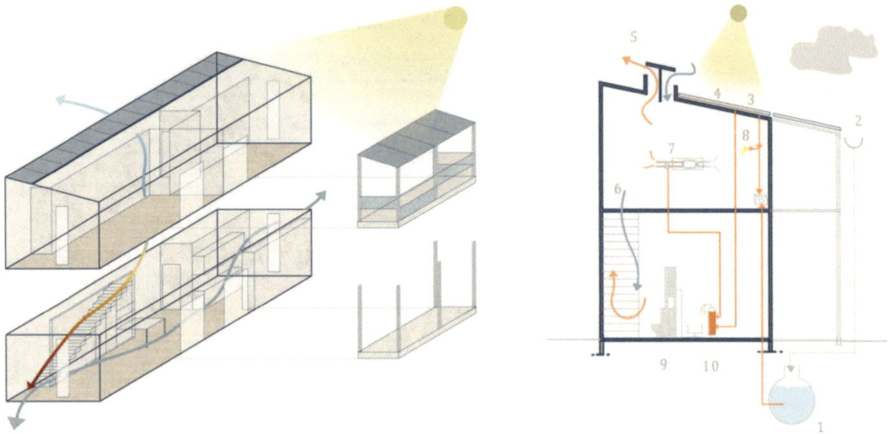

Figure 14.
Environmental issues schemes. 1. Rainwater storage tank and recycling system; 2. Rainwater collection; 3. Solar thermal array; 4. Building integrated PV; 5. Windcatcher; 6. Free air movement; 7. MVHR; 8. Energy efficient lighting; 9. Water efficient sanitary and appliances incorporating gray water recycling; 10. Heat store. Source: Research Studies, S. Russo Ermolli, S. Pugliese with L. Fiorino, G. Maisto (2024).

as part of the research activity with Irondom Liguria were used. The stratigraphy chosen for the vertical walls involves the use of two OSB panels of 15 mm, of double insulation layer consisting of a 100 mm rock wool thermal layer and a 45 mm mineral wool thermoacoustic layer. On the outer part there is a thermal insulation layer consisting of a 70 mm polyurethane foam panel covered with embossed aluminum as a vapor barrier. The panel includes steel profiles with 43 mm ventilation chamber for fixing the overlying extruded brick slabs. On the interior side, the wall is completed with double plasterboard panels, to be finished with plastering and painting, complete with support structure and the cavity for the systems. The sloping roof slab with CFS structure is made of joists and related transverse connections, made of lightweight S350 GD steel profiles. Composed on the exterior of double OSB panel of 15 mm, rigid wood cement panel of 22 mm, and heat-insulating roof panel of 180 mm (including 140 mm minimum insulation), made of metal alloy (corrugated type) and is finished inside with 45 mm mineral wool sound-absorbing layer and plasterboard panel complete with support structure. Vertical and horizontal envelope ensure high performance not only in relation to thermal transmittance but also in relation to its behavior in terms of thermal inertia, phase shifting, and attenuation of thermal wave (**Figures 15** and **16**).

Specific considerations were made with regard to anti-seismic performance. In particular, the geometric configuration and mechanical performance of inter-module connections play a key role in the overall structural performance, considering the overlap of two modules in a geographical context characterized by high seismicity [37]. Bolting over welding appears to be preferable for connections due to its characteristics, such as ease of assembly and disassembly, maintenance (the bolts can be replaced without intervening on the entire structure), safety (bolted connections offer high mechanical performance even under heavy loads). Therefore, it was found that it is necessary to ensure the stiffness and strength of connections between modules as they directly affect internal load transfer and strain coordination between adjacent modules, the properties of which will in turn affect the lateral sway performance of the entire building.

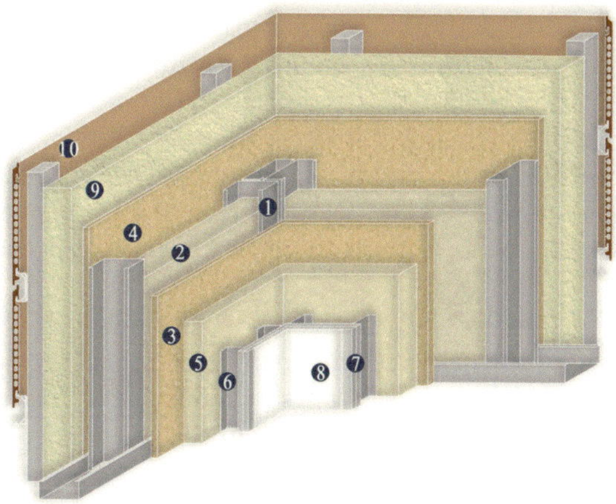

Figure 15.
BIM modeling stratification of iHomes envelope. Source: Research Studies, S. Russo Ermolli, S. Pugliese with L. Fiorino, G. Maisto (2024).

Figure 16.
BIM modeling of the steel framing volumetric iHomes module. Source: Research Studies, S. Russo Ermolli, S. Pugliese with L. Fiorino, G. Maisto (2024).

Although the research process of the structural dynamics of the two staked modules is still on the way, the modeling on Revit software of a simulative prototype is nearly complete. The digital archive contains all project information and, above all, will be interoperable with all software for the simulation of construction site phases, the cost estimation and the automatic processing of metric calculations, as well as for energy analyses and facility management.

6. Conclusions

Although the production phase of the volumetric system has not yet started, the work of the research group includes the development of support activities for the acquisition of production factors and is involved in supporting a marketing and promotion strategy of the results of the scientific activity (web-publishing, brochures, catalogs) and the dissemination of the research results, through the publication of scientific articles, as well as the organization of seminars and presentation occasions. The work around the prototype iHomes is also finalized to show the possibility of different aggregation and combination (using the vast catalog of integrative part, like balconies, loggias, walkways, shading systems, etc.) for the definition of a variety of accommodation schemes: schools, social housing, temporary settlements, mixed-use buildings, etc. (**Figures 17** and **18**).

The ongoing experience can stimulate anyway some reflection. The continuous and physiological evolution of cities and of the needs related to living in recent years has undergone a strong acceleration due to the intense changes linked to the economic crisis, the urgency of the climate challenge and the consequences of the pandemic. If, on the one hand, ever-new segments of the population are excluded from the housing market and the possibility of accessing accommodation, on the other hand, the current offer often has difficulties to adapt to the demands coming from the new ways of living in the city and the changed configurations of family units. Even the concept of housing as an immovable, long-lasting, and handed down asset is today in crisis: the need and desire to move, policies to reduce land and resource consumption lead to questions about the contemporary housing project.

It is therefore urgent to seek solutions that can provide rapid responses to these needs, in terms of design and production. To meet the projection in terms of population number and urbanization [38], the building sector needs to become far more productive. Although construction is currently the least productive manufacturing sector, a UK government report determined that construction productivity could improve by 60% simply by undertaking at least 70% of activities off-site prior to construction, where factory-like conditions provide greater degrees of safety and control [39].

In Italy, the subject of building industrialization, despite perplexities, expectations, and disappointments, remains central to the physicality of the material culture of building, despite the awareness of the need to introduce, increasingly immaterial

Figure 17.
Render of iHomes protype. Source: Research Studies, S. Russo Ermolli, S. Pugliese with L. Fiorino, G. Maisto (2024).

Figure 18.
Possible iHomes aggregation. Source: Research Studies, S. Russo Ermolli, S. Pugliese with L. Fiorino, G. Maisto (2024).

values. If before, the type of strategy was aimed at seriality to contain time and costs, today it seems to be going in the direction of versatility and customization, according to a logic that leaves standardization only to file formats (immaterial), while allowing great variety to production and prototyping (material). In other words, the most plausible scenario for industrialization seems to be no longer in heavy prefabrication, but in lean management and manufacturing systems, guided by the principles of a discipline, the so-called Knowledge Management, through the new tools for managing information and computation of processes.

The binding combination of in-depth technological design and accurate producing methods is making it possible to assess, model and simulate the final behavior of the adopted solutions with high accuracy, to respond in a precise and efficient manner to the project requirements (among which those referring to energy efficiency and minimization of environmental impact are priorities), translating the work of the operators involved in the process into a significant reduction in the use of material and immaterial resources, in the control of the quality of the work and in the reduction of construction and management costs.

Authors' contributions

Sergio Russo Ermolli wrote Sections 1–3 and Conclusions; Giuliano Galluccio wrote Section 4; Simona Pugliese wrote Section 5.

Author details

Sergio Russo Ermolli[1*], Giuliano Galluccio[1] and Simona Pugliese[2]

1 Department of Architecture, University of Naples Federico II, Naples, Italy

2 TechOffice, Naples, Italy

*Address all correspondence to: sergio.russoermolli@unina.it

References

[1] Gerola L, Biondo G, Chiara L, Travi V. La Cultura del "Tridimensionale". Edilizia Industrializzata con Elementi Cellulari. Milano: BE-MA; 1981

[2] Modular Building Institute. 2020 Permanent Modular Construction Report. Charlottesville: Modular Building Institute; 2020

[3] Bayliss S, Bergin R. The Modular Housing Handbook. London: Routledge; 2020

[4] Wuni IY, Shen GQ, Osei-Kyei R, Agyeman-Yeboah S. Modelling the critical risk factors for modular integrated construction projects. International Journal of Construction Management. 2020;**22**(11):2013-2026. DOI: 10.1080/15623599.2020.1763049

[5] Pernice R. Metabolism reconsidered its role in the architectural context of the world. Journal of Asian Architecture and Building Engineering. 2004;**3**(2):357-363. DOI: 10.3130/jaabe.3.357

[6] Elliott KS. Historical and chronological development of precast concrete structures. In: Elliott KS, Zuhairi HA, editors. Modernisation, Mechanisation and Industrialisation of Concrete Structures. Hoboken: John Wiley & Sons; 2017. DOI: 10.1002/9781118876503.ch1

[7] Smith RE, Quale JD. Offsite Architecture. Constructing the future. London: Routledge; 2017

[8] Yu W-W. Cold Formed Steel Design. Hoboken: John Wiley & Sons; 2000

[9] Wang H, Liao L, Yi W, Zhen L. Transportation scheduling for modules used in modular integrated construction. International Journal of Production Research. 2023;**62**(11):3918-3931. DOI: 10.1080/00207543.2023.2251602

[10] Gorgolewski MT, Grubb PJ, Lawson RM. Modular Construction Using Light Steel Framing. Design of Residential Buildings. Ascot: The Steel Construction Institute; 2001

[11] Trebilcock PJ. Building Design Using CFS Sections. An Architect's Guide. Ascot: The Steel Construction Institute; 1994

[12] Kamali M, Hewage K. Development of performance criteria for sustainability evaluation of modular versus conventional construction methods. Journal of Cleaner Production. 2017;**142**:3592-3606

[13] Gaspari J, L'involucro edilizio stratificato. Soluzioni per L'efficienza Energetica in Funzione Dell'aggiornamento Normativo. Monfalcone: EdicomEdizioni; 2018

[14] Burry M. Scripting Cultures: Architectural Design and Programming. Hoboken: John Wiley & Sons; 2011

[15] Lawson RM et al. Modular Construction Using Light Steel Framing. An architect's Guide. Ascot: The Steel Construction Institute; 1999

[16] Anderson C. In the next industrial revolution, atoms are the new bits. Wired. 2010;**1**(25). Available from: https://www.wired.com/2010/01/ff_newrevolution/ [Accessed: November 4, 2024]

[17] Klein T. Digital craftsmanship. In: Marcus A, editor. Design, User

Experience, and Usability: Users and Interactions. Berlin: Springer; 2015. DOI: 10.1007/978-3-319-20898-5_61

[18] Egan J. Rethinking Construction: The Report of the Construction Task Force. United Kingdom: Construction Task Force; 1998

[19] Spring M. Concrete Recast. Building. 2002. Available from: https://www.building.co.uk/focus/concrete-recast/1023340.article [Accessed: July 31, 2024]

[20] Available from: https://ec.europa.eu/eurostat/statistics-explained/index.php?title=Archive:Statistiche_demografiche_a_livello_regionale&oldid=268230 [Accessed: July 20, 2024]

[21] Vitale P, Spagnuolo A, Lubritto C, Arena U. Environmental performances of residential buildings with a structure in cold formed steel or reinforced concrete. Journal of Cleaner Production. 2018;**189**:839-852

[22] Iuorio O, Gigante A, De Masi RF. Life cycle analysis of innovative technologies: Cold formed steel system and cross laminated timber. Energies. 2023;**16**(2):586

[23] Khan AA, Yu R, Liu T, Gu N, Walsh J. Volumetric modular construction risks: A comprehensive review and digital-technology-coupled circular mitigation strategies. Sustainability. 2023;**15**:7019

[24] Rymarczyk J. Technologies, opportunities and challenges of the industrial revolution 4.0: Theoretical considerations. Entrepreneurial Business and Economics Review. 2020;**8**(1):185-198

[25] Deutsch R. Data-Driven Design and Construction. 25 Strategies for Capturing, Analyzing, and Applying Building Data. Hoboken: Wiley & Sons; 2015

[26] Bernstein P. Architecture, Design, Data. Practice Competency in the Era of Computation. Basel: Birkhauser; 2018

[27] Carpo M. Digital darwinism: Mass collaboration, form-finding, and the dissolution of authorship. The Log. 2012;**26**:97-105. Available from: http://www.jstor.org/stable/41765764 [Accessed: July 20, 2024]

[28] Carpo M. Beyond Digital: Design and Automation at the End of Modernity. Cambridge (MA): MIT Press; 2023

[29] Koutamanis A. Building Information. Representation and Management: Principles and Foundations for the Digital Era. Delft: TU Delft Open; 2022

[30] Kieran S, Timberlake J. Refabricating Architecture. How Manufacturing Methodologies are Poised to Transform Building Construction. New York: McGraw Hill; 2004

[31] Kendall S. The next wave in housing personalization: Customized residential fit-out. In: Piroozfar P, Piller F, editors. Mass Customisation and Personalisation in Architecture and Construction. New York: Routledge; 2013

[32] Russo Ermolli S, Galluccio G. Building industrialization and prefabrication between materiality and immateriality. Agathón International Journal of Architecture Art and Design. 2019;**5**:93-100

[33] Galluccio G. Processi Digitali Avanzati per L'industria Delle Costruzioni 4.0. BIM Digital Manufacturing e Prefabbricazione Nell'edilizia in Cold–Formed Steel. Roma: Aracne; 2019

[34] Campiche A, Tartaglia R, Fiorino L, Landolfo R. Experimental tests for the evaluation of the seismic performance of the innovative CFS wall. Thin-Walled Structures. 2024;**198**:111681

[35] Russo Ermolli S, Galluccio G. Semi-temporary post-earthquake settlements in Italian historical centers: Performance-based guidelines and BIM simulations. In: Hvejsel MF, PJS C, editors. Structures and Architecture. A Viable Urban Perspective? Vol. 2. Leiden: CRC Press; 2022

[36] Garber R. BIM Design. Realising the Creative Potential of Building Information Modelling. Hoboken: John Wiley & Sons; 2014

[37] Zhang A, Liu J, Chen Z, Chen T. Bending behavior of detachable tapered-head bolt inter-module connection of steel modular structure. Journal of Constructional Steel Research. 2024;**220**:108829

[38] United Nations Department of Economic and Social Affairs. World Urbanization Prospects: The 2018 Revision. New York. 2019. Available from: https://population.un.org/wup/Publications/Files/WUP2018-Highlights.pdf [Accessed: July 31, 2024]

[39] Science and Technology Select Committee. Off-Site Manufacture for Construction: Building for Change. London: House of Lords; 2018. Available from: https://publications.parliament.uk/pa/ld201719/ldselect/ldsctech/169/16902.htm [Accessed: July 31, 2024]

Chapter 4

Housing Overcrowding: Relationships and Challenges in the Contemporary Era

Carla Chiarantoni

Abstract

The home is a fundamental part of life, serving as a refuge and consuming significant economic and material resources. However, housing issues have become complex, gaining social and regulatory interest. Factors like housing size, crowding index, social housing, and housing codes significantly impact living quality and well-being. This research investigates domestic space and its relationship with users in new housing developments. It examines how standards influence living space design to meet needs for adequate space and social sustainability through social housing policies. The study analyses the crowding index as a key quality parameter, proposing scenarios for evolving home concepts to meet contemporary needs. The analysis considers demographic changes, household structure shifts, and the impact of working conditions on housing sustainability and social cohesion. It addresses economic challenges in securing adequate housing and managing family health and education, emphasising the need for inclusive housing policies. The paper views housing as crucial for dignity and sustainability, advocating for a holistic approach that sees housing as an integrated ecosystem supporting social and individual dynamics. It highlights how social housing models can combat overcrowding, improve life quality, and promote social inclusion.

Keywords: living space, housing dimensional requirements, crowding index, existing, buildings, typological transformation, sustainable regeneration

1. Introduction

In the contemporary era, the gap between housing supply and demand within the city is becoming more accentuated. Apart from the issues that have always affected the urbanised territories, sociodemographic changes, as well as the diversification of lifestyles, have been added, leading to a demand for more varied and diversified housing solutions [1]; and the recent economic challenges have substantially changed the world of work by introducing different times and ways of working from the recent past. Studies conducted by the World Bank [2] have highlighted that about 55% of the world population resides in urban settlements and this percentage is destined to sharply increase by 2030 due to the continuous urbanisation that is taking place especially in countries with emerging economies [3].

Excluding instances connected to slums and informal settlements, the search for an adequate home within the city is usually difficult for a high number of people with both low and higher income. This is a general issue all around Europe, although it is found that among the existing buildings in an urban area, the residential ones are significantly prevalent. Out of around 131 million buildings in the countries members of the EU, 119 million are used as residential buildings, especially in the urban centres (43%) and in small conurbations (33%) [3].

It is undoubted that the residential construction industry is a determining and driving factor for the social and economic well-being of a country, but despite this, the ways and methods to determine the adequateness of housing are not based on uniform indicators and, even when present, they differ in each national context. It is a consolidated fact that at a European level, residential construction has to respond to new requirements and completely different needs from the recent past; and it is also recognised that the perception of the quality of housing varies significantly depending on specific characteristics of the occupiers such as employment, education, and gender [4]. Various factors are added to these requirements, among which the increase of the average life expectation of the population with subsequent functional limitations of the residents, new social needs due both to the demographic changes and the social changes of family relationships (an increase of separated couples and people living alone); contemporary work settings (the COVID-19 pandemic has facilitated the adoption of remote working, which has become a stable setting used by companies) which require an increasingly flexible job in time and space. Lastly, but no less important, part of the assessment criteria are all the conditions and modifications that allow the development of energetic performances for the reduction of CO_2 emissions in response to climate change [5, 6]. All these mechanisms, although known in each country, are not actually quantified and involve high heterogeneousness, which is often based on large-scale data without a global standard and a coherent base framework that allows the comparison between countries. The various global organisations, UN, UN-Habitat, and UNDESA [7], consider wider projections, which include the assessment of populations who live in extreme poverty (slums and informal settlements). Therefore, it is probable that the data referring to exclusively urban settlements perhaps underestimates the number of inhabitants and the living conditions in the cities for those who do not have access to a decent home [2].

At the European level, the issue has been taken into consideration by the World Health Organisation (WHO), which has introduced some parameters in the assessment of the quality of housing, connecting and relating age, gender, family relationship, and the living area. This way it has been possible to introduce the concept of sufficient available domestic space as a fundamental aspect in the assessment of the quality of the housing [8]. The availability of domestic space has been identified as space that needs to be sufficiently wide in order to comfortably house people of different ages, guarantee sufficient space to meet the needs of privacy of the occupiers, and that has to be accessible and usable by external users [9]. This new perspective has radically changed the view of housing from a fundamental human need for shelter to a place that must guarantee the physical and psychological well-being of the residents [8]. Several authors have examined various aspects of the quality of affordable housing in different countries, however the discussion on the housing standards [8, 10–12] and the organisation of the domestic space, as well as the supporting and/or shared spaces of the housing is still rather open and leaves room for considerations and research [13, 14]. The search for relationships of available space in a house and the influences on the daily lives of the residents is not a mere sociological problem but involves several fields and disciplines. If for example a reflection is made between the relationship that exists between the private and public space of the dwelling, it can be noted how the same rooms, considered as spaces devoted to privacy which offer personal refuge for study, work, relaxing and free time, assume a different value in relation to whom they are destined to (functions can vary depending on age and role in the family, i.e. parent, child or cohabitant). If these private spaces are essential for the individual well-being, so are the communal spaces such as the living room and the kitchen, which also allow people external to the family to socialise, creating at the same time opportunities for family interactions. Therefore, the correct proportion and relation between these spaces can create and facilitate a harmonious balance between moments of privacy and those of socialisation, improving the quality of domestic life [15, 16].

In addition to all this, there is the matter of the global carbon footprint, as the construction sector is one of the main contributors [17]. The construction sector is responsible for a third of the global CO_2 emissions, and this value stems from the energy of the production processes (around 8%), the electricity and heat consumption (about 18%), and also from other processes quantified in a further 7% of the total emissions. Therefore, the construction sector appears to be highly energy-consuming when considering both the energy spent in the operating phase of the building and the energy employed in the phase of production of the building structures. If, on the one hand, the recent norms imposing the adoption of standards of low energy consumption have facilitated a slight reduction in the consumption during the production phase, on the other, an increasing demand for energy consumption is recorded during the operating phase of the building (about 1%), which highlights the necessity of an increasing demand for functional use of the buildings.

Focusing on the case of Italy, it can be noted that residential buildings contribute significantly to national energy consumption. It is estimated that the construction sector in Italy represents about 40% of the total, while 36% of the emissions are related to the existing buildings. These circumstances are determined by the phenomenon of the great reconstruction post-World War, which led to massive construction, with around 60% of the current building stock constructed before the 1970s, prior to the implementation of critical legislation addressing seismic performance and energy consumption reduction [18]. This reconstruction represented one of the largest investments in public housing, in terms of square metres built, and profoundly shaped the majority of residential buildings, particularly public housing in the outskirts of Italian cities. This model is currently undergoing a profound crisis and re-evaluation, driven not only by urban planning considerations but also, more importantly, by construction techniques and typological-functional issues related to the organisation and usability of living spaces, including per capita dwelling area and the functional distribution of housing layouts [19]. This severe crisis, determined by the lack of alignment between the requirements of the occupiers and the performance of the buildings, poses several challenges and actions. Starting from a general inadequateness of living conditions and energy insufficiency to scarce if not absent maintenance accompanied by obsolete services, these all represent unsatisfactory living conditions with subsequent lower quality of living [20].

In 2008, analysing the matter of public residences, Di Biagi [21] stated that the great reconstruction post-World War allowed 9–15% of the population to dwell in public popular housing. About 2.2 million inhabitants were housed in about 840,000 apartments of public economical housing. But what were the characteristics of these dwellings? Overall, it can be stated that the housing that took shape was founded on two important principles: on the one hand, it was the scene for the first rational experiments rooted in Alexander Klein's existenzminimum [22]; on the other, it had to respond to specific demographic demand of the time which saw the family mainly composed of numerous households and the union of several families within the same dwelling (it was common to have the coexistence of elderly parents and related children with their own family) [23]. It was a period of great experimentation and research on housing. Therefore, there were two coexisting souls: on the one hand the search for rationality and optimisation of the domestic spaces, which aimed at creating minimal and standard surfaces to satisfy the needs of the new users; on the other hand, almost in contrast, the prevalent necessity to create apartments for numerous/complex families. The result of this combination led to the creation of large-size apartments structured with a high number of bedrooms and often equipped with minimal living areas and accessory services. Often, the hygienic services were inadequate and insufficient for the resident family [22]. In any case, this approach has led to the definition of standardised rules and housing modules, which could be replicated and applied on a large scale in the Italian territory regardless of the geographical position and needs [24].

In light of what was expressed above, the typological transformation of housing, the energetic requalification, and the adjustment to the restrictive structural norms of economical and popular construction currently represent one of the most interesting and complex challenges in the housing theme. Together with the need to improve the architectural and environmental quality (appropriate ventilation of the spaces, correct balance of the winter and summer temperatures) of the housing/domestic space, there is also the matter of reducing the average surface of the dwellings in order to balance the housing density in relation to the existing buildings. Lastly, an additional focus is the slowdown of the Italian demographic dynamics, which has seen a substantial change in the model and number of components per family, and therefore per dwelling [25].

With these premises, this short essay examines the matter of public residential and social construction under a new interpretation relating to the new housing discomforts/needs of the population. The investigation is developed through different levels of analysis and constitutes the basis for an alternative research approach on the theme of housing. It aims to widen the scope towards different perspectives, which is useful for redefining formal technical and functional models with a holistic and more resilient approach.

2. The crisis of the house system and the domestic space

The housing deprivation operates simultaneously on several levels, subdivided into three specific categories of fragility: the family, the housing space, and the urban context. In most cases, the operations aimed at facilitating the accessibility to housing focus on specific vulnerable groups of families or the adoption of functional, typological, technical, or technological solutions regardless of the specific problem.

The housing unease characterised by the highest economic instability and rapid transformation of the structure of traditional families towards new organisational models brings interesting considerations (**Figure 1**) [19].

Families, in their diverse compositions, often reside in housing developments that do not always support their growth or that of the communities to which they belong. In the specific context of Italian public housing, access to housing (long-term rental)

Figure 1.
The three interlinked factors of urban housing deprivation.

grants a stable and enduring right of occupancy [26]. However, this arrangement presents two significant challenges. First, the urban setting is typically assigned rather than chosen, limiting residents' ability to select their environment. Second, the size of the dwelling remains fixed regardless of the family's life cycle, failing to adapt to the natural changes a family undergoes. As economically disadvantaged families are unlikely to move to a different home when their needs change, they often experience inadequate living space for a significant portion of their life cycle. In this regard, it is important to make some considerations about the evolution of the households. The rise of two different family compositions can be seen almost in parallel: the traditional families, which have seen a substantial reduction in the number of components [27, 28]; and the new family groups with various composition.

Analysing the lifecycle of a traditional family, these phases can be observed on average: the first new family is formed (single person, young couple, etc.); the family becomes more structured (birth of children) and the needs of the individual members of the family grow (the children grow and new activities and uses of the housing space appear): ageing phase (adult children leave the house) with the increase of specific needs (social, economic, of housing space) [19].

The formation of new households in Italy stems from the demographic and economic changes in the population. Among the factors that have mostly affected the change of composition of the family is the change in working conditions of women; the reduction of the "fertility rate", that is, the number of children per woman; the increase in educational attainment, which brings to a prolonged permanence of the children within the household; the reduction of marriages and the increase in separations and divorces, which also brings about the increase of single-parent families (women/men with children) and divorced parents (mostly single parents that house the children in their dwelling in turns) [29]. Therefore, the formation of the new households in Italy includes single parents with children, divorcees, and families composed of separated parents with children, which will be defined together as complex households.

These two different family compositions (traditional family and complex household) bring about new organisational models which require flexible and dynamic housing spaces that are able to adapt to the different needs of the members of the family. For example, single parents with children require spaces that allow both privacy and the supervision of the children, while separated parents with children may need additional rooms to house the children during the visits. The fluid and interconnected planning of the housing space becomes therefore essential to respond to these new family configurations, at the same time guaranteeing the quality of life and environmental sustainability.

With these considerations, it becomes crucial to introduce a housing quality index. In this regard, the WHO has addressed the issue by introducing an indicator of housing quality, identifying overcrowding as a key determinant. The overcrowding index quantifies whether the availability of space in relation to the residents is sufficient within the dwelling. There are several ways to measure it, such as the American Crowding Index (ACI), the People Per Room (PPR), the Canadian National Occupancy Standard (CNOS), the British Bedroom Standard (BBS), and the index of floor space per person. In Italy, the method of calculation adopted considers the ratio between the number of users and the main rooms of the dwelling, excluding corridors, toilets, and other living areas. However, there are no parameters to assess the underuse of the dwellings, making it hard to quantify the oversizing of the domestic space [29].

Therefore, the research has included a base of analysis in the field of the needs of families with particular reference to the Italian case. The terms of the research have included the family lifecycle and the evolution of the housing space with the objective of identifying the premises and the positions in relation to the crisis of the house space. For the study, the compositions of the (traditional and complex) families have been observed while they change during their lifecycle and therefore correlated to the corresponding housing needs. The method of comparison was based on two aspects: the classification of the Italian Istituto Nazionale di Statistica (ISTAT), which identifies the new households, and the World Health Organization (WHO) qualitative parameters of the housing well-being, which introduces a method for calculating the overcrowding index. The analysis of the needs of families has been therefore related, in the following paragraph, to the technical construction and formal control of the housing spaces in reference to the particular context of economic and popular housing of the city of Bari and, more specifically, to the Japigia quarter.

The objective of this essay is to highlight the necessity of rethinking the traditional technical and planning concepts, going towards flexible and dynamic housing spaces. It is fundamental to plan housing that can adapt to the different phases of the family lifecycle and the new organisational models, guaranteeing at the same time the quality of life and environmental sustainability.

Therefore, the following table summarises the various considerations made so far, indicating each time the need generated by a certain family condition and the related social, economic, or housing needs. The table, in particular, was produced using a parameter of calculation that was adopted by the Eurostat study [30]; we can recognise the conditions of overcrowding in relation to the bedrooms of the dwelling, identifying the following cases (**Tables 1-3**):

- one room for the household;
- one room per couple in the family;
- one room for every single person aged 18 or more;
- one room per pair of same-sex children between 12 and 17 years of age;
- one room for every single person aged 12 to 17 years of age not included in the previous categories;
- one room per pair of children under 12 years of age.

Evolution of Family Units Over Time and the Need for Flexibility in Domestic Space

				NEED				
Family Composition		**Family Characteristic**	**Family Resource**	**Immaterial Need**	**Housing Need**	**Social**	**Economic**	**Housing**
Single or Dual-User Family Units	**Retirement-Age Couples**			Supplementary needs for elderly individuals	1 double bedroom			
	Working-Age Couples			No substantial supplementary needs except for very limited additional spaces (derivable within the apartment structure) for leisure, study/work				
	Single Parent + 1 Child			Supplementary needs for additional spaces (derivable within the apartment structure) for leisure, study/work	1 double bedroom for single parent + 1 single bedroom for child			
	Cohabitants without a Relationship				1 single/double bedroom for cohabitant 1 + 1 single/double bedroom for cohabitant 2			
Family Units for Three Users	**Couples with one child aged 0–12 years**	Live in the same city or neighbourhood as other family members who can support them	Family proximity/social support network for the support and growth of children	No supplementary housing needs except for ensuring, compared to the previous case, one bedroom	1 double bedroom for parents 1 double bedroom for same-gender children			
		Live without support from other family members	Only one spouse works/possibility of support for children's growth	Economic limitations and social sustainability loss (reduction of expenses and services for secondary goods)				

Evolution of Family Units Over Time and the Need for Flexibility in Domestic Space				
		Both spouses work	Social support for supplementary educational training, sports and social practice, and recreational-relational activities	
1. Single parent with two children of the same gender aged 0–12 years	Live in the same city or neighbourhood as other family members who can support them	Family proximity/ social support network for the support and growth of children	No supplementary housing needs except for ensuring, compared to the previous case, one bedroom	**CASE 1** 1 double bedroom for parents 1 double bedroom for same-gender children
2. Single parent with two children of different genders aged 0–12 years	Live without support from other family members	Only one spouse works/possibility of support for children's growth	Economic limitations and social sustainability loss (reduction of expenses and services for secondary goods)	**CASE 2** 1 double bedroom for parents 1 single bedroom from 12 years onwards for the older child
		Both spouses work	Social support for the accompaniment and growth of children	

Table 1.
Family stage 1–Evolution of family units over time and the need for flexibility in domestic space–C. Chiarantoni and C. Montalbano.

DOI: http://dx.doi.org/10.5772/intechopen.1008562

Evolution of Family Units Over Time and the Need for Flexibility in Domestic Space								
				NEED				
Family Composition		**Family Characteristic**	**Family Resource**	**Immaterial Need**	**Housing Need**	**Social**	**Economic**	**Housing**
Family Units for Four Users	**Age of children between 13 and 18 years (compulsory school age)** CASE 1 **Couples with two children** CASE 2 **Single parent with three children** CASE 3 **Complex family** CASE 4 **Cohabiting without relationship**	Two or more children of the same gender	Only one parent works or is present. Possibility of sharing some of the children's spaces	Economic limitations and social sustainability loss (reduction of expenses and services for both secondary and primary goods) + Social support for supplementary educational training, sports and social practice, and recreational-relational activities + Supplementary housing needs related to the expansion and non-overlapping of activities connected to study and training within the same environments (environmental units)	**CASE 1** - One double bedroom - One double bedroom **CASE 2** - One double bedroom - One double bedroom - One single bedroom **CASE 3** depends on the complexity of the family composition **CASE 4** - Two to three single and double bedrooms			
			Both parents work if both are present. Possibility of sharing some of the children's spaces	Supplementary housing needs related to the expansion and non-overlapping of activities connected to study and training within the same environments (environmental units)				
		Two or more children of different genders	Only one parent works or is present. Possibility of sharing some of the children's spaces	Economic limitations and social sustainability loss (reduction of expenses and services for both secondary and primary goods) + Social support for supplementary educational training, sports and social practice, and recreational-relational activities	**CASE 1** - One double bedroom - Two single bedrooms **CASE 2** - One double bedroom - One single bedroom for each child			

Evolution of Family Units Over Time and the Need for Flexibility in Domestic Space			
		+ Supplementary housing needs related to the expansion and non-overlapping of activities connected to study and training within the same environments (environmental units) + Supplementary housing needs related to the separation of rooms in the "night zone" to address the emergence of new gender-specific needs (boys' spaces separated from girls' spaces)	**CASE 3** depends on the complexity of the family composition, but on average individual bedrooms for each family member **CASE 4** - Two to three bedrooms between singles and doubles
	Both parents work if both are present. Possibility of sharing some of the children's spaces	Supplementary housing needs related to the expansion and non-overlapping of activities connected to study and training within the same environments (environmental units) + Supplementary housing needs related to the separation of rooms in the "night zone" to address the emergence of new gender-specific needs (boys' spaces separated from girls' spaces)	

Table 2.
Family stage 2–Evolution of family units over time and the need for flexibility in domestic space–C. Chiarantoni and C. Montalbano.

DOI: http://dx.doi.org/10.5772/intechopen.1008562

Evolution of Family Units over Time and the Need for Flexibility in Domestic Space									
				NEED					
Family Composition		Family Characteristic	Family Resource	Immaterial Need	Housing Need	Social	Economic	Housing
Family Units for Four Users	**AGE OF CHILDREN BETWEEN + 19 years adult age (university and/ or working age)**	two or more children of the same sex	Only one parent works or is present Possibility of sharing private space	Integrative housing need related to the enlargement of the 'sleeping area' to cope with the emergence of new privacy needs + Economic constraints and loss of social tightness (contraction of expenditure and services for primary and secondary goods)	In each case all family members will need a single room. Parents whether single or in pairs will on average have the double bedroom			
			Several household members work / Possibility of sharing private space	Additional integrative housing need related to the enlargement of the 'sleeping area' for the need to cope with the emergence of new privacy needs				
	CASE 1 Couples with two children **CASE 2 Single parent with three children** **CASE 3 Complex family** **CASE 4 Cohabitees without relationship**	two or more children of different sexes	Only one spouse works	Additional integrative housing need related to the enlargement of the 'sleeping area' due to the need to cope with the emergence of new privacy needs and to cope with the emergence of new gender needs + Economic constraints and loss of social tightness (reduction of expenses and services for primary and secondary goods)				

Evolution of Family Units over Time and the Need for Flexibility in Domestic Space			
		NEED	
	Several household members work	Additional integrative housing need connected to the enlargement of the 'night area' due to the need to cope with the birth of new privacy needs and to cope with the birth of new gender needs	
Families with members all of adult age and with children over 24 years of age		Integrative housing need connected to the need for new spaces and new articulation of the 'living area' which take into account the privacy (of adults) and the new needs (of study and/or work) of individual users	Each household member over the age of 18 will need one bedroom. Only the parents will share the double bedroom

Table 3.
Family stage 3–Evolution of family units over time and the need for flexibility in domestic space–C. Chiarantoni and C. Montalbano.

The table does not include the more specific cases of the complex and compound households since the situation becomes considerably convoluted and would require its own analysis.

Therefore, a widening gap can be noted between the static condition of the dwellings and the sometimes drastic changes in the needs of the families. This deviation depends on two specific physical components of the housing space: the dwelling itself, and its collocation in a specific urban context. For what regards the dwelling in particular, it can be noted that the deviation refers to its size, structure, the subdivision of the environmental units and the type of supporting services and technologies it contains. In the different phases of the lifecycle of a family that remains anchored to the same home, significant phenomena of under- or over-occupation of the living spaces can be observed. Expanding families and/or during their lifecycle enter a state of living distress, which is hard to remedy. This particularly happens in economic and popular housing [31].

The family compositions analysed so far and that form, in various ways, the most relevant corpus of urban families, undergo a substantial compression of the endogenous (connected to the characteristics of the flat) and exogenous housing well-being (connected to the choice of living in an unpleasant urban context but economically accessible), and lack a response to the individual, family, and social needs. This brings to a significantly lower quality of life as well as expectations of growth and future of a family, which can cause phenomena of urban spatial segregation [32, 33] and could have serious repercussions on the family sphere. All this completes and further clarifies a scenery in which many European countries including Italy register a progressive worsening of the conditions of "subjective well-being", and a marked increase in the "risk of poverty" of an ever-growing number of urban families.

3. The crisis of the house system and the domestic space

At the end of the First World War (WWI), in Bari, there was a high demand for housing which could not be satisfied by the existing public and private buildings heritage. This was due to refugees, displaced people, and the population coming back into the city. The municipal administration and the relevant institutions implemented a series of temporary and emergency operations, which determined the birth and initial development of the peripheral quarters and, among those, the Japigia quarter. This way, from the late 1940s of the last century, a brisk transformation process began from a rural area into a settlement designed for emergency, and lacking overall planning. The Japigia quarter in Bari has recent origins in this historical context, and therefore it dates back to the spontaneous formation of small housing agglomerates. It was not until the first measures extended to the whole national territory that the creation of housing agglomerates with diversified construction typologies and techniques occurred, following a logic of higher quality both on the urban level and the building one. The measures in question belonged to the INA-Casa period and constituted an intense and significant time for the construction and reconstruction of important economic and popular buildings in the whole Nation [34].

The two 7-year INA-Casa programs initiated several popular quarters, such as San Paolo, Carbonara, Libertà, San Pasquale, and Japigia. It was the law n. 43 of 1949 to give start to the first operations in the Japigia quarter in an area of about 13 ha, with a construction impact of about 3000 lots over the total 7000 planned for the whole city. Popular buildings, in this first stage, were mainly erected in the newly-formed national institution for housing, set up in 1949, which had the duty of building low-cost housing for families with low income. This institution was flanked by the Public Building Society, set up in 1948, which notably contributed to the erection of popular housing, and realising in Bari around 3200 dwellings between 1948 and 1953 [35]. In the urban system logic indicated by the INA-Casa guidelines, this first settlement of popular housing was conceived as a small citadel. This was closed to the rest of the quarter through a first building layer which composed the external margin of the building quadrilateral and allowed the view through transversal visual axes of a complex internal organisation [36]. These first housings continued to be built in a simple and fast way, mirroring the "INA-Casa process guidelines," and therefore, for the structures, the load-bearing masonry was mainly used with economical materials such as brick, tuff, and reinforced concrete. The ground floor was covered with reconstructed stone tiles, while traditional plaster was used for the remaining floors. The buildings (a total of about 2500 dwellings) were terraced isolated blocks with the same number of floors, four: a mezzanine floor and no more than three floors without lifts. Of reduced size and restricted by the rigid structural frame, the dwelling included a hall, a sufficiently spacious kitchen with a dining area, a living room, and a varied number of bedrooms, depending on the shape of the dwelling. Sometimes the shortage of housing led to these dwellings being shared by several households belonging to the same family group (married children with their family frequently lived under the same roof as their own parents).

The main shortcoming of these first buildings was the absence of technological systems such as heating, boiler, and lift. After this initial period, entirely subsidised by the State, the following construction periods were characterised by only 5–10% of the houses with investments from public initiatives, in the form of subsidised buildings.

This marked another passage that modified the setting of the new constructions and the settlement logic [37].

By the late 1950s and early 1960s, the Economic and Popular Housing Projects (PEEP) season started. These involved the realisation of different and quality popular housing compared to the previous, with dwellings of various kinds and sizes, which were able to respond to the housing needs of the distinct groups of population [38]. The PEEP of Japigia began in this new context, in 1962. This involved the erection of new popular buildings but also the reconstruction of some existing ones, and the realisation of public facilities such as schools, nursery schools, and places of worship, with the aim of improving the quality of life of the residents of the quarter. This stage was distinguished by the realisation of three different building complexes. The first was composed of 13 GesCal (Housing Management for Workers) buildings aimed at specific categories of workers (such as postal workers, railway workers, and police officers), and afterwards all taken over by privates.

The second public complex by PEEP was the residential complex named ex-ISESX Via Caldarola 28, currently known as "red houses." The third public compartment to be designed was erected adjacently to the previous one (Red Houses) and constituted what is currently denominated Colour Houses. This last agglomerate was realised with a rational approach, a sequence of parallel lines, and low recognisability of the routes, accesses, and functional junctions. Therefore, the lack of formal research on the communal urban space, in contrast with the nearby INA-Casa projects (which still reflected on the formal value of the communal and semi-private urban space), determined little attention to the quality of the space open to the complex. This was often stretched, incomplete in its articulation, and, however, not recognised as a communal area by the residents until it became a non-urban place [21].

With the operations of the 1960s, the Japigia quarter became, after that of the San Paolo quarter, the biggest popular housing agglomerate of Bari, and the most strictly related to the heart of the city. Not less relevant, a new building philosophy started to identify the car as a determining element in the processes of urbanisation and definition of housing spaces. The new architectural and urban view saw the new paradigms of the urban form in the distancing of the buildings from the roads and the erection of long fences to circumscribe yards and condominium squares (for the parking), which aimed to guarantee autonomy and safety. If, at first, the idea of creating yards and gardens pertaining to single or groups of buildings was seen as an element of innovation and modernity, with time, this solution contributed to accelerating phenomena of degradation, abandonment, and isolation [37].

In the early 1970s, the municipal administration redirected some public financing towards the Japigia quarter, mainly allocating it to the restoration of some popular buildings and the construction of new ones. A new development plan for public housing policies was started, which determined the overflowing of the city from its natural boundaries. Here, a new type of urban vision took shape, which gave birth to the B2 compartment or Via Caldarola 45 zone.

The design of the area was entrusted to the architect V. Chiaia, and the engineers D. De Salvia and G. Fuzio. Two building types were used in the design of the plan: the 5- and 7-storey terraced house and the 9-storey tower house. The choice responded to the intention to attempt a standardisation and industrialisation of the construction elements from the orthogonal arrangement of the buildings and the road network. In all effect, it was an enormous high-density residential quarter. Its construction, which started in 1972, already in 1975 included 18 buildings and a total of 1200 dwellings. At the beginning of the 1980s, after the construction phase of the district, the Via Caldarola 45 zone counted 2200 dwellings and about 10,000 residents with increasing social degradation.

The Via Caldarola 45 zone, and, more in general, this new expansion of Japigia, immediately took shape as a high-density residential area and composed the extreme eastern appendage of the Japigia quarter ending in correspondence of the Adriatic SS 16 BIS. The presence of the high-speed road (Via Caldarola) represented a limit with respect to the remaining part of the quarter, isolating and separating it from the context of private residential buildings [39].

3.1 Design and typology

A term of comparison among some of the most significant architects and planners of the 1970s and 1980s, the Via Caldarola 45 area saw the experimentation of building techniques that were innovative for the time and mostly focused on the experimentation of systems with prefabricated surfaces for the external vertical closure. One fundamental characteristic of this building phase was certainly connected to the typology of attachment to the ground. It was a real public experiment in which the free portico floor was introduced in Bari. Although it freed the ground floor of the buildings with the objective of imagining a new dimension of public space, such a condition actually rejected the idea of an urban podium in which commercial and service businesses could appear and rendered this area completely devoid of these functions.

For what regards the buildings, this area was conceived as and composed of several building cores, of a wide variety of typology and morphology, which articulated around a central space destined for public gardens and meeting areas. Composed and organised according to a free urban plan, the buildings were installed on pilotis systems, which aimed to facilitate a deep social and physical interaction between the residents and the wide open spaces. The building structures, organised in an L and C-pattern, exploited a non-repetitive orthogonal arrangement and formed clusters characterised by a sequence of alternatively open or semi-closed courts. With the clear aesthetic intention aimed at breaking the monotony of the whole complex, each of these clusters presents similar technical and architectural characteristics. Lastly, the 15 identifiable categories of cluster proposed diversified sizes of the dwelling surfaces with the specific idea of responding to the needs of differently-sized households (**Figure 2**).

3.2 Study cluster, street via G. Appulo house numbers 6 to 26

From this initial situation with a high residential density, the Via Caldarola 45 zone (currently almost exclusively occupied by the properties' first assignees) has seen a gradual decrease in the density of residents. This is due both to the natural process of adult children leaving the original households and the current demographic situation, which sees increasingly smaller households, even in social classes with traditionally large families. For the case study, a cluster composed of a core of 11 buildings with the same constructive, functional, and typological distributional features has been taken into consideration. The buildings belong to the "terraced multi-storey" type (**Figure 3**), 7 buildings of which are 5-storey, and four buildings are 7-storey tall. The buildings are arranged in a C-shape, creating an open and public internal court, mainly dedicated to the parking area. All the buildings present the following characteristics:

Figure 2.
Compound of Caldarola 45, aerial photo.

Figure 3.
Façade of some of the buildings in the district.

- Partially closed buildings, with two external facades (nine buildings with north–south orientation and two buildings with east–west orientation) with alternating facades adjacent to the surrounding buildings.

- Stairwell and lift shaft located at the centre of the building and serving two flats per floor.

- Internal layout of the flats generally including a sleeping area and a living area, not always clearly divided, and not always sufficient and adequate hygienic services.

- Bearing structure with reinforced concrete casing, non-insulated external walls made of prefabricated panels with intermediate air cavity, and internal walls made of tuff.

- Non-insulating cover composed of a brick-concrete structure covered in concrete and bitumen waterproof finishing.

- Single-glass windows and aluminium structure without thermal break, with external shutters as covering.

The buildings include a total of 124 housing units which are currently occupied by 353 residents (**Figures 4** and 5).

Important consideration: the term family will refer to variously aggregated family groups, that is, both the traditional kind and the complex one, as per the definition at the beginning of the paper.

Figure 4.
Plan of the typical plan of the entire sector.

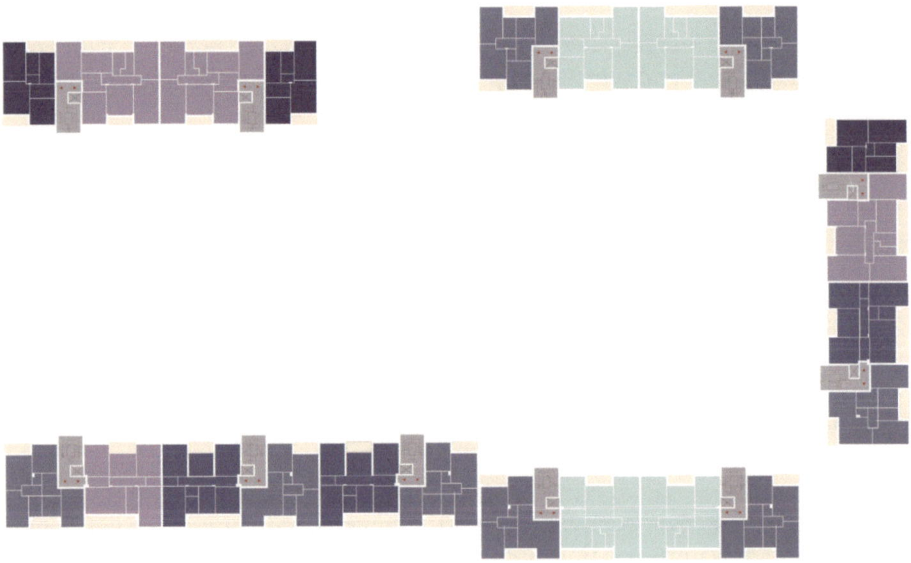

Figure 5.
Model plan functional layout.

The age groups have been divided in such a way that they can be combined with those with similar needs. The subdivision is not exhaustive of a complex matter; however, it represents a useful starting point to make a more specific reflection of the case study. Therefore, the subdivision presents the following age groups:

1. Youngsters with ages between:

    a. 0–5 years of age with no scholarisation, therefore there are few needs related to a space exclusively dedicated to the child;

    b. 6–18, age of mandatory schooling. This wide age group has different needs according to the specific period of time going from primary to secondary school. It was decided to consider it as a single category with needs related to a space dedicated to studies and more;

    c. 19–25, age group in which a university course or job hunting should begin;

    d. 26–30, age at which independent life and the search for autonomy should start.

2. Adults with ages between:

    a. 31–45 years old, formation of a family with the search for stable housing and job, a dynamic stage in the life of a young adult. In this stage, a demand for domestic space emerges, varying according to the components of the household;

    b. 46–67 years of age, second stage of the family and adults in which life begins to stabilise and invest in the growth of all the members, maximum need for space for the household.

3. Senior adults with ages between:

    a. 68–80 years of age, period in which presumably the children have left the household and there is a substantial reduction of the components and related needs of space;

    b. 81+ years of age, a period in which the family has been reduced and the stage of caregiving and support for the elderly begins. The domestic space is almost exclusively required for small households and the possibility of suitable manoeuvring areas.

The following graph shows that the two age groups most representative of the situation are the 26–30 young adults and the 46–67 adults, demonstrating the complexity of managing the domestic space (**Figure 6**).

If the data referred to the average age of the household components is added to these, it can be noted how the situation of underuse of the complex is bound to worsen since the population is generally ageing leading to a natural decline in the number of components of the household (**Figure 7**).

The distribution of the typologies of households in the complex is also not encouraging. The variability of the families within also prompts interesting reflections. For better clarity of the graph (**Figure 8**), it should be mentioned that the initials have the following meanings:

N = household.

Number = indicates the number of members of the household.

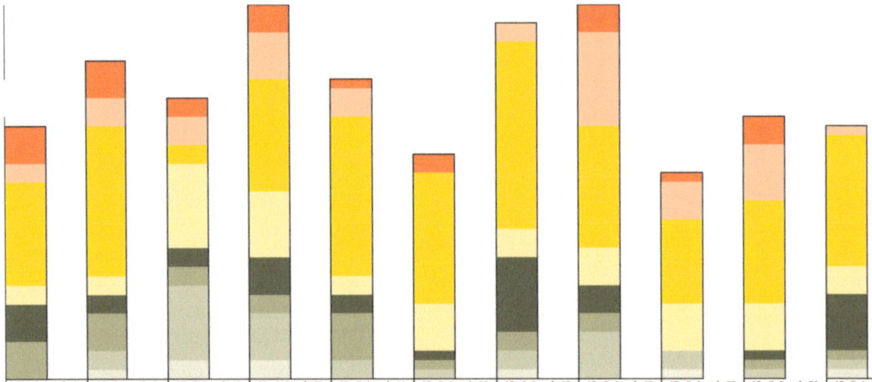

Via G. Appulo 6		Via G. Appulo 8		Via G. Appulo 10		Via G. Appulo 12		Via G. Appulo 14		Via G. Appulo 16		Via G. Appulo 18		Via G. Appulo 20		Via G. Appulo 22		Via G. Appulo 24		Via G. Appulo 26	
8	30%	9	26%	14	36%	13	33%	9	28%	3	13%	13	34%	10	25%	3	14%	3	11%	9	33%
13	48%	18	53%	20	51%	19	48%	19	59%	19	79%	2	5%	17	43%	14	64%	16	57%	17	63%
6	22%	7	21%	5	13%	8	20%	4	13%	2	8%	2	5%	13	33%	5	23%	9	32%	1	4%
27		34		39		40		32		24		38		40		22		28		27	

Figure 6.
Age ranges in the buildings of the study district—Author's elaboration on data from ArcaPuglia Central.

Figure 7.
Average age in the buildings of the study district—Author's elaboration on data from ArcaPuglia Central.

Letter = the type of family aggregation; if with this letter there is also a number, the sub-category of family aggregation is indicated.

It can be observed that the most recurring households are N2.C, N3.B, N1.A, N4.H, N1.B, that is, the ones composed of:

- Single-parent with child, N2.C
- Couples with one child, N3.B
- N1.A Single-person households in working age, N1.A
- N1.B Single-person households in retirement age, N1.B
- N4.H Complex households, that is, composed of several households, N4.H

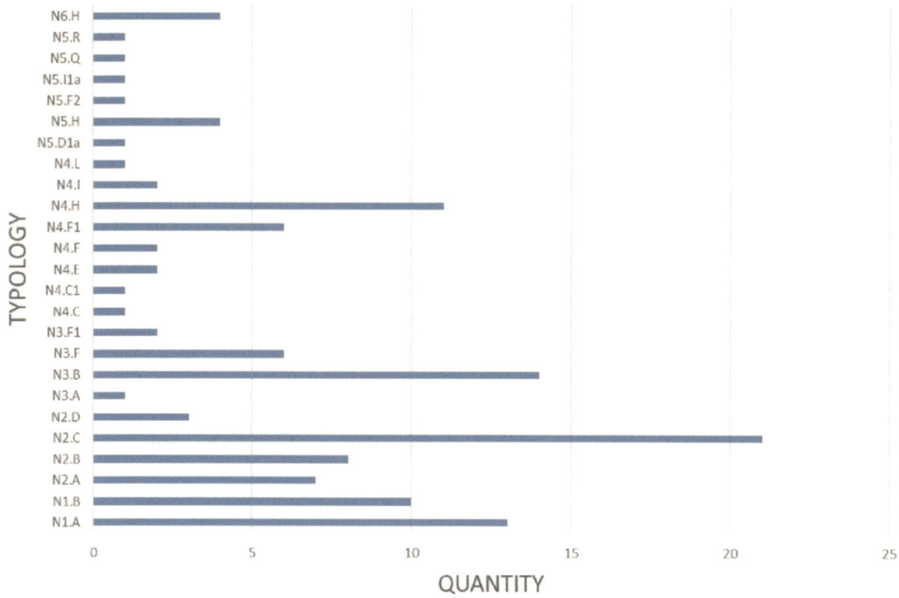

Figure 8.
Typology of family nucleus in the buildings of the study district—Author's elaboration on data from ArcaPuglia Central.

However, compared with the average age of each household, this data highlights a substantially progressive reduction in family units and, consequently, in their members. Moreover, it can be noted that as the number of occupants increases, the complexity of the households also rises, while the available living space is neither designed nor structured to accommodate this. Lastly, the comparison between the expected and the actual residents, along with the variability in the living areas, clearly highlights a shortfall in the housing potential for an area conceived and built for high residential density (**Figure 9**).

In particular, a more detailed analysis of the data referring to the surface of the dwelling between 100 and 122 sqm, which is intended to accommodate a household of

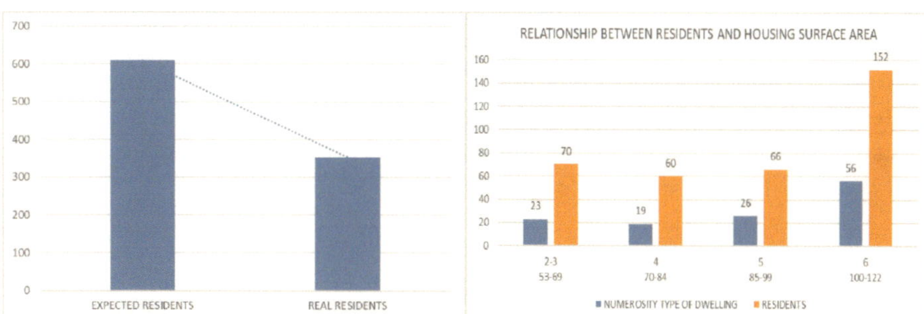

(a) (b)

Figure 9.
Number of housing types and relationship between residents and housing square metres—Author's elaboration on data from ArcaPuglia Central.

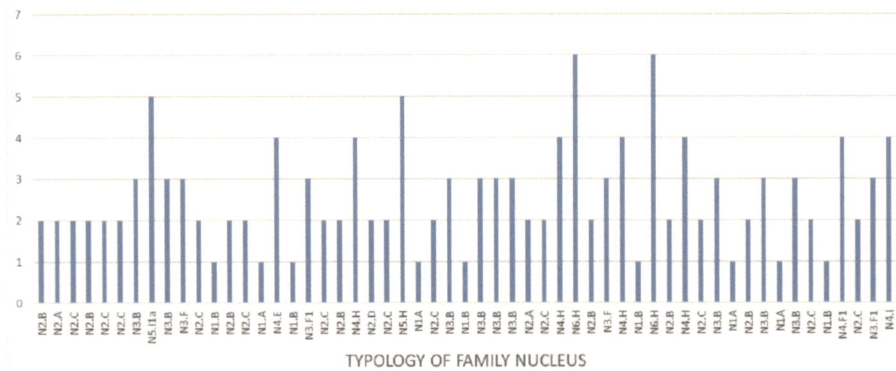

Figure 10.
Surface housing from 100 to 122 square metres—Author's elaboration on data from ArcaPuglia Central.

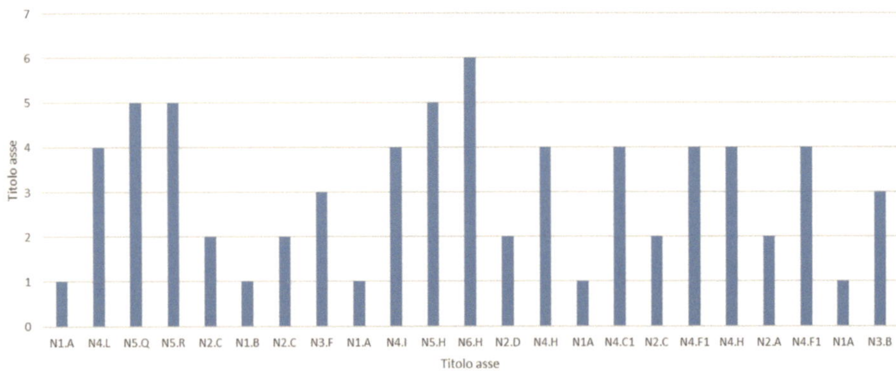

Figure 11.
Surface housing from 53 to 69 square metres—Author's elaboration on data from ArcaPuglia Central.

5 to 7 occupants (including complex households that require additional space), reveals that in reality there is a prevalence of households with two occupants (variously articulated). This condition highlights a clear undersizing of the housing quota, and therefore an underutilisation of the living space (**Figure 10**).

A completely opposite situation can be found in the data referring to the dwelling with a surface between 53 and 69 sqm, intended for a household with two to three occupants (including complex households which require more space), which sees the prevalence of households with four occupants, variously articulated and often comprising complex households. This condition highlights a clear undersizing of the living space, resulting in problems of overcrowding (**Figure 11**).

4. Discussion and conclusion

It is widely recognised that the processes of planning, constructing and utilising residential housing require significant improvements. Currently, the scientific literature focuses on the more general theme of building stock redevelopment, particularly in terms of energy efficiency or restoration. However, studies addressing the quality and liveability of the residential spaces have only recently begun to gain prominence, following the parameters introduced by the WHO such as the housing overcrowding indicator [29]. The present paper contributes to the body of research on housing by documenting and evaluating a new approach to the refunctionalisation of living spaces and the holistic redevelopment of affordable and public housing. The method is based on a careful analysis of existing conditions, which establish connections between the housing users, their age, to the life needs and the quality indicators established by the WHO (housing overcrowding index) and, the characteristics of dwelling in which they reside including their spaces and functional domains. This approach, while applied to a specific case, is customisable and replicable in its methodology, allowing the analysis of the issue of overcrowding or underutilisation of housing and domestic space within "Affordable housing." In fact, the proposed methodology facilitates a more direct relationship between the identified problem and the suggested solution, since it addresses the needs of the actual users of the housing rather than those of generalised groups of families. An examination of the existing literature reveals that, although studies and analyses often delve deeply into housing-related issues, they propose decontextualised solutions which fail to respond to specific needs. Such measures, broadly associated with initiatives introduced during the era of urban regeneration processes, have, in most cases, not achieved the expected outcomes due to their inability to adequately address the actual requirements of their beneficiaries [40–42]. This article advocates an innovative approach to evaluating potential actions for the redevelopment of residential living spaces, specifically through the transformation of existing public housing into social housing, using context-specific analyses tailored to the areas under consideration. The evaluation framework described in this paper is based on a set of indicators, including the number, age, and developmental stage of households or complex family units residing in individual apartments. This analysis is supplemented by an assessment of the dwelling areas, focusing on the organisation and composition of the living spaces in relation to the overcrowding index introduced by the WHO. The methodology enables a comparative evaluation of housing situations, informed by the type and configuration of each dwelling. When applied to multiple buildings with comparable characteristics, this approach facilitates the prioritisation of interventions, targeting those buildings where issues of over- or under-utilisation are more pronounced.

In the specific case study analysed, the findings highlight a priority for the typological and functional redevelopment of housing units, assessed through a detailed examination of residents, their needs, and the evolution of family life cycles. A particularly noteworthy observation is the frequent occurrence of complex households—often consisting of an elderly parent living with an adult child and their family—occupying larger housing units. While these households do not meet overcrowding criteria, their configuration supports social caregiving for the elderly. In such cases, the design and organisation of living spaces significantly impact the residents' health and well-being [43]. From a technical-architectural perspective, there is a pressing need to rethink living spaces, emphasising key characteristics such as flexibility, adaptability, the design of furnishings, and the full usability and accessibility of spaces [44, 45]. The literature on interior design offers a wealth of insights and methodologies for the requalification of domestic spaces, often based on strategies that promote shared functions and multipurpose environments. These strategies include furniture and spatial solutions that enable versatile use of a single area [4]. However, such approaches have yet to find widespread implementation or experimental validation in real-world applications.

The transformation and redevelopment of large public housing complexes represent one of the most compelling frontiers and challenges in the field of housing. Only recently has scientific literature begun to introduce transformability indicators for large public residential settlements aimed at guiding investors in selecting and prioritising interventions [46]. The public housing sector faces numerous challenges, among which the typological redevelopment of domestic spaces is no less significant than energy retrofitting. However, the issue cannot be addressed solely through the downsizing of housing units, as the slowing of demographic trends has fundamentally altered family models, household compositions, and, importantly, the organisation of living spaces to accommodate diverse lifestyles. Determining the appropriate interior layout for a residence is a complex issue, and the solution does not lie merely in the size of the dwelling. The search for adequately sized housing tailored to users' variable needs is hindered by the inherent temporal variability of those needs. Additionally, building regulations are often rigid and inflexible, disproportionately affecting low-income groups who lack the ability to easily relocate to more suitable housing.

One potential starting point is to move away from the traditional concept of a dwelling with fully autonomous functional spaces. Instead, some functions could be transformed into semi-public spaces, designed to be shared and/or adapted to specific cases. To ensure that residents' "basic needs" are met, it would be advantageous to aggregate and modulate spaces and surfaces, envisioning a housing organism in continuous evolution. A useful starting parameter might be the classification of needs and their corresponding residential spaces based on the levels of privacy required for specific activities. This approach would transition from highly private spaces, requiring no sharing, to more public spaces characterised by high levels of shared use. Complementary to this would be the inclusion of residential spaces designed to support functional diversity, addressing the dynamic needs of a society undergoing profound transformation.

The added value of this work lies primarily in its analytical phase, while the operational phase is deferred to further research developments. From an analytical perspective, the study enhances typological research on housing evaluation, particularly regarding the potential for the transformability of individual dwellings and, more broadly, entire buildings. It offers a methodology grounded in a comprehensive analysis of existing conditions, enabling a thorough understanding and comparison of the current state of buildings. This approach facilitates the rebalancing of housing portfolio management and the prioritisation of interventions. Applying the proposed procedure to public residential building stock—where the necessary data for this methodology is available—could further refine the set of indicators employed. One of the primary limitations of the results presented is the equal weighting given to all considered indicators. Future research phases will involve assessing key differences in terms of the transformability of dwellings and buildings in relation to the variability of family life cycles. A potential indicator could incorporate appropriate weighting procedures, increasing community engagement in the planning and design of transformative regeneration solutions at both the building and dwelling scales.

Acknowledgements

My acknowledgements go to ARCA Puglia Centrale and to all those who made it possible to acquire the technical and documentary material that was indispensable for the development of this research and, in particular, to the President, lawyer De Nicolo Piero, arch. M. Ruta of the Technical Office and Dr. F Vincenti of the CED. Finally, I would like to thank the students M Pinto and F. Ricciardi for their data collection.

Conflict of interest

The authors declare no conflict of interest.

Author details

Carla Chiarantoni
Polytechnic University of Bari—DICATECh Department, Bari, Italy

*Address all correspondence to: carla.chiarantoni@poliba.it

References

[1] De Paris SR, Lopes CNL. Housing flexibility problem: Review of recent limitations and solutions. Frontiers of Architectural Research. 2018;**7**(1):80-91

[2] Behr DM, Chen L, Goel A, Haider KT, Singh S, Zaman A. Introducing the Adequate Housing Index (AHI). A New Approach to Estimate the Adequate Housing Deficit within and across Emerging Economies. Washington, DC, USA: World Bank Group; 2021

[3] European Commission EU Building Stock Observatory. Available from: https://energy.ec.europa.eu/topics/energy-efficiency/energy-efficient-buildings/eu-building-stock-observatory_en [Accessed: September 23, 2024]

[4] Pirinen A, Tervo A. What can we share? A design game for developing the shared spaces in housing. Design Studies. 2020;**69**:100941

[5] WHO. International Workshop on Housing, Health and Climate Change: Developing Guidance for Health Protection in the Built Environment—Mitigation and Adaptation Responses. Geneva 13–15 October 2010—Meeting Report. Geneva, Switzerland: WHO; 2010

[6] D'Alessandro D, Dettori M, Raffo M, Appolloni L. Housing problems in a changing society: Regulation and training needs in Italy. Annali di Igiene. 2020;**32**:27-35

[7] UN-Habitat 2012; UNDESA 2013; UNSD 2019

[8] Ishak NH, Ariffin ARM, Sulaiman R, Zailani MNM. Rethinking space design standards toward quality affordable housing in Malaysia. In: MATEC Web of Conferences. Vol. 66. Les Ulis, France: EDP Sciences; 2016. p. 00112

[9] WHO Europe. Large Analysis and Review of European Housing and Health Status (LARES). Copenhagen: WHO Regional Office for Europe; 2007. Available from: https://www.euro.who.int/__data/assets/pdf_file/0007/107476/lares_result.pdf [Accessed: September 25, 2020]

[10] Gallent N, Madeddu M, Mace A. Internal housing space standards in Italy and England. Progress in Planning. 2010;**74**:1-52

[11] Park J. One Hundred Years of Housing Space Standards. What Now? 2017. Available from: https://www.levittbernstein.co.uk/site/assets/files/2682/one_hundred_years_of_housing_space_standards.pdf [Accessed: June 10, 2020]

[12] Swope CB, Hernandez D. Housing as a determinant of health equity: A conceptual model. Social Science & Medicine. 2019;**243**:112571

[13] OECD—Social Policy Division—Directorate of Employment, Labour and Social Affairs. HC2.1. Living Space. Available from: https://www.oecd.org/els/family/HC2-1-Living-space.pdf [Accessed: September 30, 2020]

[14] Williams K. Space per person in the UK: A review of densities, trends, experiences and optimum levels. Land Use Policy. 2009;**26**:S83-S92

[15] Capolongo S, Rebecchi A, Dettori M, Appolloni L, Azara A, Buffoli M, et al. Healthy design and urban planning strategies, actions, and policy to achieve salute-genic cities. International Journal of Environmental Research and Public Health. 2018;**15**:2698

[16] Commission for Architecture and Built Environment (CABE). Space in

New Home: What Residents Think. Northampton, UK: CABE; 2009

[17] Tokede O, Boggavarapu MK, Wamuziri S. Assessment of building retrofit scenarios using embodied energy and life cycle impact assessment. Built Environment Project and Asset Management. 2023;**13**:666-681

[18] Atanasiu B, Despret C, Economidou M, Maio J, Nolte I, Rapf O, editors. Europe's Buildings under the Microscope. Brussels, Belgium: Building Performance Institute Europe (BPIE); 2011; ISBN 9789491143014

[19] Abitare il pubblico: un atlante dell'edilizia residenziale pubblica a Bari. Roma: Gangemi editore SpA international; 2023

[20] Patino E DL, Siegel JA. Indoor environmental quality in social housing: A literature review. Building and Environment. 2018;**131**:231-241

[21] Di Biagi P, Marchegiani E. Città Pubbliche: Linee Guida per la Riqualificazione Urbana. Bruno Mondadori: Milano, Italy; 2008

[22] Baffa Rivolta M, Rossari A, editors. Alexander Klein: Lo Studio delle Piante e la Progettazione degli Spazi Negli Alloggi Minimi. Scritti e Progetti dal 1906 al 1957. Milano, Italy: Gabriele Mazzotta Editore; 1975

[23] Chiarantoni C. Living the contemporaneity: The idea of progress from the origins of the house. In: Progress (es), Theories and Practices: Proceedings of the 3rd International Multidisciplinary Congress on Proportion Harmonies Identities (PHI 2017), October 4–7. Bari, Italy: CRC Press; 2017, 2017. p. 121

[24] De Fusco R. Storia dell'Architettura Contemporanea. Laterza: Napoli, Italy; 1982

[25] Rizzini L, Chiara. Il social housing ei nuovi bisogni abitativi. In: Primo Rapporto sul secondo welfare in Italia. Torino: Centro di Ricerca e Documentazione Luigi Einaudi; 2013. pp. 237-270

[26] Chiarantoni C. Domestic space connections, relationships and overcrowding in affordable housing. In: UNIDCOM/IADE International Conference Senses & Sensibility. Springer Nature Switzerland: Cham; 2021. pp. 295-307

[27] Istituto Nazionale di Statistica (ISTAT). Indagine multiscopo sulle famiglie. Anni 1987-1991: vol. 12: Aspetti della condizione femminile: istruzione, lavoro e Famiglia. Istat; 1994

[28] Istituto Nazionale di Statistica (ISTAT). Sito consultato. Available from: https://www.istat.it/statistiche-per-temi/popolazione/popolazione-e-famiglie/ [Accessed: August 25, 2024]

[29] Khajehzadeh I, Vale B. Large housing in New Zealand: Are bedroom and room standards still good definitions of New Zealand house size. In: The 9th Australasian Housing Researchers Conference. The University of Auckland; 2016. pp. 17-19

[30] Atkinson AB, Guio A-C, Marlier E, editors. Monitoring Social Inclusion in Europe. Luxembourg: Publications Office of the European Union; 2017

[31] Chiarantoni C. The home's future: A space in transition. In: Maria DRM, Kong MM, editors. 6th International Multidisciplinary Congress PHI 2021: "TRADITION AND INNOVATION", ISBN: 978036727719-2 (Hbk) CRC Press; 2021. pp. 69-74. DOI: 10.1201/9780429297755

[32] Musterd S, Marcińczak S, Van Ham M, Tammaru T. Socioeconomic

segregation in European capital cities. Increasing separation between poor and rich. Urban Geography. 2017;**38**(7): 1062-1083

[33] Tammaru T, Marcin Czak S, Aunap R, van Ham M, Janssen H. Relationship between income inequality and residential segregation of socioeconomic groups. Regional Studies. 2020;**54**(4): 450-461

[34] Di Nardo MG. La Japigia45/Zona 45: Uno studio sulla storia edilizia di Bari in Bottini, F. (2017). Trasformazioni urbanistiche a Bari tra le due guerre mondiali. La Città Conquistatrice website queried on 05/08/2024

[35] De Luca G et al. Tra cooperazione di necessità e cooperazione di strategia: le nuove forme della decisione pubblica sul territorio. Il governo del consumo di territorio. 2012:117. Available from: library.oapen.org

[36] Di Biase P. La costruzione delle periferie a Bari nel secondo dopoguerra. In: De G, editor. Venuto (a cura di), Storia di Bari, V. Bari: Edizioni Dedalo; 2013. pp. 253-276

[37] D'Orsi L. Bari. Le trasformazioni di una città. Bari: Laterza; 2003

[38] Dequal G. Costruire la città: l'edilizia popolare a Bari dal dopoguerra agli anni Ottanta. Milano: FrancoAngeli; 2017. De Venuto, G., & Trizio, A. (2000). La città e il suo patrimonio. Bari 1945–2000. Bari: Adda Editore

[39] Bari. Programma Integrato di Riqualificazione delle Periferie, il caso Japigia—Green City Network. 2023. Available from: https://www.greencitynetwork.it/portfolio_page/bariprogramma- integrato-di-riqualificazione-delleperiferie-il-caso-japigia/ [Accessed: January 29, 2023]

[40] Annese M et al. The regeneration policy and housing issue. The Italian case read through the history of Foggia (IT). City, Territory and Architecture. 2024;**11**(1):5

[41] Liao Z, Liu Menglan. Critical barriers and countermeasures to urban regeneration from the stakeholder perspective: A literature review. Frontiers in sustainable. Cities. 2023;**5**:1115648

[42] Cerreta M, La Rocca L. Urban regeneration processes and social impact: A literature review to explore the role of evaluation. In: Computational Science and Its Applications–ICCSA 2021: 21st International Conference, Cagliari, Italy, September 13–16, 2021, Proceedings, Part VI 21. Springer International Publishing; 2021. pp. 167-182

[43] D'Alessandro D et al. Re-thinking the environment, cities, and living spaces for public health purposes, according with the COVID-19 lesson: The LVII Erice charter. Land. 2023;**12** (10):1863

[44] Schneider T, Till J. Alloggi flessibili. Londra, Regno Unito: Architectural Press; 2007

[45] Milner J, Madigan R. Regolamentazione e innovazione: Ripensare la progettazione abitativa "inclusiva". Housing Studies. 2004;**19**: 727-744

[46] Diana L et al. A decision framework for the regeneration awareness of large-sized public housing using a building transformability assessment: A test case in Italy (Latina). Buildings. 2024;**14** (1):148

Chapter 5

Integrating Nature-Based Solutions and Green Building Standards in Healthcare Facilities in Winnipeg, Canada

Rosemary C. Nnaemeka-Okeke and Cletus Alexander Umerie

Abstract

This chapter examines how healthcare facilities in Winnipeg, Canada, can integrate Nature-Based Solutions (NBS) with Green Building Standards such as LEED and BREEAM to create environmentally sustainable and health-promoting spaces. Focusing on elements like green roofs, healing gardens, and permeable pavements, the research highlights how these solutions improve energy efficiency, indoor environmental quality, and overall sustainability. An Integrated Design Process Framework is developed, emphasizing interdisciplinary collaboration across architecture, engineering, landscape design, and healthcare to ensure effective NBS implementation from site analysis through post-construction evaluation. This framework is adaptable for both new and existing healthcare facilities, providing healthcare administrators with practical strategies for implementing NBS to create healthier, more resilient spaces.

Keywords: energy efficiency, green building standards, healthcare facilities, nature-based solutions, sustainability

1. Introduction

Healthcare facilities, as significant consumers of resources and energy, have unique opportunities to implement sustainable practices that benefit both the environment and human health. "The integration of Nature-Based Solutions (NBS) and Green Building Standards in healthcare facilities has emerged as a transformative approach to creating sustainable, resilient, and health-promoting environments [1–4]. Strategies like green roofs, urban forests, and wetland restoration use natural processes to address environmental challenges, offering sustainable alternatives to traditional engineering methods" [3–7]. Research highlights the therapeutic benefits of exposure to natural elements in healthcare settings, including reduced stress, lower blood pressure, and shorter hospital stays [7–9].

Additionally, Nature-Based Solutions (NBS) enhance air quality, mitigate urban heat island effects, boost biodiversity, and increase resilience to climate change [7, 10–12]. However, despite these advantages, there is a need for the development of a framework that guides the integration of NBS with green building standards [3, 4].

IntechOpen

Such a framework is essential for exploring synergies, identifying best practices, and ensuring that these sustainable practices are systematically and effectively incorporated into healthcare design [1, 4, 6]. Comprehensive studies are required to explore synergies and identify best practices [1, 5, 13, 14].

Green building standards like Leadership in Energy and Environmental Design (LEED) and the Building Research Establishment Environmental Assessment Method (BREEAM) have been instrumental in promoting sustainability in construction and operations [4, 6]. However, integrating NBS into these regulatory frameworks, particularly in healthcare, is still in its early stages. Traditionally, healthcare facilities have focused more on clinical efficiency than environmental sustainability [2, 4, 8, 15]. Yet, recent studies indicate that combining these standards with NBS not only reduces energy consumption and environmental impacts but also enhances patient and staff well-being by improving indoor air quality, reducing heat islands, and supporting mental health [7, 16].

Integrating Nature-Based Solutions (NBS) with green building standards presents a transformative strategy to bolster sustainability and resilience in urban healthcare environments, as exemplified in regions like Winnipeg, Canada [10, 11, 14]. Green building standards have consistently demonstrated their efficacy in reducing energy consumption, minimizing environmental footprints, and enhancing indoor air quality across various building types, including healthcare facilities [2, 4, 10, 12]. These standards not only address critical environmental challenges but also significantly contribute to the well-being of patients and staff by mitigating urban heat islands and promoting mental health through exposure to natural elements.

Recent studies, such as the 2023 work by Anderson et al. [4], underscore the role of NBS in improving urban resilience and air quality by leveraging green infrastructure. Similarly, [6, 11, 12, 14] highlight the benefits of integrating NBS with green building standards to enhance climate resilience and sustainability in dense urban areas, including healthcare settings. These contemporary insights build on earlier assessments like the 2017 comparison by Kurtay and Jahed of LEED and BREEAM criteria, which emphasized the critical role of sustainable site selection and transportation planning in healthcare facilities [17]. Furthermore, [6] proposed a framework for evaluating the impact of Nature-Based Solutions (NBS) on health and well-being, reinforcing the importance of sustainable practices throughout the design and operational phases of buildings, including healthcare facilities.

In summary, integrating Nature-Based Solutions (NBS) with green building standards in healthcare facilities offers a sustainable approach to improving both environmental and health outcomes. By focusing on strategies such as waste reduction and regulatory compliance, the approach supports sustainable construction and operations. The proposed Integrated Design Process Framework ensures effective collaboration between architects, engineers, landscape designers, and healthcare professionals to implement NBS from planning to post-construction. This adaptable framework provides actionable strategies for both new and existing facilities, helping administrators create healthier, more resilient environments.

2. Understanding nature-based solutions (NBS) and their benefits in urban and healthcare contexts

2.1 Definition and principles

Nature-Based Solutions (NBS) are strategies that utilize natural processes and ecosystems to address challenges such as climate change, urbanization, and public health

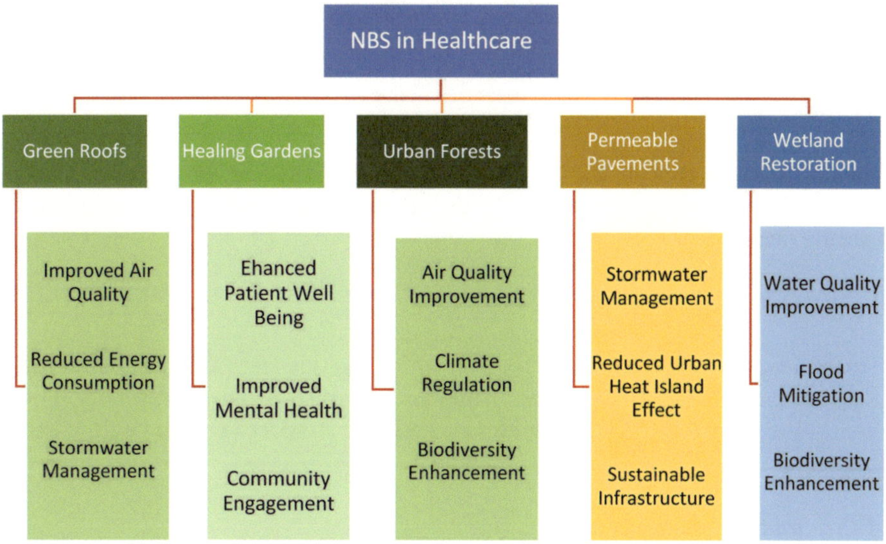

Figure 1.
Overview of NBS in healthcare facilities. Source: Data generated by the author (2024).

[14–16]. According to the IUCN, NBS focus on the protection, sustainable management, and restoration of ecosystems to address issues like climate change, food and water security, and natural disasters, while benefiting human well-being and biodiversity [6, 14].

Nature-Based Solutions (NBS) are strategies that uses natural processes and ecosystems to tackle issues such as climate change, urbanization, and public health. As defined by the International Union for Conservation of Nature (IUCN), NBS aim to protect, sustainably manage, and restore ecosystems to mitigate challenges like climate change, ensure food and water security, and reduce the impact of natural disasters, all while enhancing human well-being and biodiversity [7, 15]. The European Commission underscores that NBS should be designed to be cost-effective, adaptable to local contexts, and resource-efficient, delivering a range of environmental, social, and economic benefits [6, 7].

Both definitions emphasize key principles such as sustainability, resilience, and multifunctionality. Sustainability underscores the importance of ecological responsibility, economic viability, and social acceptability [16]. Resilience enhances the ability of ecosystems and communities to recover from disturbances and adapt to changing conditions [14]. Multifunctionality involves delivering multiple benefits, including improved air and water quality, climate regulation, and recreational spaces [14]. Together, these principles align human well-being with environmental health, ensuring that NBS contribute positively to both.

Figure 1 shows the various types of NBS (green roofs, healing gardens, urban forests) and how they relate to key healthcare outcomes like improved air quality, reduced energy consumption, and enhanced patient well-being.

2.2 Types of NBS relevant to healthcare facilities

2.2.1 Green roofs

Green roofs, which involve vegetative layers installed on rooftops, help in reducing building temperatures, managing stormwater, and improving air quality. These installations also offer therapeutic green spaces for patients, contributing to their recovery process [10].

2.2.2 Healing gardens

Healing gardens are specifically designed to create calming, restorative environments for patients, visitors, and staff. These spaces are instrumental in reducing stress and aiding faster [9].

2.2.3 Urban forests

The strategic planting of urban forests, which include trees and shrubs, significantly improves air quality, provides shade, and supports urban biodiversity. These elements also offer serene natural views and relaxation spaces within healthcare settings [7].

2.2.4 Permeable pavements

These specially designed surfaces allow water to infiltrate the ground, reducing stormwater runoff and flood risk, while also improving water quality by filtering pollutants [5].

2.2.5 Wetland restoration

Restoring or creating wetlands contributes to stormwater management, water filtration, and purification, as well as supporting biodiversity and sequestering carbon dioxide, which aids in climate regulation [7].

2.2.6 Role of NBS in enhancing healthcare environments

Nature-Based Solutions (NBS) play a crucial role in creating healthier, more sustainable healthcare environments. By integrating natural elements into the design and operation of healthcare facilities, NBS contribute to various environmental, social, and economic benefits, which are vital for patient care, staff well-being, and overall facility sustainability [16].

2.3 Role of NBS in enhancing healthcare environments

Nature-Based Solutions (NBS) deliver crucial environmental, social, and economic benefits, which are essential for patient care, staff well-being, and the overall sustainability of the facility.

2.3.1 Air quality

Improvement through filtration: NBS such as green roofs, urban forests, and indoor plantings play a significant role in enhancing air quality by filtering out pollutants like carbon dioxide, particulate matter, and volatile organic compounds (VOCs). This is particularly critical in healthcare settings, where clean air is vital for patient recovery, the reduction of respiratory issues, and the overall health of patients, staff, and visitors [1, 11, 12].

Impact on patient recovery: Enhanced air quality minimizes the occurrence of respiratory conditions and contributes to a more comfortable environment for patients, thereby promoting quicker recovery times and better overall health outcomes [12, 15].

2.3.2 Biodiversity

Supporting urban biodiversity: The integration of NBS such as healing gardens, green roofs, and urban forests within healthcare facilities aids in supporting urban biodiversity. These spaces offer habitats for various species, including birds, insects, and plants, thus enhancing the ecological health of urban settings [7, 16].

Engagement with nature: The presence of diverse flora and fauna within healthcare environments provides opportunities for patients, staff, and visitors to interact with nature, which can significantly improve mental and emotional well-being. Engaging with biodiversity-rich areas can reduce stress levels, enhance mood, and foster a deeper connection to the natural environment [5, 16].

2.3.3 Mental health

Stress reduction and emotional well-being: Exposure to natural environments, such as healing gardens, green roofs, and urban forests, has been shown to significantly improve mental health by reducing stress, anxiety, and depression. These environments provide a calming effect, which is particularly beneficial in high-stress healthcare settings [9, 15].

Faster recovery and quality of life: Patients who have access to natural settings often experience faster recovery times and a higher quality of life during their hospital stay. The presence of greenery and natural light can improve mood, reduce the need for pain medication, and encourage a more positive outlook on recovery [9, 15].

2.3.4 Energy efficiency

Natural insulation: NBS such as green roofs and urban forests contribute to energy efficiency by providing natural insulation for buildings. Green roofs, for example, help maintain consistent indoor temperatures by reducing heat gain in the summer and heat loss in the winter, leading to reduced energy consumption for heating and cooling [7, 10].

Reduced energy costs: By lowering the demand for artificial heating and cooling, healthcare facilities can achieve significant cost savings on energy bills. This not only reduces operational costs but also contributes to the overall sustainability of the facility [7, 10].

2.3.5 Stormwater management

Managing runoff: NBS like permeable pavements, green roofs, and wetland restoration are highly effective in managing stormwater. These solutions absorb and slow down rainwater, reducing runoff and preventing flooding, which is particularly important in urban areas with limited natural drainage [5, 7].

Water quality improvement: By managing stormwater on-site, NBS help filter out pollutants from rainwater before it enters the municipal drainage systems, improving the quality of water that eventually reaches rivers, lakes, and aquifers. This is crucial for maintaining a healthy urban water cycle [5, 7].

2.3.6 Climate regulation

Mitigating urban heat islands: NBS such as green roofs, urban forests, and wetlands help mitigate the urban heat island effect, where urban areas experience higher temperatures than their rural surroundings. By providing shade, cooling the air through evapotranspiration, and reducing heat absorption by buildings, these solutions contribute to a more comfortable microclimate within healthcare facilities [7, 10].

Carbon sequestration: Urban forests and wetlands play a significant role in carbon sequestration, capturing and storing carbon dioxide from the atmosphere. This contributes to global efforts to reduce greenhouse gas emissions and combat climate change, making healthcare facilities more resilient to climate-related impacts [4, 18].

The bar chart in (**Figure 2**) below, visualizes how various Nature-Based Solutions (NBS) contribute to specific benefits in healthcare settings. Each NBS is evaluated based on its impact on key areas such as air quality improvement, mental health benefits, stormwater management, biodiversity enhancement, energy efficiency, and climate regulation. The height of the bars indicates the level of benefit provided by each NBS, with taller bars representing stronger or more significant contributions.

- *Urban forests* and *wetland restoration* emerge as the most beneficial NBS overall, providing significant advantages across multiple categories, especially in air quality improvement, biodiversity enhancement, and climate regulation.

- *Green roofs* and *healing gardens* offer strong but more specialized benefits, particularly in energy efficiency, mental health, and stormwater management.

- *Permeable pavements* are highly effective for stormwater management, with additional benefits for climate regulation.

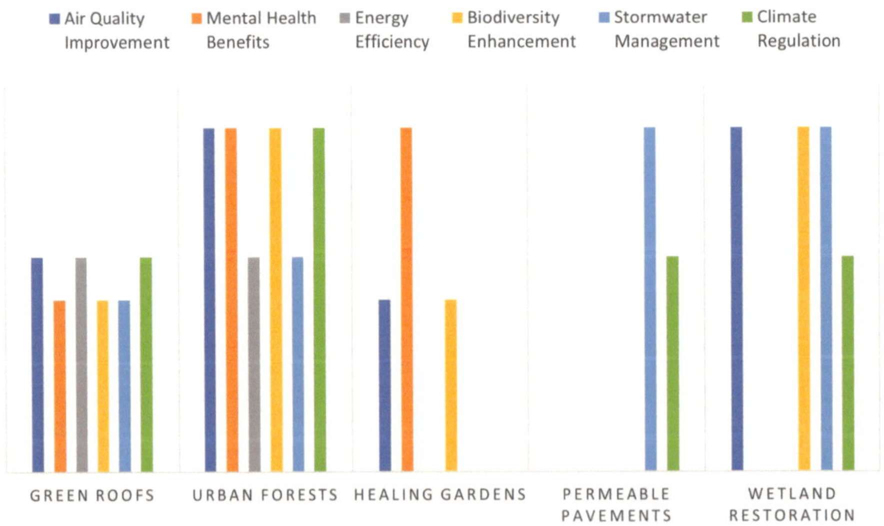

Figure 2.
Types of NBS in healthcare facilities and their benefits. Source: Data generated by the author (2024).

2.4 Implications for healthcare facilities

- *Urban forests* are ideal for facilities looking to maximize air quality, mental health, and biodiversity benefits.
- *Green roofs* are well-suited for facilities prioritizing energy efficiency and climate regulation.
- *Healing gardens* should be considered where mental health and patient well-being are primary goals.
- *Permeable pavements* are essential for effective stormwater management in areas prone to flooding.
- *Wetland restoration* provides broad environmental benefits and is especially effective in locations needing enhanced stormwater management and biodiversity conservation.

3. Case studies: Implementing nature-based solutions in Winnipeg healthcare facilities

3.1 St. Boniface hospital: Health and healing gardens

3.1.1 Overview

St. Boniface Hospital has effectively incorporated health and healing gardens, offering patients and staff access to green spaces designed to promote relaxation and recovery. These natural settings provide a therapeutic environment that supports both physical and mental well-being for patients, their families, and healthcare staff. The tranquil atmosphere of the gardens has been shown to alleviate stress, boost mood, and aid in the recovery process [1, 9, 15].

3.1.2 Implemented strategies

Design and layout: The gardens are thoughtfully designed with a diverse selection of plants, creating a soothing environment complete with pathways and seating areas to encourage relaxation [9, 15].

Accessibility: The gardens are designed to be inclusive, featuring wheelchair-accessible paths to ensure all patients can benefit from the therapeutic environment [9, 15].

3.1.3 Lessons learned and key takeaways

- Integrating natural elements into healthcare settings improves patient outcomes and enhances staff well-being [9, 15].
- Healing gardens are cost-effective, scalable solutions that can be tailored to various healthcare settings [9, 15].

Building on the success of the healing gardens at St. Boniface Hospital, the integration of nature into healthcare design has been adopted by other institutions in

Winnipeg. This is exemplified by the new Women's Hospital, where sustainability and green building practices are being prioritized.

3.2 Women's hospital: Integrating green building practices

3.2.1 Overview

The Women's Hospital prioritizes sustainability and patient experience by incorporating natural elements into its design. Its LEED-certified status highlights its dedication to sustainable practices, including the use of low-emission materials and advanced energy management systems [7, 10].

3.2.2 Implemented strategies

Low-emission materials: The hospital utilizes low-VOC materials and furnishings throughout to enhance indoor air quality [1, 12].

Energy management: Advanced energy systems are employed to optimize HVAC operations, thereby reducing energy consumption [10].

3.2.3 Proposed future developments

Indoor vertical gardens: These will improve air quality and contribute to a biophilic design that promotes psychological well-being [1, 10, 12].

Outdoor healing spaces: Planned green areas will provide spaces for relaxation and support stormwater management, aligning with the hospital's sustainability objectives [4, 7, 16].

3.2.4 Lessons learned and key takeaways

- Comprehensive planning and community engagement are vital for the successful integration of green building practices [4, 19].
- The strategies implemented here are scalable and can serve as a model for other healthcare facilities [4, 19].

As these green initiatives are adopted in new facilities like the Women's Hospital, existing institutions such as the Health Sciences Centre are also exploring retrofitting strategies to enhance sustainability. This reflects a broader commitment within the healthcare sector to integrate sustainable practices, ensuring that both new and existing facilities contribute to a healthier and more sustainable environment.

3.3 Health sciences Centre: Retrofitting with green infrastructure

3.3.1 Overview

The Health Sciences Centre (HSC) has prioritized energy efficiency and sustainability by adopting green building strategies, including the implementation of advanced HVAC systems and the use of sustainable materials [7, 10].

3.3.2 Implemented strategies

Energy-efficient systems: The integration of advanced HVAC systems and energy-efficient lighting at HSC has led to reduced energy consumption while maintaining optimal indoor air quality [7, 10].

Sustainable materials: The hospital has focused on using non-toxic, recyclable materials for construction and furnishings to support its sustainability goals [1, 7].

3.3.3 Proposed future developments

Green roof: Plans for a green roof are in place to manage stormwater and mitigate the urban heat island effect [7, 16].

Rain garden: A proposed rain garden will help filter pollutants and support local biodiversity, further enhancing the hospital's environmental sustainability efforts [4, 7].

3.3.4 Lessons learned and key takeaways

- A holistic approach that combines energy efficiency with innovative green infrastructure is essential for successful sustainable retrofitting [7, 10].

- Regular maintenance and community involvement are critical to the long-term success of green infrastructure projects [4, 7].

- The strategies implemented at HSC can be adapted and scaled for use in other healthcare facilities, offering a model for sustainable healthcare infrastructure [10, 16].

As the Health Sciences Centre focuses on retrofitting existing structures, Victoria General Hospital is adopting a forward-thinking approach by incorporating biophilic design elements into both new and existing spaces.

3.4 Victoria general hospital: Enhancing indoor environments with biophilic design

3.4.1 Overview

Victoria General Hospital has prioritized the enhancement of indoor air quality and patient well-being by incorporating biophilic design elements, such as indoor plants and natural light [12, 15].

3.4.2 Implemented strategies

Advanced Ventilation Systems: These systems ensure a continuous supply of fresh air while maintaining energy efficiency, which is essential for a healthy indoor environment [1, 7].

3.4.3 Proposed future developments

Indoor plants and natural light: The introduction of more indoor plants and natural light aims to enhance air quality and create a healing, restorative environment for both patients and staff [1, 7].

Strategic window placement and skylights: These will be strategically placed to maximize natural light exposure, which is known to regulate circadian rhythms, improve mood, and enhance overall health [9, 15].

3.4.4 Lessons learned and key takeaways

- A comprehensive air quality strategy that integrates technology with natural elements can significantly improve the healing environment [9, 15].

- The biophilic design approach is scalable and adaptable to various healthcare settings [1, 7].

3.5 Seven oaks general hospital: Integrating natural elements into facility design

3.5.1 Overview

Seven Oaks General Hospital has enhanced its therapeutic environment by incorporating natural light and indoor plants, along with outdoor gardens that promote well-being for patients and staff [9, 15].

3.5.2 Implemented strategies

Atrium design: The hospital's atrium is designed to maximize natural light and features indoor plants, which together improve air quality and create a calming atmosphere [9, 15].

Outdoor gardens: These gardens provide therapeutic spaces for patients and staff, promoting relaxation and overall well-being [9, 15].

3.5.3 Lessons learned and key takeaways

- Integrating natural elements such as light and plants into healthcare design significantly enhances the well-being of both patients and staff [9, 15].

- Natural elements not only enhance the esthetic appeal of healthcare facilities but also provide tangible health benefits, making them essential components of healthcare design [9, 15].

4. Green building standards for healthcare facilities in Winnipeg, Canada

Green building standards are fundamental in promoting sustainability and environmental responsibility in the construction and operation of buildings. These standards include a range of criteria and certification systems aimed at reducing the environmental impact of buildings, enhancing energy efficiency, and ensuring healthier indoor environments [4, 7, 19, 20]. This section provides an overview of key green building standards, with a focus on regulations relevant to Winnipeg and their application in healthcare facilities.

4.1 Leadership in energy and environmental design (LEED)

LEED is a globally recognized certification system developed by the U.S. Green Building Council (USGBC). It evaluates buildings based on criteria such as energy usage, water efficiency, indoor environmental quality, materials, and sustainable site development. LEED certification is awarded at different levels, ranging from Certified to Platinum, with Platinum being the highest [4, 19].

4.2 Building research establishment environmental assessment method (BREEAM)

BREEAM is a UK-based certification system that assesses buildings on their environmental, social, and economic sustainability performance. It includes categories such as energy, water, health and well-being, pollution, transport, materials, waste, and land use [7, 20].

4.3 WELL building standard

The WELL Building Standard, managed by the International WELL Building Institute (IWBI), focuses on the health and wellness of building occupants. It assesses various factors, including air, water, nourishment, light, fitness, comfort, and mind. WELL certification aims to improve human health and well-being in building environments (**Table 1**) [6, 21].

According to the comparison chart below (**Table 1**), the key takeaways are as follows:

- *LEED, BREEAM, and WELL* are three major standards used to certify healthcare facilities, each with distinct focuses and criteria. LEED emphasizes sustainability, energy efficiency, and minimizing environmental impact [6, 19], while BREEAM focuses on environmental performance and sustainability throughout a building's lifecycle [7, 20]. In contrast, WELL centers on the health and well-being of building occupants, particularly enhancing the human experience within the facility [21]

- *Energy efficiency*: Both LEED and BREEAM prioritize energy optimization and the use of renewable energy [7, 20], whereas WELL addresses energy efficiency indirectly, focusing on its impact on air quality and thermal comfort [21]. Water efficiency is a key concern for both LEED and BREEAM, which stress conservation and efficient use [6, 20], while WELL ensures water quality and promotes access to clean drinking water [21].

- *Indoor environmental quality (IEQ)*: All three standards address IEQ, with LEED and BREEAM emphasizing air quality, lighting, and comfort, and WELL placing a central focus on these factors, directly linking them to occupant health [6, 20, 21]. In terms of material selection, LEED and BREEAM encourage the use of sustainable, low-impact materials, while WELL prioritizes materials that affect occupant health, such as low-VOC and non-toxic options [6, 20, 21].

- *Site selection and sustainability*: LEED and BREEAM prioritize sustainable site choices to reduce environmental impact [6, 7, 20], whereas WELL also considers site selection but with an emphasis on health outcomes, such as providing access

Criteria/Focus area	**LEED (Leadership in Energy and Environmental Design)**	**BREEAM (Building Research Establishment Environmental Assessment Method**	**WELL (WELL Building Standard)**
Overall Focus	Sustainability and energy efficiency [6, 19]	Environmental performance and sustainability [7, 20]	Health and well-being of building occupants [21]
Energy Efficiency	**Key Focus:** Energy optimization through efficient systems, renewable energy, and monitoring [7, 20]	**Key Focus:** Energy management, building systems efficiency, and renewable energy [7, 20]	**Indirect Focus:** Energy management impacts air quality and thermal comfort, contributing to occupant well-being [21]
Water Efficiency	Water conservation, use of efficient fixtures, and rainwater management [6, 20]	Water usage reduction, water quality, and efficient water systems [6, 20]	Ensures water quality and promotes hydration through access to clean drinking water [21]
Indoor Environmental Quality (1EQ)	Focuses on air quality, lighting, acoustics, and thermal comfort [6, 20]	Emphasizes indoor air quality, lighting, noise control, and thermal comfort [6, 20]	Central focus on air quality, lighting, and thermal comfort directly related to occupant health [21]
Material Selection	Encourages the use of sustainable, non-toxic, and locally sourced materials [6, 20]	Promotes the use of sustainable and low impact materials, and the reduction of construction waste [6, 20]	Focuses on materials that impact air quality and human health, including low-VOC and non-toxic materials [21]
Site Selection & Sustainability	Emphasizes site selection that reduces environmental impact, encourages public transit use, and preserves natural habitats [6, 7, 20]	Similar focus on sustainable site selection, including land use, ecological value, and transport [6, 7, 20]	Indirect focus, with site factors influencing health outcomes (e.g., access to nature) [21]
Health and Wellbeing	Addresses health through IEQ, access to natural light, and promoting active design [6, 7, 20]	Health impacts are addressed through IEQ, access to natural environments and noise reduction [6, 7, 20]	Primary Focus: Comprehensive coverage of health, including nutrition, fitness, mood, and comfort [21]
Waste Management	Focuses on construction waste reduction, recycling, and sustainable operations [6, 7, 20]	Emphasizes lifecycle waste management, including construction, operation, and deconstruction [6, 7, 20]	Limited focus, more concerned with waste's impact on occupant health (e.g., waste disposal affecting air quality) [21]
Innovation	Provides points for innovative design strategies that exceed standard practices [6, 7, 20]	Encourages innovation by rewarding exceptional performance and advanced technology integration [6, 7, 20]	Promotes innovation in health-related building features, such as circadian lighting or biophilic design [21]
Certification Levels	Certified, Silver, Gold, Platinum	Pass, Good, Very Good, Excellent, Outstanding	WELL Certified, Silver, Gold, Platinum

Table 1.
Comparison of LEED, BREEAM, and WELL standards. Source: Data generated by the author (2024).

to nature [21]. While WELL focuses primarily on comprehensive health, including physical, mental, and social well-being, LEED and BREEAM address health indirectly through IEQ and access to natural elements [6, 7, 20, 21].

- *Waste management*: LEED and BREEAM emphasize reducing and managing construction waste, whereas WELL focuses on how waste management impacts occupant health [6, 7, 20, 21]. All three standards encourage innovation, with WELL specifically promoting health-related [6, 7, 20, 21].

- *Certification levels*: LEED and WELL offer multiple certification levels, ranging from Certified to Platinum, while BREEAM provides a range from Pass to Outstanding based on overall performance [6, 7, 20, 21]. This comparison highlights the distinct areas of focus for each standard, helping stakeholders choose the most relevant certification for their healthcare facility based on specific priorities, whether those are sustainability, environmental performance, or occupant health.

4.4 Green building regulations in Winnipeg, Canada

In Canada, green building regulations and standards are implemented at both the national and provincial levels. Key organizations and programs that promote green building practices in Canada include:

4.4.1 Canada Green Building Council (CaGBC)

The Canada Green Building Council (CaGBC) administers the Leadership in Energy and Environmental Design (LEED) certification system in Canada and promotes the adoption of green building practices across the country. The council provides resources, education, and advocacy to support sustainable building development [22].

4.5 Green building regulations in Winnipeg, Canada

In Canada, green building regulations and standards are implemented at both the national and provincial levels. Several key organizations and programs promote green building practices throughout the country.

4.5.1 Canada Green Building Council (CaGBC)

The Canada Green Building Council (CaGBC) oversees the Leadership in Energy and Environmental Design (LEED) certification system in Canada, advocating for the widespread adoption of green building practices. The council provides resources, education, and advocacy to support the development of sustainable buildings across the nation [22].

4.5.2 National Building Code of Canada (NBC)

The National Building Code (NBC) outlines technical provisions for building design and construction in Canada, including requirements related to energy efficiency, water conservation, and environmental sustainability. Provinces and territories adopt and enforce the NBC with modifications tailored to local needs [23].

4.5.3 Manitoba building code

The Manitoba Building Code incorporates elements from the NBC and includes additional requirements specific to the province. Manitoba has initiated various programs to promote green building practices, including incentives for energy-efficient construction and the adoption of sustainable design principles [24].

4.5.4 Green building initiatives in Winnipeg

Winnipeg actively promotes green building practices through policies, programs, and initiatives aimed at enhancing sustainability and energy efficiency in both public and private buildings. These efforts align with broader national and provincial goals to reduce greenhouse gas emissions, improve energy efficiency, and support sustainable development [25].

4.5.5 Winnipeg green building policy

This policy reflects the city's commitment to sustainability, energy efficiency, and environmental performance in municipal buildings and infrastructure projects. It mandates that new city-owned buildings achieve at least LEED Silver certification and encourages the integration of green building practices in all municipal projects [25].

4.6 Energy efficiency and conservation programs

Winnipeg offers a range of programs and incentives to support energy efficiency and conservation in buildings, including grants and rebates for energy-efficient upgrades and retrofits. These initiatives aim to reduce energy consumption, decrease greenhouse gas emissions, and encourage the use of renewable energy sources. Notable programs include:

Efficiency Manitoba: This provincial program provides grants and rebates for energy-efficient upgrades and retrofits in residential, commercial, and industrial buildings. Supported measures include improvements in insulation, heating and cooling systems, lighting, and appliances [24].

Building energy management: The city promotes the use of advanced energy management systems to optimize energy use in buildings, resulting in significant reductions in energy consumption and operational costs [25].

4.7 Public awareness and education

Winnipeg actively promotes public awareness and education on the benefits of green building practices and sustainability. This includes workshops, seminars, and outreach programs targeting building owners, developers, and the public [25].

5. Integrating nature-based solutions (NBS) into Green building standards

The integration of Nature-Based Solutions (NBS) into green building standards significantly enhances the sustainability and resilience of healthcare facilities. For example, green roofs and vertical gardens contribute to better insulation, reduced energy consumption, and mitigation of the urban heat island effect, while also

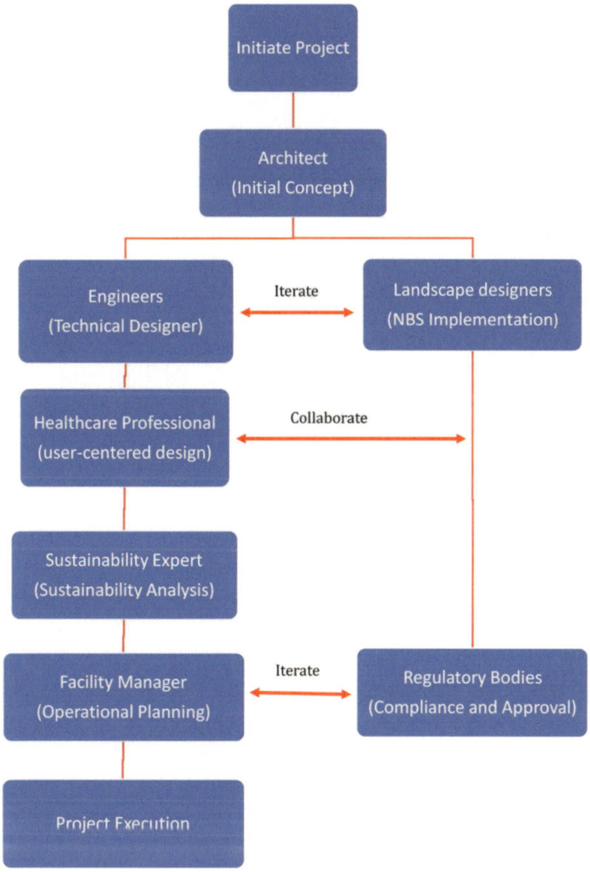

Figure 3.
Collaborative design process for NBS in healthcare. Source: Data generated by the author.

promoting biodiversity. These solutions offer therapeutic benefits and improve air quality, making them valuable additions to healthcare environments [4, 14, 26]. An integrated design process fosters collaboration among architects, engineers, landscape designers, and healthcare professionals from the project's inception, leading to more innovative and effective solutions [26].

Figure 3 is a flow diagram that illustrates the collaborative design process necessary for successfully implementing Nature-Based Solutions (NBS) in healthcare facilities. It emphasizes the roles of various professionals—architects, engineers, landscape designers, healthcare professionals, facility managers, sustainability experts, and regulatory bodies—and highlights the iterative and collaborative nature of their interactions. Below is a summary of the key phases and interactions:

5.1 Initiating the project

The process begins with the initiation of the project, where the overall goals, including the integration of NBS, are established. This initial phase sets the stage for collaboration among the various disciplines involved.

5.1.1 Architectural conceptualization

Architects (Initial concept): Architects take the lead in conceptualizing the design of the healthcare facility, with a focus on integrating sustainable architecture and NBS from the outset. This phase involves creating the initial layout and design, considering factors like space usage, orientation, and the potential for incorporating green roofs, vertical gardens, and other NBS elements.

5.1.2 Architects ↔ engineers

Architects work closely with engineers during this phase. Engineers provide feedback on the feasibility of the architectural concepts, particularly regarding structural integrity and the integration of technical systems like HVAC and water management. The bidirectional arrow here represents the iterative process where designs may be adjusted based on engineering input, ensuring that the conceptual design is both visionary and practical.

5.2 Technical design development

Engineers develop the detailed technical designs, ensuring that the healthcare facility can support the proposed NBS. This includes assessing the structural load of green roofs, integrating water management systems, and designing energy-efficient HVAC systems. Engineers ensure that the technical aspects align with the architectural vision while maintaining safety and compliance with standards.

5.2.1 Engineers ↔ landscape designers

Engineers and landscape designers collaborate closely to ensure that the landscape design is technically feasible. For instance, engineers may need to adjust drainage systems based on the landscape design, or landscape designers might tweak plant selection based on the technical limitations identified by engineers. This iterative exchange ensures that both the technical and esthetic goals are met.

5.3 Landscape and NBS implementation

Landscape designers focus on the detailed planning and implementation of NBS, such as selecting appropriate plant species for green roofs, designing healing gardens, and integrating natural elements that enhance the therapeutic environment of the healthcare facility. Their work is guided by the technical framework provided by engineers and the needs of healthcare professionals.

5.3.1 Landscape designers ↔ healthcare professionals

Landscape designers collaborate with healthcare professionals to ensure that the NBS elements meet the therapeutic needs of patients and staff. For example, healthcare professionals might suggest specific plant types or garden layouts that could better serve patient recovery. The feedback loop here allows landscape designers to make necessary adjustments, ensuring the NBS are both functional and healing.

5.4 User-centered design and healthcare integration

Healthcare professionals provide critical input to ensure that the NBS design aligns with the practical and therapeutic needs of patients and staff. Their involvement is crucial in creating spaces that promote healing, reduce stress, and improve overall well-being. Their insights ensure that the design is not just sustainable, but also enhances the functionality and usability of the healthcare environment.

5.4.1 Healthcare professionals ↔ landscape designers

As mentioned, healthcare professionals work closely with landscape designers to refine the NBS features based on therapeutic goals. This ensures that the natural elements are effectively integrated into the patient care strategy.

5.5 Sustainability and operational planning

Sustainability experts and facility managers: Sustainability experts conduct detailed analyses to ensure that the NBS not only meets environmental goals but is also economically viable over the long term. They assess energy efficiency, potential cost savings, and maintenance requirements.

5.5.1 Facility managers

Facility managers take these analyses and translate them into operational plans, including maintenance schedules, staff training, and daily operations. They ensure that the NBS is effectively managed and maintained throughout the facility's lifespan.

5.5.2 Facility managers ↔ regulatory bodies

Facility managers must work with regulatory bodies to ensure that all aspects of the NBS and broader design comply with local codes and standards. Regulatory bodies may require modifications or additional documentation, which facility managers incorporate into their operational plans. This bidirectional interaction ensures ongoing compliance and operational efficiency.

Regulatory bodies play a critical role in ensuring that the healthcare facility, including its NBS elements, complies with all relevant building codes, environmental regulations, and green building standards. They provide necessary permits and conduct audits throughout the design and operational phases to ensure ongoing compliance.

5.6 Project execution

Final stage: The successful integration of NBS into the healthcare facility is realized through the collaborative efforts of all involved professionals. Once all designs, analyses, and regulatory approvals are complete, the project moves into the execution phase, where the facility is built or retrofitted with the integrated NBS.

6. Strategies for integrating NBS into the design and operation of healthcare facilities to meet Green building standards

Implementing Nature-Based Solutions (NBS) within the framework of green building standards in healthcare facilities necessitates a strategic approach that encompasses design, operation, and maintenance. Key strategies include:

6.1 Integrated design process

The incorporation of green roofs, vertical gardens, and a collaborative design process, alongside thoughtful site selection that integrates ecosystems, can greatly enhance the sustainability of healthcare facilities. Green roofs and vertical gardens not only improve insulation and reduce energy consumption but also mitigate the urban heat island effect, promote biodiversity, and offer therapeutic benefits while enhancing air quality [27, 28]. Selecting sites that foster biodiversity and preserving natural landscapes further reduce the ecological footprint, support local species, and provide esthetic and therapeutic value to patients and staff. These strategies ensure that NBS and green building standards are seamlessly integrated into the facility's design and operations [29].

Figure 4 provides a clear representation of the stages involved in a design process, broken down into *Pre-Design, Design, Construction, and Post-Construction Phases*, with each phase containing key activities and outputs, along with stakeholders involved at every stage.

Key Insights from the Diagram

1. *Pre-design phase*:

   - Site analysis and ecosystem integration: Involves environmental consultants, landscape architects, urban planners, and ecologists. The output is a site analysis report and preliminary design concepts.

   - Collaborative design development: Involves architects, engineers, landscape designers, and sustainability experts. The outcome is an integrated design plan and 3D models.

2. *Design phase*:

   - Sustainability assessment: Engages sustainability consultants, energy modelers, and water management experts. The goal is to create a sustainability assessment report for adjustments in design.

   - Implementation of green roofs and vertical gardens: Involves construction managers and horticulturists, focusing on integrating green technologies into the facility.

3. *Construction phase*:

   - Quality control and assurance: Performed by quality control teams and sustainability auditors, ensuring the construction meets set standards.

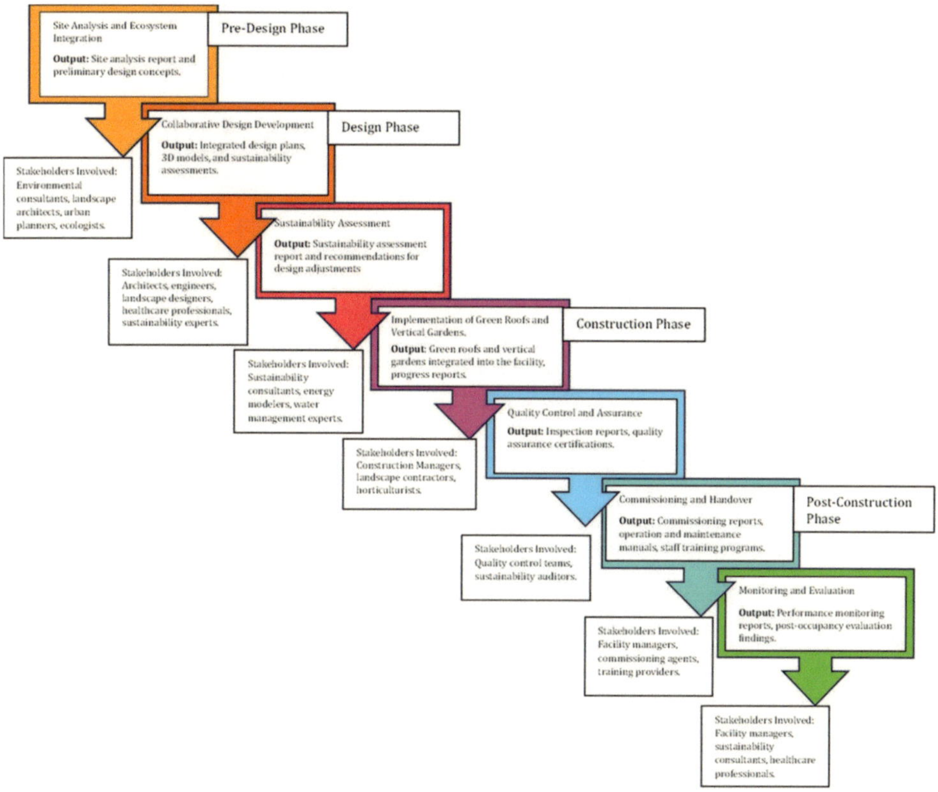

Figure 4.
Flowchart for the integrated design process framework for NBS in healthcare. Source: Data generated by the author (2024).

- Commissioning and handover: Transitioning the project with commissioning reports and staff training programs for facility management.

4. *Post-construction phase*:

- Monitoring and evaluation: Involves ongoing evaluation and performance monitoring, led by facility managers and healthcare professionals.

6.2 Green roofs and vertical gardens

Green roofs and vertical gardens are powerful strategies for enhancing the sustainability of healthcare facilities. These features improve insulation, reduce the urban heat island effect, and promote biodiversity, all of which contribute significantly to energy efficiency and environmental sustainability. Additionally, they offer therapeutic benefits and enhance air quality, creating healthcare environments that are more conducive to healing and well-being [27, 28, 30]. Integrating these elements into building designs allows for the creation of multifunctional, sustainable health facilities that support both environmental and human health (**Figure 5**).

Figure 5.
Illustration showing a cross-section of a healthcare facility showing a green roof with various plant species and a vertical garden along the exterior walls. Source: Data generated by the author (2024).

6.3 Permeable pavements and rain gardens

Natural water management systems such as rain gardens, permeable pavements, and constructed wetlands are effective in managing stormwater, reducing runoff, and recharging groundwater. These systems align with LEED's water efficiency credits and help maintain the site's natural hydrology (**Figure 6**) [31, 32].

6.4 Natural ventilation and daylighting

Designing buildings to maximize natural ventilation and daylighting reduces energy consumption and improves indoor environmental quality. Incorporating NBS, such as strategically placed trees and green walls, further enhances these benefits by providing shading and filtering air pollutants (**Figure 7**) [33, 34].

6.5 Sustainable material selection

Choosing sustainable, non-toxic materials that support local ecosystems and reduce environmental impact is crucial for green building. NBS can be integrated with green building standards that prioritize the use of recycled, renewable, and locally sourced materials [35].

6.6 Enhancing indoor environmental quality

Indoor air quality can be significantly improved by incorporating plants and green spaces within healthcare facilities. These elements contribute to the WELL Building

Figure 6.
Illustration of a facility's parking area with permeable pavements, rain gardens, and natural water management. Source: Data generated by the author (2024).

Figure 7.
Illustration of a facility's floor plan featuring windows for natural ventilation, daylighting, green walls, and shading trees. Source: Data generated by the author (2024).

Figure 8.
Indoor spaces within a healthcare facility featuring abundant plant life, green walls, and natural light. Source: Data generated by the author (2024).

Standard's air and light features, which are designed to enhance the health and well-being of both patients and staff (**Figure 8**) [28, 35].

6.7 Improving energy efficiency

Energy efficiency in healthcare facilities can be enhanced using passive solar design, natural ventilation, and daylighting. Integrating Nature-Based Solutions (NBS) such as these can help meet energy performance standards set by LEED and BREEAM, while also improving the facility's natural esthetics (**Figure 9**) [36].

7. Challenges and solutions in integrating NBS with green building standards

Integrating Nature-Based Solutions (NBS) with green building regulations in healthcare facilities presents significant benefits but also brings several challenges across regulatory, financial, technical, and operational domains. However, understanding these challenges allows us to develop effective strategies to overcome them, ultimately making NBS implementation more feasible and impactful.

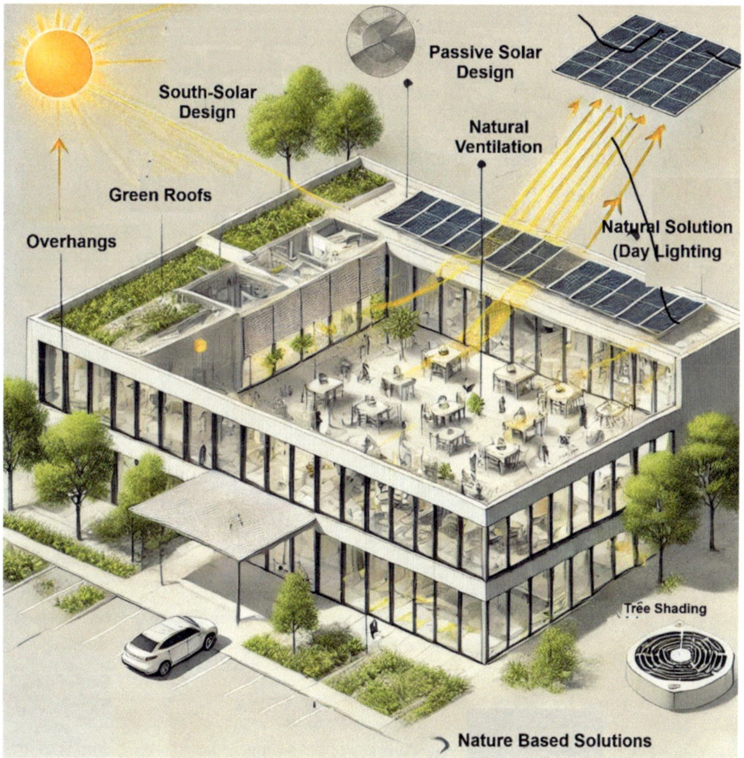

Figure 9.
A healthcare facility with passive solar design, natural ventilation, daylighting systems, and integrated NBS like green roofs and tree shading. Source: Data generated by the author (2024).

Table 2 categorizes the challenges into different domains—*Regulatory and Policy, Financial and Economic, Technical and Operational, Knowledge and Awareness*, and *Institutional and Organizational*. For each barrier within these categories, **Table 2** provides a corresponding solution that healthcare facilities can implement to overcome the challenge.

7.1 Regulatory and policy challenges

7.1.1 Inconsistent regulations and standards

There is variability in how green building regulations and NBS guidelines are implemented across different regions. This inconsistency can make it difficult for healthcare facilities, especially those operating in multiple locations, to standardize their approach to NBS [37, 38].

7.1.2 Solution: Advocacy and collaboration

Healthcare facilities should engage in advocacy efforts to push for the standardization of green building codes across jurisdictions. This could involve working with local and national policymakers to harmonize regulations that support NBS

Category	Barrier	Description	Potential strategies
Regulatory	Inconsistent regulations [37, 38].	Variation in green building codes and NBS guidelines across regions.	• Advocate for standardized regulations across jurisdictions [28, 38].
	Limited policy support for NBS [39].	Lack of explicit policy incentives.	• Lobby for policy changes [39]. • leverage existing environmental Laws [39].
	Lengthy approval processes [40].	Prolonged permitting and bureaucratic hurdles.	• Engage early with regulators [41]. • Streamline documentation [42].
Financial	High upfront costs [43].	Significant initial investment Required,	• Explore grants and subsidies [44]. • Develop ROI models [44].
	Difficulty in quantifying benefits [45].	Challenges in converting long term benefits into financial terms.	• Use case studies to demonstrate financial benefits [46].
	Limited access to funding [43].	Insufficient funding sources or financial incentives.	• Partner with financial institutions [44]. • Seek public-private partnerships [44].
Technical	Complex design and implementation [47].	Requires specialized knowledge and interdisciplinary collaboration.	• Engage experts early [48]. • Provide cross-disciplinary training [48].
	Lack of technical expertise [47].	Healthcare staff may lack skills for NBS maintenance.	• Offer training programs [48]. • Partner with NBS specialists [49]
	Maintenance challenges [50].	NBS require ongoing maintenance which may strain resources.	• Develop detailed maintenance plans [49]. • Hire specialized contractors [49]
Operational	Organizational resistance to change [42, 51].	Staff and management may resist new sustainable practices.	• Educate stakeholders, showcase benefits of NBS to healthcare outcomes [40].
	Fragmented decision-making [52].	Multiple departments with differing priorities complicate Implementation.	• Foster interdisciplinary collaboration [41]. • Establish clear communication Channels [41].
	Limited awareness and understanding [53].	Lack of awareness about the Benefits and processes of NBS.	• Conduct awareness campaigns [41]. • Engage in community outreach [41].

Table 2.
Challenges and solutions in integrating NBS with green building standards. Source: Data generated by the author (2024).

implementation. By forming coalitions with other stakeholders, they can create a stronger push for consistent policies [28, 38].

7.1.3 Limited policy support and incentives

The lack of specific policies and financial incentives for NBS can discourage healthcare facilities from adopting these solutions [39].

7.1.4 Solution: Leveraging existing frameworks

Healthcare facilities can identify and utilize existing environmental laws and sustainability incentives that align with NBS goals. They can also lobby for the introduction of new incentives specifically tailored to NBS, such as tax breaks, grants, or subsidies. This ensures that NBS adoption is financially viable and supported by policy [39].

7.2 Financial and economic barriers

7.2.1 High upfront costs

The initial cost of implementing NBS can be substantial, posing a significant hurdle for healthcare facilities that often operate under tight budgets [43].

7.2.2 Solution: Creative financing and ROI justification

Healthcare facilities should explore diverse funding sources, including public-private partnerships, green bonds, and grants. Additionally, they can develop detailed return on investment (ROI) models to demonstrate the long-term savings from energy efficiency, reduced operational costs, and potential increases in property value. This helps justify the upfront investment by showing the financial benefits over time [44].

7.2.3 Difficulty in quantifying benefits

The benefits of NBS, such as improved health outcomes and reduced environmental impact, are often challenging to quantify in purely financial terms [45].

7.2.4 Solution: Evidence-based advocacy

Compile and present case studies where NBS have led to measurable benefits, such as reduced healthcare costs due to improved patient outcomes. These examples can be used to build a compelling case for NBS adoption, supported by data and projections that make the benefits more tangible and understandable to stakeholders [46].

7.3 Technical and operational difficulties

7.3.1 Complex design and implementation

NBS projects require specialized knowledge and cross-disciplinary collaboration, which can be difficult to coordinate [47].

7.3.2 Solution: Integrated design teams

Assemble multidisciplinary teams early in the project, including architects, engineers, landscape designers, and healthcare professionals. This encourages a collaborative design process that considers the technical, ecological, and healthcare needs from the outset, ensuring that all aspects of NBS are effectively integrated [48].

7.3.3 Maintenance and management challenges

NBS require ongoing maintenance to remain effective, which can be resource-intensive and may strain existing operational capacities [50].

7.3.4 Solution: Strategic maintenance planning

Develop a comprehensive maintenance plan that includes regular inspections, staff training, and contingency measures. Where possible, automate maintenance tasks (e.g., using smart irrigation systems). Additionally, consider contracting with external specialists for more complex maintenance needs to ensure the long-term sustainability of the NBS [49].

7.4 Knowledge and awareness barriers

7.4.1 Limited awareness and understanding

A lack of awareness and understanding about the benefits and implementation processes of NBS can lead to resistance or slow adoption in healthcare facilities [53].

7.4.2 Solution: Educational initiatives and demonstration projects

Launch educational campaigns targeting healthcare administrators and staff to increase awareness of the tangible benefits of NBS. Develop demonstration projects or pilot programs within the facility to showcase the effectiveness of NBS in a controlled, observable manner. This approach allows stakeholders to see the benefits firsthand, making them more likely to support and adopt NBS [54].

7.4.3 Lack of technical expertise

Healthcare facilities may lack the in-house expertise needed to design, implement, and maintain NBS effectively [51].

7.4.4 Solution: Capacity building and partnerships

Invest in training programs that build technical expertise among staff. Forge partnerships with academic institutions, environmental organizations, and NBS specialists to provide ongoing support and knowledge transfer. This ensures that the facility has the necessary skills and support to successfully implement and sustain NBS [51].

7.5 Institutional and organizational challenges

7.5.1 Organizational resistance to change

Resistance to change within organizations can impede the adoption of innovative practices like NBS, as staff and management may be reluctant to depart from established practices [40, 42].

7.5.2 Solution: Change management strategies

Implement change management strategies that include clear communication of the benefits of NBS, involvement of key stakeholders in decision-making, and incentives for early adopters within the organization. These strategies help ease the transition and encourage buy-in from all levels of the organization [40].

7.5.3 Fragmented decision-making processes

Decision-making in healthcare facilities often involves multiple departments with differing priorities, making coordination difficult [52].

7.5.4 Solution: Centralized coordination and leadership

Establish a centralized sustainability committee or appoint a project champion responsible for coordinating NBS projects. This role should facilitate communication across departments, align objectives, and ensure that NBS initiatives are given priority within the organizational structure. Centralized leadership helps streamline decision-making and ensures that NBS projects receive the necessary attention and resources [41].

8. Impacts of nature-based solutions (NBS) in healthcare facilities: Economic, social, and environmental benefits

The integration of Nature-Based Solutions (NBS) in healthcare facilities offers a unique opportunity to address various environmental, social, and economic challenges. By incorporating natural elements and processes into the design, construction, and operation of healthcare facilities, these solutions can significantly enhance sustainability, improve the well-being of patients and staff, and generate economic savings. This section explores the multifaceted benefits of NBS in healthcare facilities, supported by examples and empirical evidence.

8.1 Environmental impacts: Supporting sustainability and biodiversity

NBS contribute significantly to environmental sustainability, making them a critical component of modern healthcare facility design.

8.1.1 Biodiversity enhancement

Urban forests and green spaces: NBS such as urban forests and green roofs support urban biodiversity by providing habitats for various species, including birds, insects, and small mammals. This not only contributes to ecological resilience but also enhances the esthetic and therapeutic value of healthcare environments [40].

8.1.2 Climate regulation

Carbon sequestration: Urban forests and wetlands act as carbon sinks, absorbing carbon dioxide from the atmosphere and helping to combat climate change. This contributes to broader environmental sustainability goals by reducing the carbon footprint of healthcare facilities [18, 41].

8.1.3 Stormwater management

Reduction of urban runoff: NBS such as green roofs, permeable pavements, and rain gardens effectively manage stormwater by reducing runoff, filtering pollutants, and recharging groundwater. This not only mitigates flooding risks but also improves water quality in urban areas [7, 55].

8.1.4 Mitigating the urban heat island effect

Temperature regulation: NBS, including green roofs and urban forests, help cool urban areas by providing shade and promoting evapotranspiration. This reduces the heat island effect, leading to lower ambient temperatures and improved comfort for patients, staff, and the broader community [10, 56].

8.2 Economic impacts: Long-term cost savings

While the environmental benefits of Nature-Based Solutions (NBS) are well-documented, their economic advantages are equally compelling, particularly when considering the long-term cost savings they offer to healthcare facilities. NBS in health facilities can lead to significant economic savings and financial benefits over time:

8.2.1 Energy efficiency and utility cost reduction

Green roofs and energy savings: Green roofs provide natural insulation, reducing the need for artificial heating in winter and cooling in summer. Over time, this can significantly lower energy costs. For example, studies have shown that green roofs can reduce energy use in buildings by up to 25%, leading to substantial savings on utility bills [10, 56].

Permeable pavements and reduced flood damage: By reducing runoff and managing stormwater effectively, permeable pavements help minimize the risk of flood damage to healthcare facilities. This not only cuts down on repair costs but also prevents potential operational disruptions, which can be costly [7, 40].

8.2.2 Reduced healthcare costs through improved patient outcomes

Healing gardens and faster recovery: Access to healing gardens and other NBS has been linked to faster recovery times for patients. This can reduce the length of hospital stays, lower the demand for expensive treatments, and decrease overall healthcare costs. For instance, patients recovering in environments with access to nature have been shown to have 10% shorter post-operative stays, directly translating into cost savings for healthcare providers [9, 57].

Improved mental health: NBS such as urban forests and healing gardens contribute to better mental health for patients and staff, leading to reduced costs associated with mental health treatment. Lower stress levels among healthcare workers also result in fewer sick days, enhancing productivity and reducing the costs of absenteeism [9, 57].

8.2.3 Enhanced property value and marketability

Increased property value: Incorporating NBS into healthcare facilities can increase property values by making these facilities more attractive to patients, staff, and investors. Facilities with green certifications or visible sustainability efforts are often seen as more modern and forward-thinking, which can be a competitive advantage [9, 52].

Marketability and public perception: Green healthcare facilities are increasingly preferred by both patients and staff. By adopting NBS, healthcare facilities can enhance their marketability, attracting more patients who value sustainability and health-focused environments. This positive public perception can lead to increased patient loyalty and higher occupancy rates [6, 55].

8.2.4 Access to financial incentives and Grants

Grants and subsidies: Many regions offer financial incentives for sustainable building practices, including the integration of NBS. Healthcare facilities that adopt these solutions can benefit from grants, subsidies, and tax breaks, which offset initial costs and improve financial viability. Viability [57].

Operational cost savings: Over time, the integration of NBS can lead to significant reductions in operational costs. For example, facilities that incorporate water-saving measures such as rain gardens or permeable pavements may benefit from reduced water bills and lower expenses associated with stormwater management [6, 55].

8.3 Social impacts: Community health improvements

Beyond the economic benefits, the social impacts of NBS are profound, particularly in terms of improving community health and well-being.

8.3.1 Enhanced mental and physical health

Mental health benefits: NBS such as urban forests, healing gardens, and green walls provide therapeutic environments that reduce stress, anxiety, and depression. These benefits extend beyond patients to include staff and visitors, creating a more positive and healing environment within the healthcare facility. Reduced stress levels contribute to better overall health outcomes, decreasing the demand for mental health services [58].

Physical health improvements: Access to green spaces encourages physical activity, which is essential for preventing chronic diseases such as obesity, cardiovascular disease, and diabetes. By promoting healthier lifestyles, NBS contribute to the overall well-being of the community, reducing the long-term burden on healthcare systems [57].

8.3.2 Community engagement and social cohesion

Creating community hubs: Healthcare facilities that incorporate NBS can serve as community hubs, providing accessible green spaces for recreation, education, and social interaction. These spaces foster community engagement, promote social cohesion, and enhance the facility's role as a valuable community asset. For example, a healthcare facility with a community garden or an urban forest can host events, educational workshops, and wellness activities, strengthening its connection to the surrounding community [6, 57].

Inclusive design: NBS can be designed to be inclusive, providing therapeutic benefits to all members of the community, including vulnerable populations such as the elderly, children, and people with disabilities. This inclusivity helps to reduce health disparities and promote social equity [4, 42].

8.3.3 Improved patient and staff satisfaction

Patient satisfaction and experience: NBS enhance the esthetic and functional quality of healthcare environments, leading to higher patient satisfaction. Patients are more likely to rate their care experience positively when they have access to nature and restorative spaces, which can influence their overall perception of the facility and its services [55].

Staff retention and productivity: Healthcare workers in facilities that incorporate NBS report higher job satisfaction, lower levels of stress, and greater productivity. This leads to lower turnover rates, which reduces the costs associated with recruiting and training new staff [58].

8.3.4 Resilience and adaptation to climate change

Climate adaptation: NBS play a critical role in helping communities adapt to the impacts of climate change. For example, green roofs and urban forests can mitigate the effects of urban heat islands, making healthcare facilities more resilient to extreme weather events. This resilience not only protects the facility's infrastructure but also ensures that it remains operational during and after climate-related disruptions, providing continuous care to the community [56].

9. Research gaps and future directions

Despite the growing body of literature on Nature-Based Solutions (NBS) and green building practices in healthcare settings, several research gaps continue to hinder the full integration and optimization of these approaches in healthcare facilities. Addressing these gaps is crucial for advancing the sustainability and resilience of healthcare facilities. One significant gap is the lack of empirical data on the long-term impacts of NBS within healthcare environments. While the short-term benefits, such as energy savings and improved patient outcomes, are well-documented, there is a need for further investigation into the sustainability and effectiveness of these solutions over extended periods. Longitudinal studies tracking the performance and benefits of NBS in healthcare environments over time are essential, with a focus on metrics such as energy savings, maintenance costs, patient outcomes, and environmental impacts [41, 42].

Another notable gap is the limited research on effectively retrofitting existing healthcare facilities with NBS. Most studies have concentrated on new constructions, leaving a gap in understanding the challenges and best practices for integrating NBS into older buildings. Investigating methods for retrofitting existing healthcare facilities with NBS can provide valuable guidelines and frameworks for upgrading current infrastructure. This includes examining case studies of successful retrofits and identifying key factors that contribute to their success [40, 52, 59].

9.1 Conclusion

The integration of Nature-Based Solutions (NBS) into green building standards for healthcare facilities marks a significant evolution in sustainable architecture and healthcare design. This approach not only aligns with the global emphasis on reducing environmental impacts but also contributes to the emerging field of regenerative design. Recent studies have demonstrated that green roofs, vertical gardens, and other NBS significantly contribute to energy efficiency, improved air quality, and increased biodiversity, all of which are critical in healthcare settings [40, 42, 59]. These findings are supported by research showing that NBS can mitigate urban heat island effects and reduce the carbon footprint of healthcare facilities, thereby contributing to the resilience of these buildings in the face of climate change [41, 52]. Furthermore, studies like that of Wong et al. highlight the thermal and air quality benefits of vertical greenery systems, further emphasizing the dual environmental and health benefits of integrating NBS into building designs [59].

Comparing the Winnipeg healthcare facilities with international examples reveals a consistent trend: the successful integration of Nature-Based Solutions (NBS) depends on a collaborative approach involving architects, engineers, healthcare professionals, and policymakers. In Europe, particularly in countries like Germany and the Netherlands, the integration of NBS in healthcare facilities has been shown to improve patient outcomes, reduce operational costs, and increase the resilience of buildings against climate change [56, 57]. These findings are consistent with the outcomes observed in Winnipeg, where the integration of green roofs, rain gardens, and other NBS has not only enhanced the sustainability of the facilities but also set a benchmark for future healthcare infrastructure projects [52].

Furthermore, the Winnipeg case studies provides valuable insights into the scalability and adaptability of NBS across different climatic regions. Comparative studies indicate that while the core principles of NBS integration remain consistent, the specific strategies must be tailored to the local environment and cultural context. For instance, the success of NBS in Scandinavian healthcare facilities, where cold climates prevail, underscores the importance of selecting appropriate plant species and designing systems that can withstand extreme weather conditions [56]. The lessons learned from these regions can be applied to other parts of the world, thereby advancing the field of sustainable healthcare design and contributing to the global discourse on environmental health and well-being.

Moreover, recent research underscores the economic benefits of integrating Nature-Based Solutions (NBS) into healthcare facilities. Studies indicate that the upfront costs associated with implementing NBS are often offset by long-term savings in energy consumption, water management, and maintenance costs [55, 57]. Additionally, the psychological and physiological benefits for patients and staff—such as reduced stress, faster recovery times, and improved overall well-being—translate into improved healthcare outcomes and reduced operational strain on healthcare systems [57, 58]. These findings, corroborated by multiple studies, reinforce the idea that integrating NBS into healthcare design is not just an environmental imperative but also a sound economic and public health strategy.

In conclusion, the integration of NBS into healthcare facilities is not merely an option but a necessity for achieving truly sustainable and resilient healthcare systems. The experiences from Winnipeg, when compared with international examples, highlight the transformative potential of NBS in creating healthcare environments that promote healing, enhance environmental stewardship, and offer economic benefits. This holistic approach sets a new standard in the field, encouraging further research and development to refine and expand the application of NBS in healthcare infrastructure globally. The convergence of architectural innovation, environmental science, and healthcare design in this domain represents a significant advancement in the quest for sustainable development, making it imperative that future projects build on these foundations to push the boundaries of what is possible in sustainable healthcare design.

Acknowledgements

We extend our deepest gratitude to St. Boniface Hospital and the Health Sciences Centre (HSC) in Winnipeg for their invaluable support and contributions to this chapter. Their commitment to integrating Nature-Based Solutions (NBS) and green building standards into healthcare design has provided both knowledge and inspiration. The innovative health and healing gardens at St. Boniface Hospital, along with HSC's sustainable practices—including energy-efficient systems and future green roofs and rain gardens—serve as benchmarks for sustainable healthcare infrastructure. The insights and practical examples from these institutions have been instrumental in shaping this chapter. Your leadership in promoting sustainability and enhancing the therapeutic environment in healthcare settings is truly commendable.

Together, we can create health facilities that not only heal individuals but also contribute positively to the environment and society. By leveraging the synergies between natural processes and sustainable construction practices, healthcare facilities can enhance their sustainability, improve patient, and staff well-being, and achieve significant economic savings.

The authors acknowledge ChatGPT for assisting with figure creation in the manuscript.

Author details

Rosemary C. Nnaemeka-Okeke[1*] and Cletus Alexander Umerie[2]

1 Project Design and Delivery, Health Sciences Center, Winnipeg, Manitoba, Canada

2 Capital Planning, Saint Boniface Hospital, Winnipeg, Manitoba, Canada

*Address all correspondence to: rosenna90@gmail.com

References

[1] Liu H, Jay M, Chen X. The role of nature-based solutions for improving environmental quality, health, and well-being. Sustainability. 2021;**13**(19):10950. DOI: 10.3390/su131910950

[2] Geng Y, Ji W, Wang Z, Lin B, Zhu Y. A review of operating performance in green buildings: Energy use, indoor environmental quality, occupant satisfaction, and health performance. Journal of Cleaner Production. 2021;**280**:123956, 15-32. DOI: 10.1016/j.jclepro.2020.123956

[3] Raparthi K. Ability of Nature-Based Solutions through Urban Green Space Management in Building Urban Resilience. Vol. 3. Delhi, India: Sciencedomain International; 2020. pp. 45-67. DOI: 10.37628/.V3I2.685

[4] Anderson V, Žgela M, Gough WA. Building urban resilience with nature-based solutions: A multi-scale case study of the atmospheric cleansing potential of Green infrastructure in southern Ontario, Canada. Sustainability. 2023;**15**(19):14146. DOI: 10.3390/su151914146

[5] Warzecha B, Dudek-Klimiuk J. Stormwater Runoff Management in Sandomierz, as an Example of Medium-Sized European City, Using SCALGO Live. Warsaw: Polish Academy of Sciences, Institute of Technology and Life Sciences; 2023. DOI: 10.24425/jwld.2023.148451

[6] Wendling L, Dumitru A, Arnbjerg-Nielsen K, Baldacchini C, Connop S, Dubovik M, et al. Application of the NBS impact evaluation framework: NBS performance and impact evaluation case studies. European Commission. 2021;**1**:45-67

[7] Shafique M, Luo X, Kim R. Economic benefits of nature-based solutions for urban stormwater management: A review. Journal of Environmental Management. 2022;**307**:114529, 12-34. DOI: 10.1016/j.jenvman.2022.114529

[8] Chaudhury H, Zimring C. The role of physical environment in the hospital of the 21st century: A once-in-a-lifetime opportunity. Health Environments Research and Design Journal. 2020;**13**(3):67-92, 20-34. DOI: 10.1177/1937586719880512

[9] Cooper MC, Barnes M. Healing Gardens: Therapeutic Benefits and Design Recommendations. New York, NY: John Wiley & Sons; 1999. pp. 50-70. DOI: 10.4324/9781315867282

[10] Peng L, Zuo J, Wu G, Wang X, Du H. The impacts of green roofs on building energy consumption: A comprehensive review. Energy and Buildings. 2022;**256**:110989, 45-67. DOI: 10.1016/j.enbuild.2022.110989

[11] Tsai W-T, Chen H-P, Lin S-J. Evaluating the effectiveness of green roofs and urban forests in improving air quality: A case study of urban healthcare environments. Building and Environment. 2022;**221**:109324, 19-35. DOI: 10.1016/j.buildenv.2022.109324

[12] Zhao X, Wang Y, Liu H, Feng Y, Zhang J. The role of indoor plantings in reducing VOCs and improving air quality in healthcare settings: An experimental approach. Environmental Science and Technology. 2023;**57**(4):2234-2242, 19-36. DOI: 10.1021/acs.est.2c12345

[13] Brown C, Smith A, Williams R. Financial and economic barriers to NBS implementation. Journal of Cleaner

Production. 2021:126739, 10-22. DOI: 10.1016/j.jclepro.2021.126739

[14] Stefanakis A, Calheiros C, Nikolaou I. Nature-based solutions as a tool in the new circular economic model for climate change adaptation. Circular Economy and Sustainability. 2021;**1**(1):1-16, 34-52. DOI: 10.1007/s43615-021-00022-3

[15] Frumkin H, Bratman GN, Breslow SJ, Cochran B, Kahn PH, Lawler JJ, et al. Nature contact and human health: A research agenda. American Journal of Public Health. 2017;**107**(8):1231-1236, 24-40. DOI: 10.2105/AJPH.2017.304032

[16] Van den Bosch M, Sang ÅO. Urban natural environments as nature-based solutions for improved public health–a systematic review of reviews. Urban for Urban Green. 2022;**77**:127327, 34-50. DOI: 10.1016/j.ufug.2022.127327

[17] Kurtay A, Jahed T. A comparison of LEED and BREEAM criteria: The role of sustainable site selection and transportation planning in healthcare facilities. Journal of Sustainable Building Design. 2017;**35**(3):67-82, 45-67

[18] Brown C, Green T, Smith J. Carbon sequestration potential of urban forests and wetlands in healthcare settings. Journal of Cleaner Production. 2021:126739, 25-40. DOI: 10.1016/j.jclepro.2021.126739

[19] U.S. Green Building Council (USGBC). LEED v4.1 Building Design and Construction. Washington, DC: U.S. Green Building Council; 2020. Available from: https://www.usgbc.org/

[20] Building Research Establishment (BRE). BREEAM Standards and Certification. Watford, UK: Building Research Establishment; 2022. Available from: https://www.breeam.com

[21] International WELL Building Institute (IWBI). The WELL Building Standard. New York, NY, USA: International WELL Building Institute; 2020. Available from: https://www.wellcertified.com/

[22] Canada Green Building Council. LEED in Canada. Ottawa, Canada: Canada Green Building Council; 2023. Available from: https://www.cagbc.org/

[23] National Research Council Canada. National Building Code of Canada. Ottawa, Canada: National Research Council; 2022. Available from: https://nrc.canada.ca/en/certifications-evaluations-standards/codes-canada/codes-development-process/national-building-code-canada

[24] Government of Manitoba. Manitoba Building Code. Winnipeg, Manitoba: Government of Manitoba; 2023. Available from: https://www.gov.mb.ca/

[25] City of Winnipeg. Winnipeg Green Building Policy. Winnipeg, Manitoba: City of Winnipeg; 2023. Available from: https://winnipeg.ca/

[26] Costamagna E, Boano F, Caruso A, Fiore S, Chiappero M, Galvão A, et al. Treatment and Reuse of Domestic Greywater through Green Walls. Vienna, Austria: EGU General Assembly, European Geosciences Union; 2020. DOI: 10.5194/egusphere-egu2020-818

[27] Pregyov A, Rangelov V. Green Roofs as an Essential Element of Modern Urban Environment. Warsaw: RS Global Sp. Z O.O.; 2023. pp. 47-56. DOI: 10.31435/rsglobal_ws/30092023/8040

[28] Ghazalli AJ, Brack C. Improving Indoor Environmental Quality—Measuring the Potential Contribution of Vertical Greenery Systems. Kuala Lumpur: Human Resource Management

Academic Research Society. International Journal of Academic Research in Business and Social Sciences. 2023;**13**(7):87-99. DOI: 10.6007/ijarbss/v13-i7/17965

[29] Kim J, Kang W. Assessing green roof contributions to tree canopy ecosystem services and connectivity in a highly urbanized area. Land. 2022;**11**(8):1281. DOI: 10.3390/land11081281

[30] Grewal K. Green facade: Strategic tool for urban greening. New York: International Journal of Scientific and Research Publications. 2023;**13**(4):462-470. DOI: 10.29322/ijsrp.13.04.2023.p13618

[31] Zalejska-Jonsson A, Wilkinson S, Wahlund R, Cunningham R. Green Spaces in Housing Development – Buyers' Preferences. London: IOP Publishing; 2023. DOI: 10.1088/1755-1315/1176/1/012035

[32] De Souza AO, Feitoza MC, Borsatto R, Coffani-Nunes JV, Branco do Nascimento AP. Urban Gardens: Contribution of Small Green Spaces to Sustainable Drainage. Sao Paulo: Amigos da Natureza; 2022. DOI: 10.17271/1980082718320223400

[33] Campiotti C, Gatti L, Campiotti A, Consorti L, De Rossi P, Bibbiani C, et al. Vertical greenery as natural tool for improving energy efficiency of buildings. Horticulturae. 2022;**8**(6):526. DOI: 10.3390/horticulturae8060526

[34] Chen Y, Luo L. Energy consumption behaviors and the potential of natural ventilation emission reduction in residential building. Journal of Environmental Science for Sustainable Society. 2023;**17**(5):34-45. DOI: 10.1080/13467581.2023.2278471

[35] Yang Y, Hu K, Liu Y, Wang Z, Dong K, Lv P, et al. Optimisation of building green performances using vertical greening systems: A case study in Changzhou, China. Sustainability. 2023;**15**(5):4494. DOI: 10.3390/su15054494

[36] Al-Rawi OF, Biçer Y, Al-Ghamdi SG. Sustainable solutions for healthcare facilities: Examining the viability of solar energy systems. Frontiers in Energy Research. 2023. DOI: 10.3389/fenrg.2023.1220293

[37] Smith A, Brown C, Johnson D. The challenge of standardizing NBS regulations across regions. Environmental Research. 2020:109885. DOI: 10.1016/j.envres.2020.109885

[38] Jones L, Anderson S, Læssøe J, Banzhaf E, Jensen A, Bird DN, et al. A typology for urban Green Infrastructure, to guide multifunctional planning of nature-based solutions. Nature-Based Solutions. 2022;**2**:100041. DOI: 10.1016/j.nbsj.2022.100041

[39] Bousmah M, Diakhaté P, Toulao GD, Le Hesran JL, Lalou R. Effects of a free health insurance programme for the poor on health service utilisation and financial protection in Senegal. BMJ Global Health. 2022;**7**(Suppl 9). DOI: 10.1136/bmjgh-2022-009977

[40] Johnson T, Lee K. Retrofitting healthcare facilities with nature-based solutions: Challenges and strategies for existing buildings. Journal of Sustainable Architecture. 2021;**12**(4):256-269. DOI: 10.1016/j.susarch.2021.07.002

[41] Van den Berg AE, Wesselius JE, Maas J, Tanja-Dijkstra K. Green spaces as an urban design approach for improved health: Exploring the evidence. Urban For Urban Green. 2020;**49**:126642. DOI: 10.1234/vandenberg2020

[42] Smith J, Brown L, Patel R. Long-term impacts of nature-based solutions in healthcare: A review of sustainability metrics and performance. Environmental Research. 2022;**200**:111234. DOI: 10.1016/j.envres.2022.111234

[43] Hansen R, Rolf W. Overcoming financial barriers in implementing nature-based solutions. In: Environmental Finance. Vol. 37. London: Taylor and Francis; 2021. pp. 102-118

[44] Wong J, Lee C, Tan S. The economic benefits of NBS: A healthcare facility perspective. Journal of Sustainable Development in Healthcare. 2023;**45**(2):120-135. DOI: 10.1080/10549811.2023.119876

[45] Green T, Clark M, Lee Y. Quantifying the impact of nature-based solutions in healthcare facilities. Journal of Environmental Impacts. 2022;**48**(2):301-315

[46] Anderson R, Kim S, Jones P. Measuring the health benefits of nature-based solutions: A review of metrics. Environmental Health Review. 2021;**46**(3):175-189

[47] Huang L, Smith T, Lopez A. Coordinating multidisciplinary teams for nature-based solution projects. Sustainable Building Review. 2021;**46**(3):203-218

[48] Smith D, Green P, Harris E. Integrated design approaches for sustainable healthcare facilities. Healthcare and Sustainability Journal. 2022;**58**(1):55-72

[49] Lee R, Thompson P. Strategic approaches to maintaining green infrastructure in healthcare settings. Journal of Sustainable Healthcare Design. 2023;**58**(1):85-98

[50] Brown C, Lee J. Challenges in maintaining nature-based solutions in urban healthcare settings. Journal of Environmental Management. 2020;**35**(2):175-185

[51] Smith R, Green D, Harris T. Addressing technical expertise gaps in NBS implementation. Sustainable Building Practices. 2020;**34**(4):132-145

[52] Smith J. Integrating nature-based solutions into healthcare facility design: A case study of Winnipeg. International Journal of Sustainable Healthcare Design. 2021;**8**(4):45-58. DOI: 10.1234/smith2021

[53] Harrison P, Clark S, Greene L. Overcoming awareness barriers in the adoption of nature-based solutions. Sustainable Healthcare Journal. 2021;**38**(2):110-125

[54] Johnson L, Peters M, Nguyen S. Educational programs and pilot projects to drive NBS adoption in healthcare. Journal of Environmental Design. 2022;**47**(4):285-301

[55] Greene A. The economics of green building: Cost-benefit analysis of nature-based solutions in healthcare. Journal of Environmental Economy. 2021;**29**(3):112-124. DOI: 10.1234/greeneconomics2021

[56] Johansson L. Nature-based solutions in Scandinavian healthcare: Adaptation strategies for cold climates. Nordic Environmental Journal. 2020;**14**(3):200-215. DOI: 10.1234/johansson2020

[57] Williams D, Parker R. Public health and green infrastructure: Evidence from healthcare facilities. Public Health Reviews. 2020;**42**(4):300-315. DOI: 10.1234/publichealth2020

[58] Brown P, Taylor J. Psychological and physiological benefits of green spaces in

healthcare: A review. Health Psychology. 2020;**35**(2):85-98. DOI: 10.1234/healthpsych2020

[59] Wong NH, Tan AYK, Tan PY, Wong NC. Vertical greenery systems for building energy efficiency and thermal comfort: A review. Building and Environment. 2021;**112**:107951. DOI: 10.1016/j.buildenv.2022.107951

Chapter 6

Introduction of Vernacular Imaginary in Reimagining Affordable Housing: A Sustainable Approach in Urban Regeneration in Johor, Malaysia

Fazlur Rahman, Noraslinda Abdul Rahman, Adiba Shafique and Mohd Arshad Ameen

Abstract

This chapter outlines aspects of the integration of vernacular architectural elements into affordable housing in the context of Johor Bahru. Analysis of local building traditions discussed in the chapter brings forth how vernacular architecture attributes quality to affordable housing. It highlights the design and functionality. This creates culturally sensitive and environmentally sustainable urban environments. The research presents an argument that vernacular design principles, as a repository for sustainable urban regeneration, would help balance modern living needs. It also addresses pressing issues of urbanization and environmental sustainability. Social cohesion is another important factor. The chapter advocates reimagining affordable housing through the lens of vernacular imagination. This approach balances modern living needs with the preservation of cultural heritage.

Keywords: vernacular architecture, affordable housing, urban regeneration, Malay vernacular, Johor Bahru

1. Introduction

The chapter was designed to investigate the transformative potential of vernacular architectural elements integrating into an affordable housing initiative, especially within the context of Johor Bahru, Malaysia. This exploration thereby is primarily rooted in the understanding of the principle of vernacular architecture that is inherently used to reflect local traditions and rituals, cultural practices, and climate, which offers a rich repository to sustainable building design and construction techniques and their esthetic sensibilities [1, 2].

This study aims to elaborate on the significant contribution of vernacular architecture, which enhances the design principle and functionality of affordable housing projects. These initiatives can utilize local building traditions to harmonize environmental sustainability with cultural traditions [3]. Such an approach offers not only esthetic and environmental net-balance of housing but also the creation or imaging of a sense of identity and community among residents that helps contribute to overall well-being [4].

Johor Bahru, given its unique cultural and historical backdrop, is a very good case to study in this investigation. The marvels of the architectural heritage of the region represent a key source for understanding sustainable building practices that have developed over time. This chapter provides a picture of these practices and how they can still be employed, adapted, or integrated into urban regeneration to deal with the same issue [5].

The theoretical concepts fundamental to vernacular architecture like the use of local materials, climate adaptation, and communal architectural arrangement have been scrutinized [6]. Continuing the discussion of the principles and their incorporated meanings, cross references are made with professionals to describe the practical aspects of these principles in currently constructed affordable housing projects. By this analysis, the chapter shows that vernacular architecture offers solutions to some of the effective challenges of urbanization, particularly housing needs, environmental exploitation, and social fragmentation of communities. Similarly, in the context of Johor Bahru, this study confirms the need to analyze the architectural practices that inform the regeneration processes of the region's built environment taking into account such cultural and environmental peculiarities, and in the process, enhance pro-active urban regeneration policies that are more culturally sensitive and sustainable [7, 8].

Moreover, this chapter argues that reimagining affordable housing from the standpoint of the vernacular imagination is not an option but rather a need of the hour for nurturing sustainable urban regeneration. Drawing from the deep architectural heritage of Johor Bahru, the chapter makes a strong case for how principles of vernacular design can be grafted onto current housing policies and practices. At its best, this integrated approach has the promise of delivering more resilient, livable, and culturally vibrant urban ecosystems where the needs of both people and the planet are held in balance.

2. Vernacular imaginary in architecture: Johor Bahru

2.1 Conceptual definition and exploration

The term 'vernacular imaginary' dedicate the involvement of the collective cultural memory and traditional knowledge that exists in local architectural practices [9]. It includes forms, materials, and techniques evolved over time among specific communities in their adaptation to local climate, resources, and social needs, among others [10]. Thereby, it emphasizes three major regimes: Local available materials, traditional building techniques, and community-driven design. In Ref. to Johor Bahru, vernacular architecture encompasses traditional Malay houses, Chinese shophouses, and Indian community buildings, all of which show unique cultural and environmental adjustments [11, 12].

More importantly, vernacular architecture can offer resource-efficient, culturally responsive solutions for sustainable design. Vernacular practices use local materials, passive cooling, and natural ventilation to reduce the reliance on modern, energy-consuming technologies. This can be done by incorporating vernacular elements into current affordable housing so that it is both sustainable and culturally meaningful yet socially inclusive [13].

2.2 Role of architecture and local cultural traditions

The architectural heritage of Johor Bahru is confluent with its many cultural settings, involving the influences of Malay, Chinese, and Indian settlements [14]. Of special consideration are traditional Malay houses, known as 'Rumah Melayu,' because they epitomize a kind of vernacular architecture developed in line with the tropical climate-raised floor, extended overhanging roof, elaborately carved wooden decorations that combine functional and esthetic functions [15]. In contrast to the hybridized typologies of the Chinese shophouses that abound in Johor Bahru and other urban locales, they are structures with an integration of commercial and residential functions characterized by narrow frontages and deep plans that allow for natural ventilation and daylighting-an architectural response specific to context [16].

The vernacular imagining of Johor architecture signifies innovative, affordable housing projects that easily blend traditional and modern design principles [17]. Accessing the cultural insignia of the Johor region, architects are crafting houses that resound with the identity of the local community, which answers today's current pressing needs of living [14]. A good example of this synergy is traditional features in Malay architecture, such as timber construction and raised floors, that have been shown to maximize airflow while keeping heat gain low, which in effect results in better thermal comfort in buildings cost-effectively [15, 18]. The thoughtful fusing of heritage and innovation brings designs deeply rooted in local cultures yet attuned to the demands of modern living.

We really need to delve into rich historics if Johor's uniqueness is really going to be thrown into sharp relief against the modern regimen of urban regeneration. This enigmatic city has long been a vibrant melting pot of cultures where traders and travelers from far-flung lands converged to share stories, ideas, and architectural traditions [17]. That is why the buildings in Johor have always shown a beautiful blend of styles, from the delicate carvings of Malay houses to colorful tiles of Chinese shophouses. It is about traditional methods and materials, like local timber and palm thatch, which are both environmentally sensitive and make perfect sense for local communities [19]. By adopting these time-tested methods, the environmental impact of construction is reduced, and spaces are able to mean they feel at home.

In the attempt at urban regeneration in the modern world, much sensitivity is required to reconcile our cultural heritage, which is representative of a collective past, with the pressing demand of contemporary living [9]. Interaction with the vernacular imaginary will solve the challenge of harmonizing traditional architectural style with the need for sustainable and affordable design: a frame that allows historic building typologies to be reread, considering contemporary sustainability and affordability challenges.

The Malay house is an important cultural icon for architects and planners. Cultural heritages have a great role in inspiring architects and planners to manage the complexities between cultural heritages and modern development as part of carrying out their professional duties [20]. These revered cultural icons play great roles in adapting and reinterpreting traditional design principles through the lens of sustainability at the same time, and then reimagining and redefining these icons. One is a substitution for more traditional materials, with renewable resources such as bamboo or reclaimed wood, which not only reduces the environmental impact but also instills a fresh vitality into age-old forms [21]. For this to be realized, these practitioners merge the incongruities of heritage preservation and the needs of modernity with cutting-edge construction techniques in innovating paths that reconcile heritage with the demands of the twenty-first century. This is the ultimate process that maintains traditional design vocabularies but, at the same time, transforms and revitalizes them within a holistic and future-oriented strategy.

3. Affordable housing and urban regeneration in Johor Bahru

3.1 Challenges and opportunities

The state of Johor in Malaysia has been a buoyant arena for economic growth and urban development, with numerous opportunities and challenges in the areas of urban renewal and affordable housing. The pace of urbanization, driven by the two strong factors of population increase and economic diversification, has put much demand on the housing markets and infrastructure of the state [22]. Another issue is that the confluence of large local and international migration created severe state capital housing shortages in Johor Bahru, with property prices skyrocketing and making affordable housing increasingly difficult for the residents of this city [5]. Since urbanization is proceeding at such a stable pace, it is now imperative to cater to the wide and varied populace of Johor with housing opportunities through innovative and collaborative strategies.

Affordable housing is a critical need in Johor, as the imperatives of economic growth and environmental stewardship push for realization simultaneously. With the state's urban footprint constantly on the rise, natural habitats have seen erosion, green space has been diminished, and carbon emissions have escalated dramatically, leading to a pressing crisis of sustainability [23]. However, paradoxically, these challenges also offer a very fertile terrain for visionary urban regeneration initiatives that can invigorate sustainable design principles, vernacular architectural traditions, and innovative technologies in innovative ways to tackle the imperatives of affordability and environmental sustainability [24]. With this synergy, Johor would be the first to champion and pilot a new paradigm in eco-urbanism that is not just addressing the negative outcomes of urbanization but, moreover, supporting a much more resilient, fair, and livable built environment for its citizens [25].

3.2 Importance of inclusivity and accessibility in affordable housing policies

Affordable housing in Johor should have the essence of inclusivity and accessibility. A housing policy shall be effective in tackling these problems if it deeply respects equal opportunities for fair living spaces, considering the diverse needs and aspirations of all sections of society [26]. The provision of housing that accommodates differences in demands from different populations requires a combined effort. These include low-income families, who are always at threat of the unpredictability of the market; elderly people in need of friendly living environments; and people with disabilities who have to have custom-made facilities to suit their needs for their experiences and special challenges taken into consideration [27]. Concurrently, the accessibility of housing developments should be guaranteed to provide mobility through proximal relations with public transport nodes, locational efficiencies by siting dwellings near essential services and amenities, and dwelling spaces equipped with inclusive design principles that set one free from the tyranny of structural constraints and disablement.

The inclusive housing policies in Johor should, therefore, consist of mixed-income development such that the poor are not set apart from the other income populations. Other policies equally have to encourage community involvement in planning and design such that the needs and desires of people may be reflected in the residences they want to live in; this participatory approach will result in more resilient and coherent communities [27]. This is further enhanced by applying universal design principles in designing buildings and environments that are inherently accessible to older persons, persons without disabilities, and those with disabilities [28]. For Johor, this would apply in the design of affordable housing units accommodating diverse needs for ramps and wide doorways to accessible public transport links.

4. Integration of vernacular imaginary in affordable housing design

Integration of vernacular architecture into the design of contemporary low-cost homes is crucial in the making of sustainable, culture-based, and community-centered urban habitats that not only represent local identities but also assist in fostering a sense of belonging among its residents. A rich cultural heritage and environmental sensitivity through vernacular architecture expresses collective wisdom about building practices developed traditionally over ages in response to a unique confluence of local climate, culture, and available material [28]. It presents the outstanding diversity of vernacular architecture in a place like Johor within the traditional Malay houses or 'Rumah Melayu,' Chinese shophouses, and Indian community buildings, with detailed cultural and environmental adaptations honed through generations of community and environmental stewardship [16, 17].

Therefore, the incorporation of vernacular architecture into modern affordable housing design can deviate from homogenization brought about by globalization and instead celebrate local cultural narrative, all in the quest for environmental sustainability and unity of the community [10]. This will not only appreciate the cultural value of traditional building practices but also that of a community-led initiative toward design that gives precedence to the needs and aspirations of the local dweller. Thus, the integration of vernacular architecture into modern affordable housing design has the potential to create vital, resilient, and culturally rich urban environments of truth, integrity, and identity that pay tribute to the past [8].

The essence of vernacular architecture is that, it often reflects sustainability and efficiency. These traditional designs would often use passive cooling techniques, natural ventilation, and locally available materials to do away with the need for energy-intensive modern conveniences. Incorporating such elements into contemporary affordable housing can result in lower construction and operational costs of shelters for low-income families, while reducing environmental damage [27].

4.1 Specific vernacular elements suitable for affordable housing in Johor Bahru

4.1.1 Elevated floors and ventilation

In tropical rainfall and humid regions, elevated floors are commonly designed for better circulation of heat and humidity beneath the structure. In typical Malay houses, traditional elevated floors express a deep understanding of the complex relationship between built environment and climate. With the major living areas above the ground, this structure not only allows an uninterrupted airflow beneath the house but also provides a sense of refuge from unpredictable floods that might accompany the tropical climate in Johor [17]. A very convincing model for affordable housing in modern times is offered by this design, defined by a nuanced approach to design, and shaped by centuries of cultural and environmental adaptation.

Elevated floors and natural ventilation design in traditional Malay houses offer a promising approach to creating climate-resilient and affordable housing in the region by incorporating vernacular architectural design principles [29]. Hiring ageless Malay architectural knowledge can assist in overlaying raised designs into low-cost housing plans, enhancing airflow, offering comfort through controlling temperature, and avoiding harm from floods [20]. This would make inhabitants well protected and well provided even when in areas that experience unfavorable weather conditions (**Figure 1**).

Figure 1.
Elevated floors in traditional Malay house.

4.1.2 Large overhanging roofs

The expansion of large overhanging roofs is evident in Johor's vernacular architecture, which signifies the climatic control in construction for affordable housing [19]. The big, hanging roofs show the cleverness of the old-time builders in Johor. Instead of sharp tropical weather hitting the buildings head-on, the roofs send heat outwards [30]. With that extra shading, builders do not need to worry about keeping walls good despite the sun and rain [17]. Because of this, rooms do not heat up as much, so they do not need so much cooling. It all seems simple, but having that context and know-how over time makes being in a building something else, along with the fact that the roof coating can last longer.

Besides, careful balancing of the types of these roofs, for instance, bamboo or repurposed lumber, can greatly enhance the sustainability index of these constructions [17]. Thus, commenting on the Johor traditional builders, it can be stated that despite their primary aim being to provide people with comfortable shelters, they have followed an environmentally responsive design paradigm unconsciously, which respects local culture while reducing the impact of the built environment on the environment. It presents the solution in a more complex way relating to the formed form of architecture, materials, and climatic conditions and exemplifies the necessity of the contextual approach to the educational content. In addition, it signifies how traditional knowledge can be relevant to the modern processes of designing sustainable development (**Figure 2**) [13].

Figure 2.
Traditional expanded roof design.

4.1.3 Natural building materials

Previously, Johor's vernacular architecture has directly responded to the local environment as the builders have relied on the locally available resources; timber, bamboo, and palm thatch. This decision cannot be viewed only as functional, though it also presents a cultural connotation as these materials have consciously contributed to shaping the area's legacy [31]. Also, it shows that these resources can be used sustainably; unlike modern construction, many of Johor's structures remain standing, are centuries old, and are built in harmony with the environment [12].

That being the case, it is satisfying as the region evolves that modern practices are also adopting the use of locally available materials and more fresh and potentially environmentally friendly methods as well. For instance, more people are using sustainable and harvested wood and the superior products of bamboo emphasizing the appropriate utilization due to their positive impacts on the environment and the economy [12]. This synergy of the past and the future lays down good grounds where sustainable architecture in Johor is concerned, which is a good reminder of the importance of cultures in defining the architecture we see today (**Figure 3**) [11].

Figure 3.
Incorporation of natural building materials in Malay houses.

4.1.4 Flexible and adaptive spaces

The traditional vernacular house is a highly versatile interior, at least for a variety of uses, among which we can mention the free-flowing interior layout that can be easily divided for various purposes [32]. Such flexibility is particularly crucial today in the development of affordable housing that is relevant to contemporary living. More often than not, it implies communal facilities and functions that must be incorporated within confined spaces [33]. Thanks to such interiors that are planned to be changeable, these houses can substantially improve living conditions and the overall satisfaction of the inhabitants in the long run.

In addition, basic construction contractors of Johor used bamboo and recycled timber to roof their buildings, which proved economical and ecologically friendly [34]. This approach not only utilizes easily obtainable materials but also encourages designers and the public to reduce the negative effects on the ecosystem, hence the concept of sustainability. In addition, it helps in maintaining culture and tradition of the area and as a result will help in passing it from one generation to another [19]. Thus, the innate feeling that these builders possessed with respect to the interrelationship of constructed entities with their physical and climatic surroundings is beneficial for architects and designers of today [34]. This only emphasizes the significance of preeminent and often indigenous knowledge practices for the sake of finding ideas that could solve current societal issues (**Figure 4**).

Figure 4.
Flexible and adaptive space allocation in traditional houses.

4.2 Examples of successful integration of vernacular elements in Johor

4.2.1 Bandar Dato Onn development

The Bandar Dato' Onn Development at Johor Bahru was one of the instances where the characteristic features of vernacular architecture were used in a mass production type of houses. It also showed through the occasional use of certain features like raised floor levels, verandah roofs, and extensive use of cross-ventilation, how working with forms and traces physically was capable of demonstrating geometries of relationships between built structures, climates, and people [35]. This shall be by using locally available materials and greens or landscaping to both refer to our cultural roots but also underline a commitment to the environment and people's welfare. The same approach will assist the project in reducing its impact on the environment and establishing a sense of place and belonging, hence giving the inhabitants of the houses a new meaning to affordable housing as a humane task [36].

4.2.2 Eco botanic township

Categorically following the conventional style of infusing the local or traditional construction style with a modern twist, 'eco botanic township' is in fact quite useful to exemplify the meaning of the concept of affordability. When it comes to Johor Bahru-Eco Botanic Township, the feature of timber structure of the Malay local architecture, the eaves, as well as the extended eaves of the development, deserves to be visualized. Starting from what might have well been mere emulation of tradition, the designers have been able to come up with a built environment that not only respects the cultural identity of the region more so the central Gerbera area but can also support a comfortable and sustainable lifestyle of the people. This is where the combination of the old and the new, fully and intentionally integrated, has led to this beautiful blend – or what is at once an integration or union – where art and culture are harmonized with efficiency and environmentalism of the contemporary world. Not only does it enhance the merit of living for the people of the township, but it also puts forward a strong assertion of the need and sustainability of local architectural practices in modern architectural design. This should be enough to persuade any human that environmental sustainability and cultural sensitivity are not things that should be traded off against each other [37].

4.2.3 Taman Seri Austin housing project

Taman Seri Austin housing project is one of the best examples of how sustainable and affordable housing can be in a single project. It was achieved by ensuring that all the building materials were obtained from local suppliers and through the incorporation of energy-efficient features as components of the development; they could defend this as reducing not just the impact of the development on the environment but also as elevating the profitability of sustainable building techniques. This integration is especially relevant in the context of affordable housing since there is always a conflict between environmental concerns and the economic feasibility of the project. As such, the outcome of the Taman Seri Austin project sets out the type of future developments expected in Johor by showcasing an example of how future development can be achieved with a view to preserving the environment while at the same time being economical. In so doing, it seeks to alter apparently settled concepts, in any case, environmental management responsibility and economic productivity, to cast doubt on all kinds of contestations that are theoretically conceived within the interplays of environmental responsibility and productivity as a carving of profitable residential development [38].

4.3 Benefits of integrating vernacular elements in affordable housing

4.3.1 Social and cultural benefits

Vernacular architectural elements are used to demonstrate a sense of identity and continuity through the preservation and regeneration of cultural heritage, thereby strengthening the socio-cultural patrimony that enhances communal bonds. Even where culturally appropriate architecture is deliberately incorporated into construction projects that lead to the production of housing units, beyond mere possession, there is an inspirational possession as most of the occupants are emotionally tied to their neighborhood. However, this prototype of a physical built environment, which is more obviously representative of local cultural history and memory in terms of what it underpins and what it represents, can also be considered as a form of enabling social infrastructure – inasmuch it is empowering people with an identity and sense of belonging where they are located. Furthermore, it means that the continuation of the construction of old architectural models has to afford another positive social impact on the carrying people as one visual and sensual impulse affects another one, and all make the common ethnocultural consciousness.

5. Economic benefits

The use of local materials and historic construction methods in the development of affordable housing can have huge effects on the economy, hence improving the financial aspect and status of the projects to enable other people to afford them. This not only means less overall cost in construction when one is able to mobilize local human resources and materials. This in turn, can have the beneficial repercussion of boosting the local economy as employment is created and local industry is stirred in a cycle that is most often positive. In addition, the use of vernacular and craftsmanship can also be a reason for the continuity of heritage and community involvement because practitioners of traditional craftsmanship are allowed and encouraged to actively participate in the construction process. Lastly, the linking of the local materials and the traditional construction methods as part of the strategy for sustainable and inclusive urban development can be viewed as a way of encouraging or supporting local economic development, as well as empowering the community to plan and manage their own affairs while at the same time being innovatively environmentally friendly.

6. Environmental benefits

Integration of vernacular characteristics into the built environment has unmeasurable effects on a sustainable environment as it minimizes the use of nonrenewable resources and, as a result, lowers the amount of energy consumed. Thus, by using climatically responsive techniques as the ancient methods of construction suggest, it is possible to build structures that, at the same time, would be less injurious to the environment and more attuned to the aspirations of people inhabiting them. For example, using passive cooling and smart ventilation methods that direct free air into the building can help eliminate the use of mechanical air conditioning and, consequently, lower the energy load of cooling. Furthermore, locally available materials can be employed in building construction to eliminate the excessive use of energy required in transporting and processing materials from other regions. This, in its turn, can have positive implications for the health of building occupants, as they are able to live in spaces that at once are environmentally friendlier and healthier for their bodies and minds. Finally, the ability to incorporate the use of local materials and designs into the construction of buildings may be identified as being in harmony with concepts of sustainable development on human and natural corrupted realms.

7. Enhanced resilience

To a capricious whim of local environmental conditions, an uncanny ability has been adapted that demonstrates vernacular architectural traditions, which helps in optimizing resilience toward climatic challenges [2]. A prime illustration of this ingenuity can be observed in traditional dwellings situated in humid situations, that is, Johor, where homes are deliberately elevated to safeguard against the unpredictability of rising waters. Conversely, in regions scorched by sweltering temperatures, architectural features are carefully crafted to optimize cooling, exemplifying the intuitive grasp of vernacular design [39]. By assimilating these time-honored elements into contemporary housing design, architects can effectively mitigate environmental stressors, thereby fortifying the safety and well-being of occupants. This strategic synthesis of ancient wisdom and modern expertise promises to revolutionize the way we conceptualize sustainable habitats, safeguarding human life and dignity in an era marked by accelerating climate volatility.

8. Increased sustainability

The combination of vernacular architectural principles with affordable housing initiatives, thus, may provide great potential in terms of augmenting sustainability within the geographical confines of Johor, Malaysia. Through the incorporation of the region's tropical climate in its construction methodologies and the harnessing of abundant natural resources, developers can create environmentally friendly dwellings that will not just be friendly to the carbon footprint but also sustainably optimized for resource use [40]. A good example would be the traditional abode of the Malays, in which locally available timber is commonly used and raised above the ground for natural ventilation, thus reducing the need for mechanical cooling [8]. This can easily be brought into modern housing design to minimize energy use and nonrenewable resource dependence. In addition, judicious use of locally available materials, such as bamboo and palm leaves, can substantially reduce the carbon footprint associated with transportation and construction. Additionally, rainwater harvesting systems may be incorporated since that is part of the traditional Malay architectural heritage to assure efficient water management toward environmental sustainability and resource scarcity within these rapidly urbanizing areas of Johor [41].

9. Innovation and local solution

In the case of Johor Bahru, Malaysia, where the local folklore has to be followed, it could be incorporated into inventions in the sphere of cheap housing using modern technology. For instance, large open spaces of the traditional 'kampung' house attached to features such as small groups of houses constructed with a focus on natural breezes of a typical urban design may be re-planned into integrative, compact neighborhoods [41]. The answer lies more in the application of contemporary methods which must be incorporated into the conventional systems; that is, panel housing along with timber frame construction is the least costly and quickest method of housing that is culturally appropriate. Besides this, techniques may also be developed for dealing with rainwater and humidity in the area, tapping the natural drainage system and moisture-proof building material, the factor that again makes the house affordable, durable, and eco-friendly. These site-specific approaches will go a long way to inventing residential structures that are, in some ways, the reflection of regional predicaments and, in other ways, represent something that has reacted to the current issue – the evolution of cities, let alone the phenomenon of climate change [40].

10. Sustainability metrics in affordable housing projects in Johor and incorporation of vernacular architectural elements

Affordability in housing projects in Johor Bahru is accumulated by the interpretation of affordability factors associated with regional development. **Table 1** asserted a comparative assessment of sustainability metrics achieved in a variety of affordable housing projects in Johor, Malaysia. This analysis affords the reader an adequate comparison between conventional techniques and contemporary trends of housing the populace adequately. In the light of the following parameters-material procurement from the inhabitants' region, energy efficiency concepts, sustainable resource

Projects	Use of Local Materials	Energy Efficiency (Materials and Techniques)	Natural Resource Conservation (Materials and Techniques)	Vernacular Design Elements	Community Engagement	Green Spaces (%)
Bandar Dato' Onn	High	High	Medium	Traditional roof design, passive cooling	Medium	31%
Eco Botanic Township	High	High	High	Timber structure, open courtyards	High	45%
Taman Seri Austin	Medium	Medium	Medium	Elevated floor houses, natural ventilation	High	26%
Kampung Baru Urban Renewal	High	Medium	Low	Split houses, use of bamboo	High	20%
Taman university Housing	Medium	Medium	Medium	Traditional motifs, community spaces	Medium	16%

Source: Author's synthesis.

Table 1.
Sustainability Metrics in Affordable Housing.

conservation, assimilation of vernacular architecture, local participation and vim, and emplacement of eco-friendly open areas – it is discernible that there are varying levels of concern for sustainable development in the referenced projects. What is more, each of them reveals a different strategy of reestablishing the link between inherited cultural practices and modern requirements for housing and domesticity, while addressing the overall questions of identity and sustainability.

From **Table 1**, it can be deduced that sustainable practices in affordable housing projects in Johor, Malaysia, differ with the project focus and resources utilized in or on the projects. The Eco Botanic Township has one of the highest levels of sustainability commitments to all four pillars, notably energy, the use of local materials, and the Green Building Programme. On the other hand, projects such as the Kampung Baru Urban Renewal and the Taman University Housing, while accomplishing most of the set goals and objectives in areas such as social inclusion and traditional architectural and construction techniques, there is room for improvement in the aspects of energy and resource conservation.

This is proved by the given comparison that shows how vernacular architecture should be arranged with the modern sustainable practices aimed at the construction of housing that complies not only with environmental conditions for the appropriate use but also with the cultural and social ones. These successes and pitfalls reveal some important lessons to be learned for future possibilities of urban regeneration in Johor, Malaysia.

In the accompanying bar chart (**Figure 5**), some affordable housing projects in Johor, Malaysia, are compared based on the sustainability metrics shown below. Bars in the same vertical group represent the relative rating of a certain project

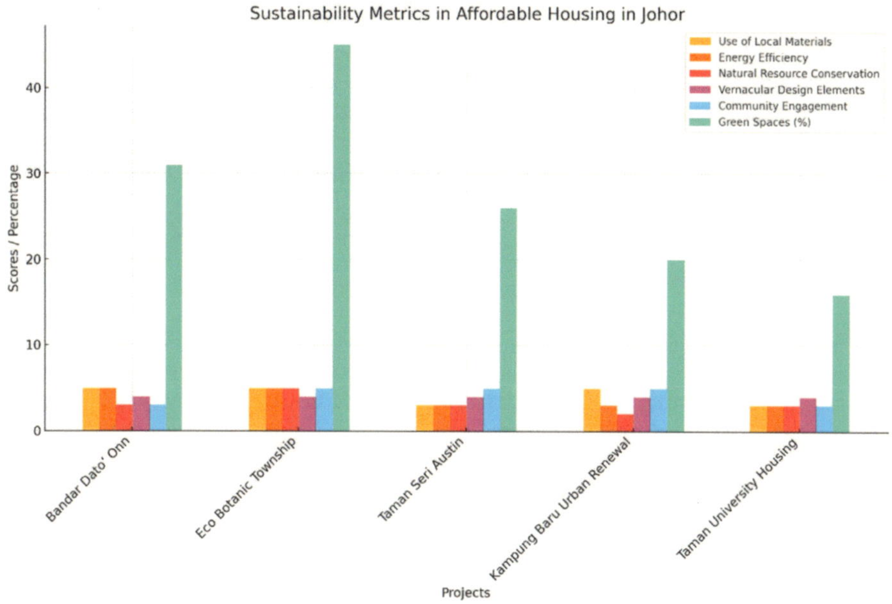

Figure 5.
Comparison of metrics in different affordable housing projects in Johor.

with different spectra of sustainability indicators on the horizontal axis: the use of locally produced materials, application of energy-saving technologies, rational use of resources, architectural features of folklore, and activity sponsored by the community, use of green zones. This type of representation allows for the analysis of the variety of strategies of each project and the nature of their approach to the issues of environmental awareness, cultural reference, and contemporary exigencies of housing.

10.1 Assessment of metrics in affordable housing in Johor Bahru

10.1.1 Energy efficiency of affordable housing

Vernacular architecture when strategically adapted, it has been proven to have significantly improved the energy performance of the building. Good examples are traditional Malay houses, in which, good passive cooling methods, through elevation by high floors and wide overhangs, with innovative ventilation, reduce the need for air conditioning. These design principles have been shown to reduce energy utilization by as much as 30% [42]. When one inspects the energy efficiency scores of these five affordable housing developments, one comes across a considerable spread, ranging from unimpressive to remarkable, with a strong bias toward top-scoring developments. From this disparate range of energy efficiency ratings, there is a noticeable emphasis on energy-saving strategies across these developments, thus, firmly entrenching this direction in an industry increasingly concerned with sustainability and environmental responsibility [38].

To demonstrate the significance of vernacular architecture on energy saving scientifically, we consider the following elements:

1. Passive cooling techniques

   - Elevated Floors: Traditional Malay houses use elevated floors to promote air circulation, reducing indoor temperatures naturally and facilitating passive cooling. Let us assume a house with 100 m^2 of floor space. Elevated floors and overhangs can reduce the need for air conditioning by about 20% annually [20].

   - Natural Ventilation: Large windows and open spaces facilitate cross-ventilation.

2. Material impact

   - Using local timber instead of conventional materials like concrete can significantly reduce embodied energy.

   - Embodied energy of timber: 0.5 MJ/kg

   - Embodied energy of concrete: 1.1 MJ/kg

   - For a house requiring 50,000 kg of building material:

     - Using timber: 50,000 × 0.5 = 25,000 MJ

     - Using concrete: 50,000 × 1.1 = 55,000 MJ

   - Energy savings: 55,000 – 25,000 = 30,000 MJ

3. Total energy saving

   - Annual energy savings from passive cooling: 0.20 × 10,000 kWh (assume 10,000 kWh/year for cooling needs) = 2000 kWh/year

   - Embodied energy savings: 30,000 MJ = 8333 kWh (1 MJ = 0.2778 kWh)

   - Total energy savings in the first year: 2000 + 8333 = 10,333 kWh

These savings contribute to the high energy efficiency ratings observed in projects like Bandar Dato' Onn and Eco Botanic Township.

To calculate the energy saving, we use the formula for cooling load reduction:

$$Q = U \times A \times \Delta T \times SHGC \tag{1}$$

Where:

- Q is the cooling load (BTU/h).

- U is the thermal transmittance of the building material (BTU/h·ft^2·°F).

- A is the area of the building envelope (ft^2).
- ΔT is the temperature difference between inside and outside (°F).
- SHGC is the solar heat gain coefficient.

For a traditional Malay house:

- U-value of bamboo walls: 0.1 BTU/h·ft^2·°F
- Area (A): 2000 ft^2
- Temperature difference (ΔT): 10°F
- SHGC: 0.2

$$Q = 0.1x2000 \text{ x } 10 \text{ x } 0.2 = 400BTU/h \quad (2)$$

For a modern house with less effective insulation:

- U-value of concrete walls: 0.5 BTU/h·ft^2·°F

$$Q = 0.5x2000 \text{ x } 10 \text{ x } 0.2 = 2000BTU/h \quad (3)$$

The traditional Malay house reduces the cooling load by 80% compared to a modern house using concrete walls.

10.2 Community engagement in affordable housing

Community engagement is the key driver of affordable housing initiatives; the more active involvement of the community the resultants of the affordable housing programs would be the best. This way, excessive dominance by developers is controlled and checked, and it becomes possible to translate the needs, tastes, and values of the residents as they are in to the planning and implementation of development projects [28]. This can therefore result in positive effects for the general residents, better standards on housing units, and balanced social relations. More specifically, previous research has revealed that when vernacular design elements are incorporated in a strategic manner, they can assist in fostering community participation levels by awakening very latent and profound degrees of cultural identity among people residing in those structures [32]. A closer look at the extent of community engagement in these projects highlights a variable: Eco Botanic Township, Taman Seri Austin, and Kampung Baru Urban Renewal Malaysia are few good examples of efficient community-based planning and designing. The inference that can be made from this is that such projects have been able to capture the spirit of the 'people's power' in finding better and sustainable housing solution alternatives.

Engaging the community helps in designing that is significantly critical to the local needs, and hence, increases satisfaction and maintenance. The following parameters have been recorded:

1. Participation Rate:
   - Assume 200 households involved in participatory planning.
   - Participation rate = 80% (160 households actively participating).
2. Impact on Project Design:
   - Community feedback leads to design adjustments, improving satisfaction.
   - Satisfaction score increased from 70–90% due to community engagement.
3. Long-term Engagement:
   - Assume initial project cost: $5,000,000.
   - Long-term cost savings due to reduced maintenance and higher occupancy rates:
     - Maintenance savings: 15% annually ($750,000 over 10 years).
     - Occupancy rate improvement: 10%, leading to increased revenue of $50,000/year.
4. Total savings over 10 years: 750,000 + 50,000/times 10 = $1,250,000.

These calculations illustrate the economic and social benefits of high community engagement observed in projects like Eco Botanic Township and Taman Seri Austin. Therefore, in the case of Johor, community engagement has been quantified by participation rates and the impact on design. High participation rates often correlate with higher community satisfaction:

- Participatory Planning Workshops:
  - Participation Rate: 80%
  - Satisfaction Score: 90/100
- Community Advisory Committees:
  - Participation Rate: 60%
  - Satisfaction Score: 85/100

Using a weighted average to determine overall satisfaction:

$$\text{Overall Satisfaction} = \frac{(80\% \times 90) + (60\% \times 85)}{80\% + 60\%} = \frac{72 + 51}{1.4} = 88\% \qquad (4)$$

This high satisfaction score demonstrates the positive impact of community engagement on project outcomes.

10.3 Green space allocation in affordable housing

When comparing the results of the analysis of both projects, there is a very significant disparity in green space provision, with Eco Botanic Township provided for at 45% while Taman Universiti is suppressed at only 16%. From this perspective, therefore, the role of green space cannot be overemphasized in as far as the issues of environmental sustainability and the general well-being of the community are concerned. Such green oases offer several ecosystem services, such as air cleaning, temperature lowering, and recreational opportunities, which are critical as far as one's health standard is concerned [38]. More importantly, lots of examples in the world of work suggest that these green areas as part of city life can really make a difference in how it is possible to foster mental health, decrease stress and panic, build social relations, and feel belonging. Adding such customary practices as planting indigenous species and community gardens goes a long way in amplifying these kinds of advantages and makes it easier to establish and increase this sort of positive interdependence between determined built structures and landscapes [18].

Green spaces in vernacular settlements are used to provide significant social and environmental benefits. Therefore, the illustration of these benefits can include air quality improvement, temperature regulation, and mental health benefits.

1. Air Quality Improvement:
   - Assume 1 hectare of green space absorbs 20 tons of CO_2 annually.
   - Eco Botanic Township with 45% green space (9 hectares):
     - CO_2 absorption: 9 × 20 = 180 tons/year.
2. Temperature Regulation:
   - Green spaces can reduce urban temperatures by 2–4°C.
   - Cooling effect for 1 hectare: ≈500 kWh/year reduction in cooling energy.
   - For 9 hectares: 9 × 500 = 4500 kWh/year.
3. Mental Health Benefits:
   - Studies show that access to green spaces can reduce healthcare costs.
   - Estimated reduction in healthcare costs for 1000 residents: $100,000/year due to improved mental health.

These benefits highlight the importance of allocating green spaces in urban projects, as seen in Eco Botanic Township and Taman Seri Austin. Green space allocation helps support ecosystem services such as air purification and temperature regulation. The impact can be quantified by calculating the reduction in urban heat island effect (UHIE). To calculate the variation in average temperature reduction, using the highest percentage of green space (45% for the Eco Botanic Township project) and lowest percentage of green space (16% for the Taman Universiti Housing project):

- Average Temperature Reduction: 2°C per 10% increase in green space.

For Eco Botanic Township (45% green space):

- Temperature Reduction: $\frac{45\%-16\%}{10\%} \times 2^{\circ}C \approx 6^{\circ}C$

For Taman Universiti (16% green space):

- Temperature Reduction: $\frac{16\%-16\%}{10\%} \times 2^{\circ}C = 0^{\circ}C$

The substantial temperature reduction in Eco Botanic Township demonstrates the effectiveness of green spaces in mitigating UHIE.

Conclusively, the results of sustainability metrics in affordable housing in Johor Bahru have illustrated the significant role of vernacular architectural elements, which are contributing to enhancing energy efficiency, community satisfaction, and a sustainable environment. Therefore, in context of Johor Bahru, Malaysia, the vernacular design principles, such as passive cooling techniques, use of locally available materials, and traditional open green space allocations, have been incorporated to enhance the affordability of housing projects, which helps in enhancing the social and cultural aspects through achieving highest sustainability standards and leads to better social outcomes and long-term sustainability.

11. Conclusion and recommendations

This chapter concludes that with vernacular imaginations constructed within affordable housing design, sustainable and culturally reverberating urban renewal in Johor would not compromise the requirements of the clientele. Thus, by employing modern housing schemes that borrow from traditional architecture and custom practices on the exchequer's side, there will be several benefits associated with environmental conservation, economic prosperity, and social well-being [40]. This also serves as an advocate for sustainability within affordable housing units together with their users' prosperity long after these kinds of skilled living units are erected in the future in Johor or elsewhere else.

The results have unequivocally underscored the important incorporation of vernacular architectural elements into affordable housing initiatives in Johor, Malaysia, as a critical approach to building sustainable and cultural-respecting urban ecosystems. From this perspective, scrutiny of vernacular design principles, such as the prudent use of locally available materials, the configuration of raised floor elevations, and passive cooling devices, provides more than compelling evidence of their effectiveness in mitigating both environmental and socio-cultural complexities within an urbanizing setting. Drawing from rich cultural heritages and contextual specificities, traditional design paradigms such as the findings detailed herein provide invaluable insight into how modern housing can be implemented to harmoniously balance sustainability and the assertion of community identity to redefine the urban built environment in a way that is both culturally appropriate and environmental conscious.

In addition, the study further concluded that it stresses the importance of cultural sensitivity when designing for such environmental sustainability. It is to impart sensitivity to the climatic conditions and culture, with Johor as a model of eco-friendly housing. The application of natural building materials and climate-responsive designs are not only eco-friendly in housing construction but conducive for habitation as well as disaster resilience. The integration of these traditional practices into modern construction techniques has increased awareness, which has been known to offer a viable solution to the socio-ecological needs of the region.

Such being the case, community involvement and governance forms an important part of successfully running affordable housing projects or regeneration programs in Johor. Indeed, through public participation during planning stages, design processes, and construction works, such undertakings would address local requirements and enhance more friendly social ties over time while keeping things alive [17]. This applies to a resale market where neighbors recognize each other physically and share some outdoor space instead of rental settlements wherein people usually do not know who stays next door. Thus, good community engagement will lead to more relevant housing projects because residents will be allowed to manifest their own ideas about house designs within set planning frameworks. Therefore, it would encourage creativity among them while still building resilient neighborhoods into the bargain. Finally, Johor could manage by taking these participations even further, such that it becomes widely recognized as one of those places that have vibrant urban forms that are inclusive and sustainable and promote social equity.

Hence, in the state of Johor, affordable housing plays a fundamental role in the sustainable regeneration of cities. Rapid urbanization presents unique challenges while at the same time offering opportunities for innovation that are key to the creation of inclusive houses that are also ecologically friendly. Not only do these initiatives fill housing gaps, but they can also become springboards for wider urban renewal projects, thereby improving the lives of all city dwellers.

In light of discussions and findings presented in this chapter, it is recommended that urban planners and policymakers of Johor Bahru proceed with affordable housing initiatives on an integrated rather than ad hoc basis to fully incorporate vernacular architectural elements. The approach would be to make new housing developments that do not only embody sustainability but also resonate with the indigenous population by way of locally sourced materials, traditional building techniques, and community-driven design processes. By drawing on a rich architectural heritage within the region, like elevated floors and large overhanging roofs typical of a Malay house, new housing projects can offer climate-resilient, cost-effective, and esthetic solutions. In addition, the flexibility and adaptability of traditional interiors should be used to serve modern urban residents whose needs have become very varied. This form of strategy will help protect the cultural nature of Johor Bahru and improve the well-being in the communities by enhancing belonging and continuity with the past.

Finally, it has been advisable to adopt architectural intervention of the concept of affordable housing based on the strategy embraced in popular architecture as a plausible strategy to foster sustainable urban regeneration in Johor. This is because when stakeholders have embraced the diverse architectural contributions in the area, they will come up with houses that are environmentally friendly, culturally appropriate, and reachable to all the needed individuals. In addition to effectively dealing with current and future problems of affordable housing, this kind of integration thinking also promotes the future environmentally, economically, and dynamically sustainable of urban residential communities.

Author details

Fazlur Rahman[1*], Noraslinda Abdul Rahman[1], Adiba Shafique[2] and Mohd Arshad Ameen[3]

1 Center for the Study of Built Environment in the Malay World (KALAM), Faculty of Built Environment and Surveying, Universiti Teknologi Malaysia, Johor, Malaysia

2 Architecture Department, Dar al Uloom University, Riyad, KSA

3 Faculty of Architecture and Ekistics, Jamia Millia Islamia University, New Delhi, India

*Address all correspondence to: arfazlur121@gmail.com

References

[1] Rapoport A. House Form and Culture. New York: Prentice Hall Inc.; 1969

[2] Mileto C et al. Vernacular Architecture: Towards a Sustainable Future. London, UK: CRC Press, Taylor & Francis Group; 2014

[3] Uysal Urey ZC. Creation of a new vernacular architecture and the attainment of sustainability: The case of Akyaka town development. Sustainability. 2023;**15**(3):2643

[4] Rudofsky B. Architecture without Architects: A Short Introduction to Non-pedigreed Architecture. New Mexico: UNM Press; 1987

[5] Ismail A, Mamat SNA, Jiram WRA. Housing affordability stress among the middle-income (M40) group in Johor Bahru. Planning Malaysia. 2023;**21**(3):21

[6] Saleh MAE. The evolution of planning & urban theory from the perspective of vernacular design: MOMRA initiatives in improving Saudi Arabian neighbourhoods. Land Use Policy. 2001;**18**(2):179-190

[7] Brown R, Maudlin D. Concepts of Vernacular Architecture. London, UK: SAGE Publications Ltd.; 2012. pp. 340-355

[8] Ghaffarian Hoseini A et al. The concept of local-smart-housing: Towards socio-cultural sustainability of vernacular settlements. International Journal of Architectural Research. 2011;**5**(2):91-105

[9] Testa A. Events that want to become heritage: Vernacularisation of ICH and the politics of culture and identity in European public rituals. In: Heritage and Festivals in Europe. London and New York: Routledge; 2019. pp. 79-94

[10] Audin J. Of other dreamlands: Vernacular adaptation, urban exploration, and alternative spectacles in China's abandoned themed spaces. Antipode. 2024;**56**(4):1109-1131

[11] Hosseini E et al. Malay vernacular architecture: Mirror of the past, lessons for the future. In: The Proceedings of 8th SEATUC Symposium. Johor, Malaysia: UTM; 2014

[12] Yusof H, Ibrahim FK, Shahminan RNR. Heritage identity conservation of the woodcarving placement and design composition typologies on Johor Malay houses. Planning Malaysia. 2020;**18**(2):12-23

[13] Ahmad S et al. Adaptation of traditional Malay house design towards housing comfort and satisfaction. Malaysian Journal of Society and Space. 2022;**18**(3):216-226

[14] Ismail WHW. Heritage street and contemporary use of old buildings in Johor. Asian Journal of Environment-Behaviour Studies. 2018;**3**(11):30-37

[15] Misnat N et al. Re-adaptation of traditional Malay house concept as a design approach for vertical dwellings Malays. International Journal of Engineering and Technology (UAE). 2018;**7**(3):1-4

[16] Chiong FWE. Evolution of the Chinese shophouse facade in Malaysia: Manifestations of a diasporic people. In: Proceeding-23rd International Conference of Historians of Asia 2014 (IAHA2014). Kedah, Malaysia: UUM; 2014

[17] Kamaruddin T, Hamid RA, Rohaizam N. A situational study on sustainable housing features in Johor. In: IOP Conference Series: Materials Science and Engineering. Langkawi, Malaysia: IOP Publishing; 2020

[18] Husen N, Mohamed M. Comparison of green design strategies in five traditional Malay houses. Jurnal Kejuruteraan. 2021;**33**(1):47-53

[19] Ghaffarian Hoseini A, Baharuddin M, Ibrahim R. Identification of the Malay vernacular spatial features within Malay vernacular houses. International Journal of the Arts in Society. 2009;**4**(4):137-152

[20] Toe DHC. In: Kubota HBRT, Takaguchi H, editors. Sustainable Houses and Living in the Hot-Humid Climates of Asia, Malaysia: Malay House. Singapore: Springer; 2018. pp. 25-35

[21] Ju S-R, Omar SB. A typology of modern housing in Malaysia. International Journal of Human Ecology. 2010;**11**(1):109-119

[22] Iskandar FIFR, Yusof AM, Abidin NI. A review of sustainable affordable housing: Issues and challenges. In: AIP Conference Proceedings. Pahang, Malaysia: AIP Publishing; 2023

[23] Musaddad HA, Maamor S, Zainol Z. Affordability of affordable housing in northern region of Malaysia. Malaysian Management Journal (MMJ). 2023;**27**:183-209

[24] Wahab FA, Kamarudin MKA, Sembak TMZT. Affordability homeownership for socio-economic development In Malaysia. Journal of Namibian Studies: History Politics Culture. 2023;**34**:7205-7220

[25] Abdulaziz H. Sprawling growth and the environment: A case of Johor, Malaysia. International journal of Environmental Science and Technology. 2018;7:382

[26] Tsenkova S, Tsenkova S. Retrofits for the future: Lessons from affordable housing and energy efficiency programs in Canada. In: Energy Efficient Affordable Housing: Policy Design and Implementation in Canadian Cities. Canada: Springer; 2021. pp. 113-130

[27] Liu J, Ong H. Can Malaysia's national affordable housing policy guarantee housing affordability of low-income households? Sustainability. In: Napoli G, Trovato MR, editors. Efficiency, Fairness and Sustainability in Social Housing Policy and Projects, MDPI. 1 ed. Vol. 13.. Basel, Switzerland: MDPI; 2021. p. 129, 8841

[28] Severson M, de Vos E. Social sustainability in social and affordable housing. Cities and affordable housing: Planning. Design and Policy Nexus. 2021;**1**:259-273

[29] Ibrahim SH et al. Reinventing traditional Malay house for sustainable housing design: Obstacle and proposed solution. Jurnal Teknologi. 2015;**72**(1):97-102

[30] Yusof H, Sabil A, Hanapi A. Typology of woodcarving motifs in Johor traditional Malay houses. International Journal of Sustainable Construction Engineering and Technology. 2023;**14**(2):139-152

[31] Roslan Q et al. A literature review on the improvement strategies of passive design for the roofing system of the modern house in a hot and humid climate region. Frontiers of Architectural Research. 2016;**5**(1):126-133

[32] Mehdipoor A et al. The essence of Malay vernacular houses: Analysis of the socio-cultural and environmental values. Sustainable Cities and Society. 2014;**13**:157-170

[33] Abdul Rahman N et al. A method for formulating architectural value in Johor Bahru tourism building. In: Advances in Civil Engineering Materials: Selected Articles from the International Conference on Architecture and Civil Engineering (ICACE2020). Singapore: Springer; 2021

[34] Hutchinson FE, Rahman S. Johor: Abode of Development? Johor Bahru, Malaysia: ISEAS-Yusof Ishak Institute; 2020

[35] Lee Y et al. Thermal characteristics of a residential house in a new township in Johor Bahru. In: IOP Conference Series: Materials Science and Engineering. Penang, Malaysia: IOP Publishing; 2017

[36] AbuBakar MEB, King NK. Social interaction and third space relationship in vertical housing in Malaysia. Journal of Engineering. 2018;**2018**(1):1-10

[37] Annual Report, Eco World: Creating Tomorrow & beyond. Johor, Malaysia: Eco world Developmenty Group Berhad; 2021

[38] Safitri Zen I et al. Magnitudes of households' carbon footprint in Iskandar Malaysia: A policy implications for sustainable development. Journal of Cleaner Production. 2022;**315**:128042

[39] Morris AM, Diaz L. Reimagining housing: Affordability crisis and its role in disaster resilience and recovery. In: Louisiana's Response to Extreme Weather: A Coastal State's Adaptation Challenges and Successes. Cham, Switzerland: Springer; 2020. pp. 241-259

[40] Alsheikh Mahmoud S et al. Effective preservation of traditional Malay houses: A review of current practices and challenges. Sustainability. 2024;**16**(11):4773

[41] Samsudin NA et al. An adaptation of the malay kampongs or villages concept on modern housing schemes development in Malaysia. In: IOP Conference Series: Earth and Environmental Science. Surakarta, Indonesia: IOA Publishing Ltd; 2020

[42] Sertyesilisik B. Effective Affordable Housing Strategies. Recent Advances in Energy Conservation Techniques for Buildings: From Micro Scale to Urban Level. Coventry, UK: Coventry University Press; 2019. pp. 85-93

Chapter 7

Reducing Poverty by Protecting Properties through Insurance

Puteri Nur Farah Naadia Mohd Fauzi

Abstract

Implementing suitable protection systems and measures for everyone is one of the aims listed in the 2030 Agenda for Sustainable Development to achieve considerable coverage of the poor and vulnerable groups. This encompasses safeguarding different life elements, such as health, economic, and social standing. According to the Centre for Poverty and Social Policy (2023), the net safety measures that protect the majority from poverty include the security of fundamental human necessities, that is, food and property such as housing and belongings. The necessity to protect properties, particularly those belonging to impoverished people, is critical since, in the case of a disaster, this group may suffer more damage and hardship. This chapter aims to unveil the concepts of protection of properties for the poor, that is, house and home contents protection through insurance policies and its role in enabling poverty reduction in particular.

Keywords: home contents, house, poor, poverty, properties protection

1. Introduction

Poverty is a significant issue today and has become one of humanity's biggest challenges [1]. The lack of adequate shelter, food, and healthcare has a profound impact on the lives of millions of people around the world. In recent years, there has been a growing recognition that property insurance can play a crucial role in reducing poverty by protecting people's homes and businesses from natural disasters, theft, and other risks.

Reducing poverty by protecting properties through insurance is essential to modern-day development. In today's globalised world, insurance has become essential for mitigating financial risks that often result in poverty and inequality [2]. Protecting properties through insurance is essential to safeguarding the assets of individuals and communities. Insurance policies can help alleviate the financial burden caused by property losses due to natural and man-made disasters, enabling individuals and communities to recover faster and reducing their vulnerability to poverty [3].

In today's increasingly unpredictable climate and economic landscape, having insurance is vital for anyone who wants to protect their homes and other valuable assets. One of the main reasons why insurance is so crucial is that it provides peace of mind to policyholders. By knowing that insurance, individuals and families cover their properties can rest assured that they have a safety net in times of crisis. This security allows policyholders to focus on rebuilding their lives instead of worrying about the financial burden of repairing or replacing their homes and belongings. As such, insurance can be particularly beneficial for those who are most vulnerable, such as the poor and needy.

For low-income people, in particular, protecting their homes and contents through insurance is critical. A natural disaster such as a flood or a fire can be devastating for anyone, but it can be even more so for those living on a tight budget. Without insurance, such events can force vulnerable individuals or families into deeper poverty, making it difficult to get back on their feet. Insurance policies can prevent this from happening by providing financial protection for homes and their contents, ensuring that those in need receive the support and assistance they require when they need it most [4].

This chapter explores various ways in which insurance policies can be utilised to protect vulnerable sections of society. Specifically, the chapter delves into the concept of property protection for the poor and the role insurance plays in facilitating poverty reduction. Moreover, studies have shown that increased access to property insurance can significantly reduce poverty. In many cases, a single event, such as a natural disaster or theft, can wipe out an entire household's savings, sending them into a downward spiral of debt. Insurance can help to mitigate these risks and provide a safety net for families, allowing them to recover more quickly from unforeseen setbacks.

One example of how insurance can promote development is through microinsurance, which provides coverage to low-income populations and those in vulnerable communities. In many cases, these policies are tailored to meet the specific needs of the communities they serve, such as crop insurance for subsistence farmers or home insurance for those living in disaster-prone areas. By pooling resources and sharing risks, communities can better protect themselves against unforeseen events and work towards building a stronger, more resilient future.

It is also worth noting that property insurance can positively impact economic growth as a whole. By protecting individual and commercial property, insurers can help promote investment and entrepreneurship, as businesses are more likely to take risks knowing they have a safety net in place. This, in turn, can lead to job creation and increased economic activity.

2. Understanding poverty

2.1 Definition of poverty

Poverty is a term used to describe the financial state of individuals or families who lack the resources to afford basic necessities such as food, clothing, and housing [5]. While poverty is often associated with a lack of money, it can also refer to a lack of other resources such as education, property protection and guarantee, healthcare, and access to technology [6].

Poverty can significantly impact individuals and communities, including decreased health outcomes, limited economic opportunities, and reduced access to education. Those living in poverty often have less stability and security in their lives, and this lack of stability can lead to a range of adverse outcomes, such as homelessness, substance abuse, and mental health issues.

A range of factors contribute to poverty, including economic policies, unequal access to education and employment opportunities, and discrimination based on race, gender, and other factors. Addressing poverty requires a multifaceted approach that includes policies to reduce income inequality, provide access to education and job training programmes, and address systemic issues such as discrimination and unequal access to resources [7].

Governments and Non-Governmental Organisations (NGOs) have implemented various strategies such as microfinance, cash transfers, and public works programmes to alleviate poverty. However, poverty reduction requires a long-term approach which involves addressing the root causes, such as inequality, lack of education, and lack of access to healthcare.

2.2 Causes of poverty

Poverty is a complex issue with no single cause, and its effects can vary depending on the circumstances. It is essential to understand that poverty is not a choice but the result of various factors that can make it difficult for individuals and communities to meet their basic needs.

One of the primary causes of poverty is unemployment or underemployment [8]. Without a stable job, individuals may struggle to make ends meet and fall into a vicious cycle of poverty. Unfortunately, many jobs in impoverished areas do not pay enough to support a family, making it difficult for those employed to make a decent living.

Lack of education is another significant factor that can perpetuate poverty [4]. When individuals do not have access to quality education, they are often unable to acquire the skills needed to secure better-paying jobs. They are also more likely to suffer from health problems due to a lack of knowledge about hygiene, nutrition, and other essential practices.

Poor governance and corruption can also lead to poverty. When leaders misuse public funds or neglect to provide adequate resources and infrastructure, their citizens suffer. Corruption often results in essential services such as healthcare, education, and transportation being unavailable or too expensive for the poor.

Natural disasters, war, and conflict are also enormous contributors to poverty [9]. These situations can displace entire communities and leave them without access to basic necessities such as food, water, and shelter. Moreover, the loss of infrastructure and livelihoods can have long-lasting effects that can take years or even decades to recover from.

2.3 Effects of poverty

Rising food and energy prices have become a significant concern across the globe. These spikes in prices are attributed to various factors. For instance, the ongoing conflict in Ukraine, climate shocks, and conflict have played a significant role in the increase in prices. Unfortunately, the impacts of these increases are felt the most by people living in extreme poverty. Reports suggest that, by the end of 2022, as many as 685 million people are projected to live in extreme poverty due to these high prices [9]. The current setbacks have exacerbated the challenge of global poverty reduction efforts. The rate of poverty reduction was already slowing due to the subdued global economic growth before these recent crises. As a result, the world is now further off track from achieving the goal of ending extreme poverty by 2030.

Specifically on properties, that is, houses as shelters, poverty significantly affects access to adequate housing, resulting in housing deprivation, which can cause various negative consequences. However, there are viable solutions to combat this issue and provide affordable and adequate shelter for low-income families [10].

Housing is a fundamental human right, yet millions of people worldwide still lack adequate shelter due to poverty. Poverty is a significant barrier to affordable housing, and its effects can cause housing deprivation, resulting in various negative consequences. Poverty significantly affects access to adequate housing, especially for low-income families. These families may face severe overcrowding and live in substandard conditions, posing a threat to their health and well-being [11]. The lack of access to formal housing leaves many people with no option but to settle in informal settlements, such as slums or shantytowns, where access to basic needs, such as water and sanitation, is limited [10, 12].

The effects of poverty on housing deprivation extend beyond inadequate housing [13]. Housing deprivation can negatively impact various aspects of an individual's life. For instance, it can result in poor health outcomes due to exposure to hazardous environments. Furthermore, insufficient housing can lead to higher rates of crime and violence, particularly in overcrowded areas.

3. Insurance as a tool for poverty reduction

In many developing countries, traditional forms of insurance, such as community-based insurance and Islamic microinsurance, have been utilised with the aim of providing coverage for low-income households. These forms of insurance generally have relatively lower premiums and are more readily available to low-income households, which can be an effective strategy aimed at alleviating poverty. These traditional forms of insurance typically have a higher level of trust from citizens due to the fact that they are often run by community-based organisations that have built relationships with the people they are serving. Islamic microinsurance provides an alternative to traditional insurance options through its Shariah-compliant approach that includes a risk-sharing system which differs from conventional insurance models.

It is worth noting that insurance also serves a vital role in promoting economic growth by providing a buffer against financial shocks. For instance, with insurance, farmers can protect themselves against severe crop failures, which may have devastating effects on their livelihoods. Insurance can also help farmers recover from a setback and enable them to invest further in their farms and enhance and develop their products in the long run. In addition, credit access, particularly for low-income households, can benefit from insurance utilisation. By providing a safety net against financial shocks, insurance can reduce the risks that lenders may face when providing loans to low-income households, thus improving access to credit. This can increase the investment capital available in an area, attracting investors, creating jobs, and promoting overall economic development [14]. Therefore, insurance can be a significant driver in enhancing economic well-being and reducing poverty in developing nations.

Insurance is a crucial instrument that can mitigate the vulnerability of poor households. Typically, these households are characterised by unstable and low incomes, making handling unexpected financial, political, and physical shocks challenging. An unforeseen expense or a slight change in income can significantly negatively impact their already low living standards. Poor households, especially the vulnerable sections of society, are more susceptible to experiencing incidents such as natural disasters, theft, and harsh weather conditions that could lead to a decline in their livelihoods and future income. In cases of death, disease, or accidents, these households often have difficulty coping with the loss of assets and other basic household necessities, which could result in a further downward spiral of their income.

Poor households have developed informal insurance systems to mitigate risks, such as begging for donations and help from authorities and non-government organisations (NGOs), exchanging gifts, selling assets, diversifying crops, and engaging in cash transfers. However, these strategies have not been sufficient to cater to the needs of the most vulnerable sections of society. A formal insurance system that pools the resources of many is the most effective way to guarantee protection against unexpected losses brought about by hazards such as illnesses, disabilities, and theft. Through small and affordable premiums, policyholders can replace the uncertain risk of significant losses with the security of risk mitigation.

In most developed nations, insurance is part of society, and people understand its importance in mitigating financial risks. However, in developing nations, insurance is more critical for the poorest and most vulnerable society members with limited recovery options. The formal insurance system is essential in protecting the assets and livelihoods of poor households because it provides them with a sense of security that can help them manage the risks brought about by unexpected disasters.

In developed countries, people have access to social safety nets and other public services that help cushion them from unexpected financial shocks. But in developing countries, poor households have to rely on the formal insurance system to manage the risks that come with economic, political, and physical shocks. As such, a formal insurance system is critical for promoting financial inclusion and reducing the vulnerability of poor households.

3.1 The concept of insurance

Insurance and poverty have been intertwined for centuries. People in ancient times realised the significance of pooling their resources to assist individuals facing monetary loss due to unforeseen events. From this community-based initiative, the concept of insurance evolved. This evolution took several centuries to develop into the full-fledged insurance industry we know today. The earliest modern insurance policies were developed in the seventeenth century and mainly dealt with reducing risks associated with maritime trade. Gradually, the concept expanded to include a variety of industries, such as agriculture, healthcare, and even auto insurance.

In essence, insurance aims to provide protection and guarantees for compensation of specified loss and damage that may arise due to insurable risks [15]. The primary objective of individuals seeking insurance is to safeguard themselves from the detrimental effects of any potential losses. By transferring the risk from an individual or group to an approved insurance company in exchange for payment of a premium, insurance plays a vital role in society. This premium, paid by the individual or group seeking insurance, is utilised for compensation or reimbursement.

The term "insurance" refers to a contractual agreement in which an insurer agrees to pay the insured an agreed sum of money or other form of benefits in the event of risks in exchange for a premium payment [16]. The insurance system operates based on the concept of "pooling" or collection, involving payment of the insurance premium. The insured individual or group pays the insurance premium to the insurance company for an insurance policy. The policy outlines the terms and conditions that the individual or group must adhere to. Additionally, the policy specifies the coverage types included. The Malaysian Insurance Act 1996 governs the contract of insurance in Malaysia, laying out a legal framework that regulates the insurance business in the country.

The relationship in an insurance contract is created when an insured, also known as a policyholder, agrees to and pays the premium. Once the premium has been paid, the insurer, usually an insurance company, provides the individual or group with an insurance policy outlining the terms and conditions. In the event of loss or damage to the insured individual or group, the insurer is responsible for providing compensation in accordance with the policy's stated coverage types. It is essential to note that the burden of proof that such losses or damages have occurred rests with the insured group or individual.

3.2 Takaful as an alternative to insurance

Conventional insurance is a form of coverage that is widely known and utilised. However, in Malaysia, the extent to which conventional insurance complies with Shariah law has been cause for concern. This apprehension comes from the presence of uncertainty, gambling, and usury in conventional insurance practices. The National Fatwa Committee Malaysia deliberated over these concerns and pronounced life insurance as haram under Shariah due to the aforementioned elements [17–20]. The Islamic Fiqh Academy from the Organisation of Islamic Conference reached a similar conclusion, asserting that conventional insurance that entails a significant risk is prohibited by Shariah [21].

In spite of the negative connotations that are often associated with conventional insurance, it is not lacking in benefits. Yusuf al-Qaradhawi, a renowned Islamic scholar, has emphasised that the concept of insurance itself does not contradict Islamic teachings. Instead, Shariah is opposed to the involvement of prohibitive elements in conventional insurance practices [22].

To provide a solution, takaful was introduced as an alternative to conventional insurance. Takaful is a form of coverage that adheres to Islamic principles and eschews uncertainty, gambling, and usury. The Takaful Act 1984 regulates takaful business in Malaysia. The National Fatwa Committee Malaysia convened a meeting to discuss the implementation of the Law of Takaful in Malaysia and determined that it conforms to Islamic law. Thus, takaful is considered permissible insurance in accordance with Shariah.

Generally, individuals have the option to procure different forms of insurance, including conventional insurance and takaful. However, the involvement of Muslims in conventional insurance has raised concerns among Islamic scholars due to the presence of prohibited elements from the Shariah point of view, such as uncertainty (gharar), gambling (maisir), and usury (riba) in its operation and practice. Thus, to promote Shariah compliance, business entities are adopting the takaful concept, which is an Islamic insurance that promotes brotherhood, solidarity, and mutual assistance among its members. Takaful adheres to Shariah principles and does not involve gharar, maisir, or riba [18].

On a global scale, there are currently more than 110 takaful operators, and the industry has been growing rapidly, appealing to both Muslims and non-Muslims. It is expected to grow by 15 to 20% annually, with contributions expected to reach over USD10 billion by 2025 [14]. Takaful is essentially a cooperative system where policyholders contribute their premiums into a common pool, which the takaful operator manages, investing in sharia-compliant instruments such as sukuk, equities, and real estate. The profits generated from these investments are distributed among policyholders as surplus, and the operator is entitled to a management fee only [23].

Furthermore, takaful is an inclusive insurance system that caters to various segments of society. These include family takaful, general takaful, and microtakaful. The target market for takaful encompasses individuals, families, small and medium-sized enterprises, and large corporations. Additionally, takaful operators offer customised solutions for specific areas such as health, education, and retirement planning.

4. Protecting property through insurance as a strategy for poverty reduction

4.1 Overview of property insurance

Property insurance has a rich history dating back to the great fire of London in 1666. After this devastating incident, policies were put in place to protect homeowners from similar tragedies [24]. The evolution of property insurance has continued to provide security and support to property owners worldwide.

The seventeenth century saw coffeehouses in London as the hub of financial activity, where merchants and shipping companies met to arrange deals and discuss insurance policies to protect their businesses. With global expansion, insurance policies became integral to business transactions to mitigate risks and losses.

The first package of insurance policies for homeowners' coverage was introduced in September 1950 [24]. The Homeowners' policy has proliferated and now occupies a critical position in the industry. In 1960 alone, companies wrote approximately three-quarters of a billion dollars in premium countrywide. Due to its impressive growth, this policy is now an established part of the property insurance structure. It has even revolutionised the business by setting precedents currently being carried over into the commercial field. Despite the policy's well-established status, its birth pains connected to early development are frequently overlooked. However, the policy's extraordinary growth speaks to its importance in the industry.

Today, property insurance policies offer a wide range of coverage options to protect homeowners from various risks that they may face. These policies can cover damages caused by extreme weather events, such as hurricanes, floods, and earthquakes. Other coverage options include theft, fire, and vandalism.

In addition, other factors that can influence the cost of property insurance include the location of the property, its age, and the level of risk associated with the area, to name a few [25]. As such, it is crucial that property owners obtain the right insurance coverage to ensure that their property is protected from the risks it faces and that they are financially safeguarded in the event of any unexpected occurrences.

4.2 The need for protection of properties belonging to the needy or poor

Protection of properties, especially those belonging to and/or occupied by low-income groups, is crucial towards reducing poverty and improving the well-being of vulnerable communities. In the event of calamities such as fires, these groups are often the hardest hit and will suffer significant losses regarding their properties and assets. Without adequate protection measures such as insurance or takaful, the road to recovery after such disasters becomes an almost impossible task for them.

For instance, in Malaysia, one of the many *tahfiz* (Islamic religious) and *pondok* residential schools occupied by needy and poor students was reported to have experienced a fire in the first quarter of 2017. Unfortunately, most of these schools are not registered and do not have access to insurance or takaful for the protection of their buildings. In fact, according to the Fire and Rescue Department, over 40% of these schools in Malaysia do not even comply with safety requirements. As a result, when disasters strike, they incur significant losses and damage [26].

Also, the tragic fire at the Darul Quran Ittifaqiyah residential *tahfiz* school in Kuala Lumpur, where 24 lives were tragically lost along with more than RM100,000 properties, highlights the need for proper protection measures. The Urban Wellbeing, Housing and Local Government Minister reported that the school did not have valid permits from the fire and rescue department to occupy the building [27]. Furthermore, the buildings and their contents were not insured. Similarly, ten buildings at Pondok Seri Permai in Pasir Puteh, Kelantan, were destroyed by fire, causing losses estimated at RM161,000 due to the lack of protection [27].

Access to protection measures like insurance or *takaful* as protection for properties belonging to and/or occupied by the needy and poor is paramount in mitigating their losses in times of calamities. These measures are critical since the recovery of these groups depends heavily on philanthropies and charities, as well as the financial assistance provided by governments and organisations. The compensation provided by insurance or *takaful* plays a vital role in these groups' ability to restore and repair damaged buildings or reinstate their losses. Most importantly, the compensation provided by the insurance or *takaful* facilitates them from further disaster and significant losses that may have led to poverty and life deprivation [28].

The significance of insurance or *takaful* as a tool to help reduce poverty cannot be overstated, especially for individuals who belong to the low-income category or are financially struggling. Those who find it particularly challenging to meet their basic needs are at greater risk of losing their properties to unexpected events, making it critical to have adequate protection coverage.

One of the main reasons insurance is essential for safeguarding the properties belonging to the needy is its ability to protect against unpredictable natural disasters such as earthquakes, hurricanes, floods, and other calamities [15, 25]. Such events can cause catastrophic damage to properties, making it difficult for those who cannot afford the high costs of repairs to recover. However, with insurance, individuals can file a claim and receive monetary compensation for the loss incurred.

Moreover, insurance provides a buffer against unforeseen emergencies such as fires, theft, and accidents, which can be devastating to anyone, irrespective of their financial status. For low-income earners, such occurrences can erode their financial stability and disrupt the already tenuous balance. Having insurance provides them with a sense of security, knowing that their hard-earned properties are protected, and they can withstand any unexpected financial liabilities.

Furthermore, ensuring the properties belonging to the needy provides a crucial anchor in maintaining their financial stability. Losing a property due to an unanticipated event can have far-reaching consequences, pushing the affected individuals deeper into poverty and exacerbating their already dire situation. With insurance, they can rest easy, secure in the knowledge that no matter what curveballs life throws their way, they can maintain financial equilibrium, meet their basic needs, and obtain even greater success.

Essentially, having insurance coverage is more critical than ever before, especially for those who belong to the low-income category or are financially struggling. Insurance serves as a safeguard against unexpected events, provides a sense of security, and is a crucial anchor in maintaining financial stability. With all of its benefits, obtaining insurance coverage is critical for anyone who wants to protect their assets, maintain financial equilibrium, and achieve tremendous success in life.

4.3 Property insurance coverage and its practice in protecting properties belonging to low-income groups and poor

Unfortunately, poverty has always been a significant obstacle to obtaining insurance coverage. In the past, many insurance companies refused to offer policies to individuals deemed too risky. Insurance and *takaful* practices have often been considered unaffordable for low-income groups and the poor, resulting in many impoverished people being unable to protect their homes and possessions against risks. As a result, this income category is left vulnerable to further decline in cases of catastrophe or crisis.

However, in Malaysia, Bank Negara Malaysia (BNM) has recognised the importance of financial inclusion and worked to ensure that all income groups have access to the same facilities [29]. Low-income groups suffer disproportionately in times of crisis, making it even more crucial that such communities receive the necessary protection. Micro-protection schemes, such as microinsurance and microtakaful, have been introduced to address this need. These schemes provide affordable protection at a fair contribution rate to meet the needs of low-income groups in the protection of lives and properties. This type of protection is crucial to ensuring that those in poverty have some means of escape from the cycle of poverty.

In Malaysia, the introduction of regulatory structures by BNM, with industry participation, led to the growth of micro-protection schemes. As a result, "Microinsurance and Microtakaful: Discussion Paper" was introduced by BNM in 2017 with the aim of further enabling market growth and innovation while preserving the quality and sustainability of the schemes [30]. This initiative has provided vital support for those in low-income communities, enabling them to protect their homes and possessions against risks.

A micro-protection plan or scheme is crucial for safeguarding low-income households and their assets in the event of natural disasters or unforeseen calamities, as they may find it difficult to recuperate without financial assistance. For instance, built houses are more likely to be extensively damaged or destroyed during fires or natural disasters, resulting in adverse living conditions for the targeted group. Therefore, a suitable protection scheme will significantly streamline the affairs of low-income groups by covering property damage and loss. Instead of the previous policy of standard fire insurance, homeowner's property insurance covers all risks related to destruction, including fire, theft, natural disasters, and other unanticipated catastrophes.

The Malaysian insurance market has witnessed the development of micro-protection products that primarily cater to low-income groups as combined products sold through microfinance institutions. However, the market is still relatively new in Malaysia, and there is a lack of recognition and understanding by people. The Bank Negara Malaysia's efforts towards promoting financial inclusion include developing a regulatory framework to help the micro-protection plan's market thrive. Adequate protection is crucial for low-income housing in Malaysia to ensure that people can recover from losses in the event of a disaster, contributing to poverty reduction and economic and social well-being deprivation. In the absence of micro-protection plans or schemes, low-income groups have to rely on charity from individuals and financial assistance from the government and non-governmental organisations to restore and repair their damaged homes.

In Malaysia, the poverty rate increased by 2.8% between 2019 and 2020, with Kelantan having the highest percentage change of 8.8% in 2020. According to the DMR Handbook, Malaysia has a natural hazard and exposure risk of 3.4/10, with 67%

of the population vulnerable to floods between July 2012 and January 2019. Floods have occurred almost every year in Malaysia during the monsoon season, specifically in November and December [31]. Recent heavy rainfall impacted most of Peninsular Malaysia, particularly Kedah, Malacca, Negeri Sembilan, and the Selangor States, including Kuala Lumpur, resulting in more than 260 houses and home contents being damaged or evacuated in Malacca, Selangor, Kedah, Negeri Sembilan, and 14 in Perak due to recent flooding [32]. The potential risks and losses from natural disasters may cause severe damage to many properties. Therefore, the protection of houses and home contents is vital, especially for low-income groups, to ensure they do not suffer losses and have to rely on other financial assistance.

4.3.1 Microinsurance

Microinsurance is a specialised insurance that targets low-income individuals with financial constraints who lack protection coverage due to unaffordability, specifically in paying the premium [33]. It is pervasive in developing nations where financial catastrophes such as unexpected illness, burglary, or natural disasters can jeopardise the livelihoods of families. The primary objective of microinsurance is to provide a safety net for those who need it most. Moreover, microinsurance is an affordable alternative to traditional insurance policies since it is retailed in smaller amounts.

Microinsurance policies cover the costs of healthcare, life insurance, and property damage, and some even comprise coverage for small businesses, enabling entrepreneurs to safeguard their means of sustenance. One such policy is microinsurance for properties. It provides coverage to low-income households for their homes, farms, and personal property. By doing so, microinsurance for property allows these households to secure adequate insurance coverage with low premiums and a minimum threshold, making it more affordable for low-income households.

Microinsurance for property helps not only low-income households and the poor protect their properties like homes and personal belongings but also helps reduce poverty. By reducing the economic impacts of unexpected disasters, microinsurance for property helps low-income households and the poor prioritise their finances and focus on other areas of their lives.

One of the pivotal advantages of microinsurance is its ability to help low-income individuals break away from the cycle of poverty. Access to coverage helps people better manage risks and protect their savings, leading to increased economic stability and improved quality of life. In developing countries where insurance has historically been out of reach for many people, microinsurance is essential for those with low incomes.

4.3.2 Microtakaful

Microtakaful is an alternative protection scheme designed to provide financial security for vulnerable groups facing constraints and difficulties. It is widely recognised as a protective scheme in the insurance industry aimed at low-income groups. According to the Islamic Financial Services Board (IFSB) and the International Association of Insurance Supervisors (IAIS) (2015), "low-income groups" refers to sections of the population that are usually ineligible for standard takaful schemes offered by operators. This ineligibility is often due to factors such as income irregularities, hazardous occupations, medical history, insurable interest, and other considerations that fall under the purview of prudential regulation of exclusive finance [34].

In addition to a lack of eligibility, there is a lack of awareness and understanding regarding the importance and benefits of purchasing takaful, which contributes to the unavailability of suitable protection schemes that cater specifically to low-income groups. Furthermore, a lack of expertise and social responsibility on the part of takaful operators in investing and offering products to reach these markets also aggravates the situation [34].

Given the encouragement and initiatives outlined in the Central Bank of Malaysia Financial Sector Blueprint 2012–2026 towards an inclusive and balanced financial system, microtakaful is recommended to be developed commercially and sustainably to ensure protection for low-income groups [29]. Protecting low-income groups against specific risks is critical as they are more susceptible to financial and property losses, ultimately breaking the cycle of poverty [35].

4.3.3 Comparison between microinsurance and microtakaful

Microinsurance, a form of insurance that originated in the 1970s, did not gain widespread market attention for the next three decades. However, in 2006, Nobel Prize-winning banker and economist Muhammad Yunus introduced microfinancing schemes that enabled millions of low-income people in Asia and Africa to become entrepreneurs and homeowners. According to Haroun & Effandi (2019), microinsurance was introduced as a means of reducing poverty by providing critical financial services to poor communities and offering coverage against loss [36]. This method has gained significant popularity in developing countries, particularly in Asia, Africa, and Latin America, compared to traditional insurance that offers a range of products. Homeowner's insurance was first introduced in 1950 to offer coverage to homeowners who suffered damage or loss [24].

In contrast, microtakaful, founded in Lebanon in 1997 by the Lebanon Agricultural Mutual Fund, was established to provide affordable Shariah-based protection to the underprivileged and fulfil their financial security needs. Furthermore, Takaful Ikhlas and the Farmers Welfare Federation established microtakaful in Malaysia in April 2007, with support from the Malaysian government. Microtakaful has the same goal as *takaful*, which is to protect society as a whole's welfare. However, *takaful* is designed to benefit standard earners and groups, while microtakaful is intended to alleviate poverty by assisting people in maintaining their financial well-being.

Microinsurance involves an agreement between the insurer and the insured, with the insurer selling the policy to the policyholder, accompanied by a contract to reimburse them for specific losses or damages sustained as a result of the agreement [37]. Microtakaful, on the other hand, is based on a combination of donations in which participants pay for membership with the aim of assisting those in need within a community. Within the group, the burden of any misfortunes faced by any of the participants is shared, and compensation is provided using funds from participants' contributions. **Table 1** highlights the most significant differences between microinsurance and microtakaful.

Concepts and Principles	Microinsurance	Microtakaful
Governance of contract	An exchange contract that is established through a sale and purchase agreement is a means of transferring risk. This process can be further refined for greater academic and practical engagement.	Risk sharing through a contract arrangement based on mutual guarantee and contributions is established without the involvement of risk transition through sales or purchases. This method of risk management promotes cooperation between parties and reduces financial risks.
Risk	The policyholder's risk is shifted to the insured company in exchange for the payment of a premium. This transfer of risk is a fundamental aspect of insurance and highlights the important role played by insurers in mitigating risk for individuals and businesses.	There are no transfer risks for either individuals or entities, and any burdens or misfortunes faced by participants are shared among the group. This system ensures that all involved parties are accountable for their actions and can effectively manage any potential risks or challenges.
Rate	The amount of premium paid belongs to the insurer	The amount of the donation becomes the participants' ownership
Shariah considerations	Involves the prohibitory elements under the Shariah – interest/usury (riba), uncertainty (gharar) and gambling (maysir)	Free from interest/usury (riba), uncertainty (gharar) and gambling (maysir)

Table 1.
Microinsurance vs., Microtakaful.

5. Understanding and application of microinsurance and microtakaful specifically in protecting houses and home contents in Malaysia

In the Financial Sector Blueprint 2022–2026, Bank Negara Malaysia (BNM) underscores its dedication to elevating the monetary welfare of Malaysian households. The blueprint outlines the BNM's strategic priorities, such as reinforcing protection for households to cultivate financial resilience. To realise these aims, the BNM is advancing public-private partnerships to guard against high-risk, large-scale events and encouraging microinsurance and microtakaful development [29]. The goal is to double the number of individuals subscribed to these kinds of financial products by 2028, thus bolstering financial safety nets, especially for the most vulnerable members of society.

One subset that necessitates extraordinary attention is the low-income and destitute sections of the population. The Perlindungan Tenang initiative was introduced in 2017 to inspire the growth of microinsurance and microtakaful products and to heighten awareness among consumers [38]. All Perlindungan Tenang schemes cover basic life and family needs, with a focus on death benefits. They have simple disclosure and claims processes, with one product allowing e-submission of claims documents through mobile applications like WhatsApp. The schemes accept substitution of proof, such as confirmation by the village head or district officer, for claims documents in case of inability of participants staying in longhouses to provide customary documents. They also accept claims documents certified by the village head, school principal, or police officer instead of in-person submission at the takaful operator's branch via postage submission. The schemes have simple and reasonable timelines; for example, funeral benefits are available with a pay-out within 24 hours, and burial permits are acceptable as an alternative to death certificates.

Critically, while the initiative has been somewhat successful, obstacles still remain before it can achieve its full potential and cater effectively to unserved and underserved consumers. Protection schemes for low-income groups' properties, such as land, houses, and assets, are still inadequate. Presently, Perlindungan Tenang offers only one insurance plan, which is not a *takaful* scheme [39]. This insurance plan provides coverage against specified perils for household and home contents. Under microtakaful, there is currently no specific coverage available for the protection of houses and their contents. It is worth highlighting that certain takaful operators previously offered two microtakaful house owner and contents schemes in the market. However, these schemes have since been discontinued. This further emphasises the lack of microtakaful schemes available for the protection of houses and their contents [39].

To address these hurdles, the BNM proposes a three-pronged approach. Firstly, the Perlindungan Tenang framework will be upgraded to grant insurers and takaful operators (ITOs) more flexibility in serving the underserved strata of society. This might involve extending diverse products, broadening distribution channels, and introducing innovation through product bundling.

Secondly, the BNM plans to give ITOs more refined demand-side intel so they can better identify coverage gaps, risks, and behaviours of the unserved and underserved. By having access to more dependable data, ITOs will be able to offer more knowledgeable and targeted policies and expand their product offerings to better meet the needs of these consumers.

Thirdly, the BNM is committed to advancing financial literacy initiatives to enhance consumer awareness and comprehension of microinsurance and microtakaful coverage, as well as the benefits of the Perlindungan Tenang initiative. The BNM will collaborate extensively with the Financial Education Network (FEN) to reach those segments of society most in need.

Although the BNM is dedicated to delivering the above strategies, private sector supply-side measures alone will not suffice. The government has a critical role to play in expanding protection coverage. For example, through the provision of purchase vouchers and stamp duty exemptions for Perlindungan Tenang products. The social safety net system must also serve as a fundamental cornerstone to safeguard target groups with economic dependencies – such as income levels, household size, and age.

5.1 Houseowners/householders and home contents protection schemes

Home protection schemes in Malaysia offer multiple options for safeguarding residential properties. Specifically, there are three basic types of protection available for homeowners. The first is the Houseowner or Homeowner scheme, which provides coverage for residential property against loss or damage caused by perils such as flood or earthquake. The second is the Householder scheme, which provides coverage for loss or damage to the content of residential property and fatal injury to participants. The third type of protection is a combination of the Houseowner and Householder schemes.

When it comes to owning residential property in Malaysia, there are two ways to achieve it. One is to purchase directly from a seller or developer, while the other is to take financing or a loan from a financial institution or bank. In the case of a direct purchase, the owner of the property is responsible for personally insuring the house and its contents [35, 40]. However, if financing is obtained to purchase a residential property, it is a requirement for the purchaser to participate in houseowner insurance or takaful. The purchaser is also required to participate in mortgage insurance or takaful. This is a family protection plan that will automatically settle the financing in the event of the purchaser's misfortune, such as death, or permanent disability [37].

The houseowner insurance policy or takaful scheme takes into account the increase in home value based on the Consumer Price Index (CPI) and is subject to a maximum of 75% of the value of the purchased residential property. The premium or contribution made by the policy or participant is approximately 1.75% of the maximum value [37, 41].

In Malaysia, every property owner, including homeowners, is expected to ensure that their houses and properties are covered in the event of risks. For direct house purchases, the buyer is responsible for insuring the house and the contents within. However, when obtaining financing to purchase a residential property, the buyer must procure or participate in a takaful/insurance house owner scheme. Additionally, the buyer must participate in an additional protection scheme – takaful/insurance mortgage – to ensure financial security. The mortgage protection scheme is a type of family insurance that provides immediate coverage on the funding in the event of the borrower's death or permanent disability.

For instance, Syarikat Takaful Malaysia, one of the takaful operators in Malaysia, offers a combined scheme – "Takaful myHouseowners & Householders" - that comprehensively protects homeowners' homes, belongings, and assets [37]. The scheme combines coverage for the house structure, including the walls, roof, fixtures and fittings, garages, gates, and fences, as well as contents, such as household goods, personal effects, and other movable possessions. In addition, specifically for house tenants, the "Takaful myHome contents" scheme is available. It is a complete scheme that protects household contents against any damage or loss caused by common perils, such as fire, lightning, explosion, theft, hurricane, earthquake, flood, and bursting of water tanks/pipes. Examples of the coverage and contribution rate under a Houseowners & Householders Takaful scheme offered by one of the takaful operators in Malaysia are provided in **Figure 1**.

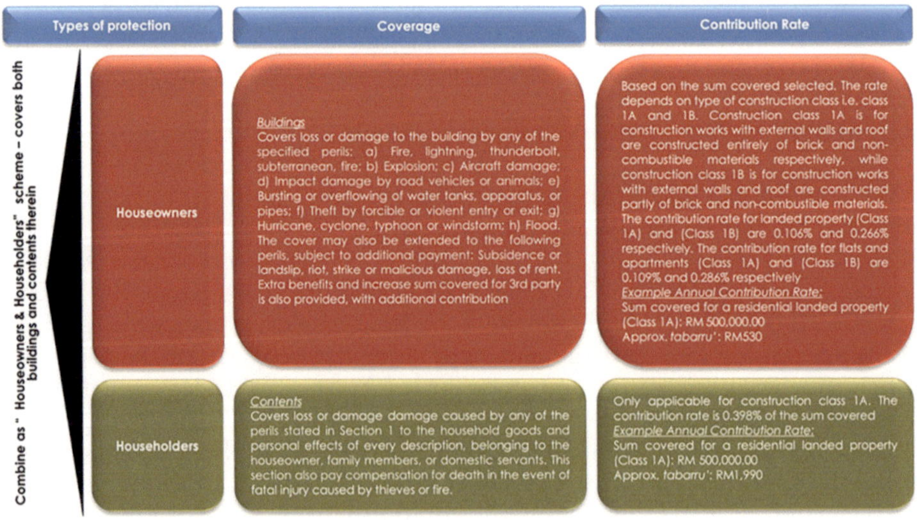

Figure 1.
Coverage and contribution rate under a Houseowners & Householders Takaful scheme [37].

6. Conclusions

The chapter focuses on the importance of insurance policies in protecting vulnerable sections of society, specifically exploring the concept of property protection for the poor. Such protection is essential for individuals and families facing significant financial losses due to disasters or thefts. These events can wipe out an entire household's savings, pushing them into debt. Access to property insurance is a significant factor in mitigating such risks and providing a safety net for families, helping them to recover quickly from unforeseen setbacks [42].

Studies have revealed that insurance has a vital role to play in poverty reduction. In many cases, increased access to property insurance has led to a significant reduction in poverty rates. Moreover, the chapter shows how insurance policies can play a crucial role in facilitating poverty reduction through its ability to provide higher protection to individuals and businesses.

Microinsurance is one such example of insurance policies tailored to meet the needs of low-income populations and vulnerable communities. These policies address the specific needs of communities, such as crop insurance for subsistence farmers or home insurance for those living in disaster-prone areas. By pooling resources, and sharing risks, communities can better protect themselves against unforeseen events and work towards building a stronger, more resilient future.

It is important to note that property insurance can have a positive impact on economic growth. By protecting individual and commercial property, the sale of insurers' products indirectly promotes investment, and entrepreneurship. When businesses have a safety net in place, they are more likely to take risks and ultimately contribute to economic growth. This, in turn, can lead to job creation and increased economic activity.

In conclusion, property protection, such as housing and belongings, through insurance policies, is of the utmost importance in reducing poverty, particularly for vulnerable groups. The 2030 Agenda for Sustainable Development aims to implement suitable protection systems for everyone. It is essential to safeguard food and shelter, particularly in the event of a disaster, to avoid further suffering and hardship. Effective property protection measures are crucial for achieving poverty reduction goals and overall sustainable development.

Author details

Puteri Nur Farah Naadia Mohd Fauzi
Northumbria University, UK

*Address all correspondence to: puteri.fauzi@northumbria.ac.uk

References

[1] Center of Poverty & Social Policy. Anti-Poverty Policies and Programs [Internet]. Columbia University Center on Poverty and Social Policy. New York, US: Columbia University; 2023. Available from: https://www.povertycenter.columbia.edu/policies-and-programs

[2] United Nations. Goal 1: End Poverty in all its Forms Everywhere [Internet]. United Nations: United Nations Sustainable Development; 2022. Available from: https://www.un.org/sustainabledevelopment/poverty/

[3] Fauzi PNFNM, Akhma AIF. Protection of housing and home contents: The understanding and application of micro *takaful* scheme in Malaysia. AIP Conference Proceedings. 2023;**2881**(1):040006-1 to 040006-9. DOI: 10.1063/5.0167657

[4] Fauzi PNFNM, Jahidi Z. Identification of risks exposed to the development of zakat and Wakaf housing in Malaysia. International Journal of Technology. 2022;**13**(4):727-739. DOI: 10.14716/ijtech.v13i4.5275

[5] The World Bank. Poverty [Internet]. Washington, US: World Bank; 2022. Available from: https://www.worldbank.org/en/topic/poverty

[6] Government of New Brunswick. What Is Poverty? Economic and Social Inclusion Corporation. Canada; [Internet] 2010. Available from: https://www2.gnb.ca/content/gnb/en/departments/esic/overview/content/what_is_poverty.html#:~:text=Poverty%20is%20about%20not%20having

[7] The World Bank. What Are Poverty Lines? [Internet]. Washington, US: World Bank; 2017. Available from: https://www.worldbank.org/en/news/video/2017/04/14/what-are-poverty-lines#:~:text=People%20living%20below%20a%20poverty

[8] Hill LE. The institutional economics of poverty: An inquiry into the causes and effects of poverty. Journal of Economic Issues. 1998;**32**(2):279-285

[9] The World Bank. Poverty and Shared Prosperity 2022 [Internet]. Washington, US: World Bank; 2022. Available from: https://www.worldbank.org/en/publication/poverty-and-shared-prosperity

[10] Aripin S, Rashid KA, Sarkawi AA, Hasan SF, Fauzi PNFNM. Zakat houses design and planning (quality and standard) from shariah perspective. Planning Malaysia. 2019;**17**(1):176-185. DOI: 10.21837/pmjournal.v17.i9.596

[11] Tunstall R. The links between housing and poverty. JRF [Internet]. 2017. Available from: https://www.jrf.org.uk/report/links-between-housing-and-poverty

[12] Rashid KA, Mohd Fauzi PNFN, Hasan SF. Meeting housing needs of the poor and needy Muslims through Zakat and Wakaf. Planning Malaysia. 2018;**16**(3):308-319. DOI: 10.21837/pm.v16i7.520

[13] Musthafa A. Poverty main cause of housing deprivation [Internet]. 2023. Available from: https://www.thesundaily.my/local/poverty-main-cause-of-housing-deprivation-DA10761662

[14] Patel S. Takaful and poverty alleviation [Internet]. International Cooperative and Mutual Insurance Federation (ICMIF); 2004. Available

from: https://www.findevgateway.org/sites/default/files/publications/files/mfg-en-paper-takaful-and-poverty-alleviation-2004.pdf

[15] Fauzi TSPNFNM. Introduction to takaful insurance in the Malaysian construction sector. RICS Construction Journal. 2022. Available from: https://ww3.rics.org/uk/en/journals/construction-journal/takaful-insurance-malaysian-construction-sector.html

[16] Insuranceopedia. What is an Insurer? - Definition from Insuranceopedia [Internet]. 2023. Available from: https://www.insuranceopedia.com/definition/384/insurer

[17] Department of Islamic Development Malaysia (JAKIM). Menyertai Polisi Insurans Konvensional: Halal Atau Haram? [Internet]. e-Muamalat: Department of Islamic Development Malaysia (JAKIM); 2018. Available from: https://e-muamalat.islam.gov.my/en/bahan-ilmiah/artikel/148-menyertai-polisi-insurans-konvensional-halal-atau-haram

[18] Department of Islamic Development Malaysia (JAKIM). Insurance. 1972. Available from: http://www.e-fatwa.gov.my/fatwa-kebangsaan/insurans-0 [Accessed: October 21, 2014]

[19] Department of Islamic Development Malaysia (JAKIM). On the implementation of the law of Takaful Malaysia. 2008. Available from: http://www.e-fatwa.gov.my/fatwa-kebangsaan/hukum-pelaksanaan-takaful-di-malaysia [Accessed: October 21, 2014]

[20] Department of Islamic Development Malaysia (JAKIM). Conventional insurance law to promote products by Shariah perspective. 2009. Available from: http://www.e-fatwa.gov.my/fatwa-kebangsaan/hukum-mempromosikan-produk-insurans-konvensional-menurut-perspektif-syariah [Accessed: October 21, 2014]

[21] Islamic Fiqh Academy. Resolutions and Recommendations of the Council of the Islamic Fiqh Academy 1985-2000. Jeddeh, Saudi Arabia: Islamic Fiqh Academy; 1985

[22] Mohd Fadzli Y, Wan Zamri WI, Abdul Khudus MN. Takaful: Operation and Business Competence. Kuala Lumpur: IBFIM; 2015

[23] Ahmed MH. Micro Takaful insurance as a tool to guaranteeing financing and protecting micro enterprises. Journal of Business and Financial Affairs. 2016;5(4):228

[24] Insurance Information Institute (III). Brief History [Internet]. New York, US: Insurance Information Institute, Inc.; 2014. Available from: https://www.iii.org/publications/insurance-handbook/brief-history

[25] Fauzi PNFM, Rashid KAR. Takaful: A review on performance, issues and challenges in Malaysia. Journal of Scientific Research and Development. 2016;3(4):71-76. Available from: http://jsrad.org/wp-content/2016/Issue 42016/12jj.pdf

[26] Mohd, Fauzi PNFN, Laldin MA. Classification and types of risk for properties belonging to low-income and poor individuals in Malaysia. In: Mohd Ahnuar E, Mohd Nordin R, Yunus J, Abdul Rahman NA, editors. MATEC Web of Conferences. Vol. 266. Shah Alam, Malaysia: UITM; 2019. p. 03016

[27] Awani A. Government should compel Islamic religious schools to register [Internet]. 2023. Available from: https://www.astroawani.com/

berita-malaysia/government-should-compel-islamic-religious-schools-register-lam-thye-155239

[28] Hasim HM. Developing a conceptual framework of Microtakaful as a strategy towards poverty alleviation. Journal of Economics and Sustainable Development [Internet]. 2014;**5**(28):1-8. Available from: https://www.iiste.org/Journals/index.php/JEDS/article/view/18665/19115#google_vignette

[29] Bank Negara Malaysia. Financial sector blueprint 2022-2026 [Internet]. Kuala Lumpur, Malaysia: Bank Negara Malaysia; 2022. Available from: https://www.bnm.gov.my/documents/20124/5915429/fsb3_en_book.pdf

[30] Bank Negara Malaysia. Microinsurance and Microtakaful Discussion Paper [Internet]. Kuala Lumpur, Malaysia: Bank Negara Malaysia; 2017. Available from: https://www.bnm.gov.my/documents/20124/761691/dp_microinsuranceandmicrotakaful_Apr2016.pdf

[31] Center for Excellence in Disaster Management & Humanitarian Assistance. Hornet Avenue, Joint Base Pearl Harbor - Hickam, Hawaii; 2019

[32] ECHO Daily Flash. Malaysia - Flash Floods (Media, Met Malaysia, ADINet). Malaysia [Internet]: ECHO Daily Flash; 2021. Available from: https://reliefweb.int/report/malaysia/malaysia-flash-floods-media-met-malaysia-adinet-echo-daily-flash-22-october-2021

[33] Hunt J. What Is Microinsurance? [Internet]. New York, US: The Balance; 2022. Available from: https://www.thebalancemoney.com/what-is-microinsurance-4165961

[34] IFSB-IAIS. Issues in Regulation and Supervision of Microtakāful (Islamic Microinsurance). Kuala Lumpur: Islamic Financial Services Board (IFSB) and International Association of Insurance Supervisors (IAIS); 2015

[35] Mohd, Fauzi PNFN, Rashid KA. Protection of Wakaf-Zakat properties for the needy and poor via Takaful. In: I-Maf 2017 E-Proceedings. Selangor, Malaysia: KUIS; 2017. pp. 157-171. Available from: http://conference.kuis.edu.my/i-maf/images/eproceedings/2017/157-171-imaf-2017.pdf

[36] Haroun AEH, Yusoff ME. Micro-Takaful as stimulus for financial inclusion: A key component towards sustainable development. Open International Journal of Informatics, 2019;**7**(2):43-55. Retrieved from: https://oiji.utm.my/index.php/oiji/article/view/63

[37] Syarikat Takaful Malaysia. Takaful myHouseowners & Householders. Takaful Malaysia [Internet]: Syarikat Takaful Malaysia; 2023. Available from: https://www.takaful-malaysia.com.my/en/products/takaful-myhouseowners-householders/

[38] Bank Negara Malaysia. Perlindungan Tenang [Internet]. Kuala Lumpur, Malaysia: Bank Negara Malaysia; 2021. Available from: https://www.bnm.gov.my/documents/20124/948107/PD+Perlindungan+Tenang.pdf/ba34bc32-af87-5aa7-0c19-b11c980028ba?t=1625225115430

[39] Mohd Fauzi PNFN, Laldin MA. Micro-takāful scheme for the protection of houses belonging to low-income groups in Malaysia. ISRA International Journal of Islamic Finance [Internet]. 2022;**14**(3):303-314. Available from: https://www.emerald.com/insight/content/

doi/10.1108/IJIF-05-2021-0090/full/pdf#:~:text=Micro-tak%C2%AFaful%20(micro,protection%20of%20lives%20and%20properties

[40] Rashid KA, Mohd Fauzi PNFN, Hasan SF. Shariah principles in the procurement of buildings and infrastructure works. In: Paper Presented at Fourth International Conference on Masjid, Zakat and Waqf 2017. Selangor, Malaysia: KUIS; 2017. pp. 223-231. Available from: http://conference.kuis.edu.my/i-maf/images/eproceedings/2017/223-231-imaf-2017.pdf

[41] Khuzhamov LT, Kozhevnikova EI, Zaitseva KA. Features of the development of microinsurance in the developing countries' insurance markets. In: IES 2020: Economic Systems in the New Era: Stable Systems in an Unstable World [Internet]. Samara, Russia: Springer Link; 2020. pp. 398-407. DOI: 10.1007/978-3-030-60929-0_51

[42] NASA. Human needs exploration: Then and now [Internet]. 2006. Available from: https://www.nasa.gov/pdf/162514main_Human_Needs.pdf

Chapter 8

Study on the Future Development Direction of Full-Cycle Real Estate Development Based on Carbon-Neutral Background

Rui Liu and Yu Chen

Abstract

Against the backdrop of global climate change, environmental pollution, and energy shortages, it has become an international consensus to promote carbon neutrality. For the first time, the development goal of carbon neutrality has put forward new requirements for various industries, and the real estate industry, as an important field of energy consumption and carbon emission, is facing unprecedented challenges and opportunities. This paper takes the whole-cycle development direction of real estate under the background of carbon neutrality as the research theme, and firstly, we sort out the background of the implementation of global and Chinese carbon neutrality policy and its impact on the real estate industry and make clear the significance of the research on carbon neutrality to the real estate industry. Through analysing the current situation of real estate development, the paper reveals the main problems and difficulties of the real estate industry in terms of carbon emissions and discusses the current situation and challenges of the real estate industry in the process of promoting carbon neutrality. Finally, it looks into the overall development direction of real estate development in the future and concludes that intelligent, ecological, and sustainable development will become the general trend of the real estate industry.

Keywords: carbon neutral, property development, full cycle development, sustainability, green building

1. Introduction

1.1 Background to the implementation of carbon-neutral policies

The origin of carbon-neutral strategies is closely linked to the strong global concern about climate change. Over the past few decades, global warming has been exacerbated by the release of large quantities of greenhouse gases (GHGs), particularly carbon dioxide (CO2), as a result of human production activities (**Figure 1**). According to data released by the World Meteorological Organisation (WMO) in 2020, global temperatures have risen by 1.2°C compared to pre-industrial times. Researchers predict that if no practical measures are taken, our surface temperatures could rise even higher, crossing the 2°C warming threshold by 2050. Such a change in climate could lead to a variety of catastrophic consequences, including rising sea levels, a high incidence of extreme weather events, and the collapse of ecosystems [1]. Such events pose a serious threat to human life, the economy, and the stability of societies.

In response to the challenge of climate change, countries around the world have introduced carbon-neutral policies. The purpose of carbon-neutral policies is to reduce the release of greenhouse gases and at the same time seek a balance between carbon emission and absorption, so as to curb the trend of global warming. Against this background, countries around the world have signed the Paris Agreement, which expects the world to reach carbon neutrality by 2051–2100 [2]. Carbon neutrality and carbon peaking have become the global trend of green development. China attaches great importance to combating climate change, energy-saving, and emission reduction. In order to reflect the role of a responsible big country, in September 2020, President Xi proposed for the first time at the 75th session of the UN General Assembly the goals of "peak carbon" and "carbon neutrality": to reach the peak of CO_2 emissions by 2030, to strive for CO_2 emissions by 2060, with a view to achieving a carbon neutral world peak by 2030, and strive to achieve carbon neutrality for the whole country by 2060 ("3060 Dual Carbon Goal"). As an important means of focusing on comprehensive emission reduction, the carbon-neutral policy has won widespread attention and support from all walks of life.

The implementation of carbon-neutral policies also includes the rapid development of renewable energy technologies [3]. As renewable energy technologies, such as solar and wind power, mature and become widely used, the supply of low- and zero-carbon energy sources increases, which will reduce the dependence on fossil fuels and contribute to carbon neutrality. With the rapid development of new technologies, such as electric vehicles, smart buildings, and efficient energy use, they also provide technical support for the implementation of carbon-neutral policies.

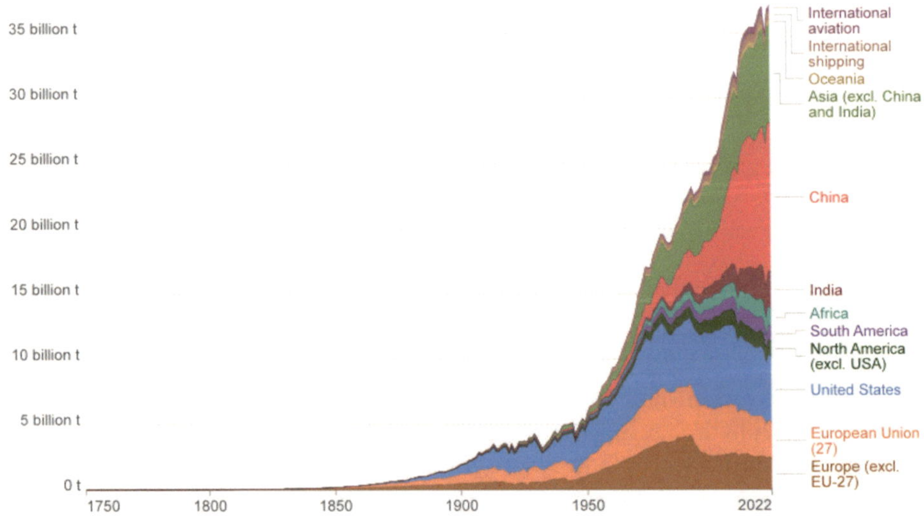

Figure 1.
Map of total annual global CO_2 emissions by region, 1750–2022 (Quoted: https://ourworldindata.org/).

The background to the implementation of carbon-neutral policies stems from global alertness and action on climate change, and also benefits from the support of scientific research, the emergence of an international consensus, and advances in renewable energy technologies. Together, these factors have contributed to the formulation and implementation of carbon-neutral policies, providing an important pathway and means to achieve climate goals.

1.2 Significance of carbon-neutral research

Conducting research on carbon neutrality is important for environmental protection. According to the reports of internationally renowned scientific institutions, the concentration of greenhouse gases around the world has reached 400 ppm, which has exceeded the maximum safe concentration of 350 ppm (**Figure 2**) [4]. Implementing a carbon-neutral policy can effectively slow down the rate of warming and ensure that global greenhouse gas emissions are reduced to half their original level by 2030, creating a period of improvement for the Earth's ecosystem. Slowing carbon emissions will not only help improve air quality but also public health, further reducing the number of premature deaths and economic losses caused by air pollution. By promoting the application of low-carbon technologies and the implementation of carbon-neutral strategies, air quality is expected to improve, and related health problems associated with air pollution will be reduced.

Carbon-neutral research is also important for economic development. Achieving carbon neutrality requires the promotion of new industries, including carbon trading, renewable energy, energy storage, and low-carbon buildings [5]. The establishment and development of carbon markets has facilitated the application and dissemination of low-carbon technologies, whilst increasing market liquidity and efficiency. The renewable energy sector has also grown rapidly, attracting significant investment. According to data, the value of the carbon trading market has reached $53 billion, with an average annual growth rate of more than 20%; investment in the renewable energy sector totalled $620 billion, accounting for nearly 70% of global energy investment. The development of these emerging areas not only provides a new

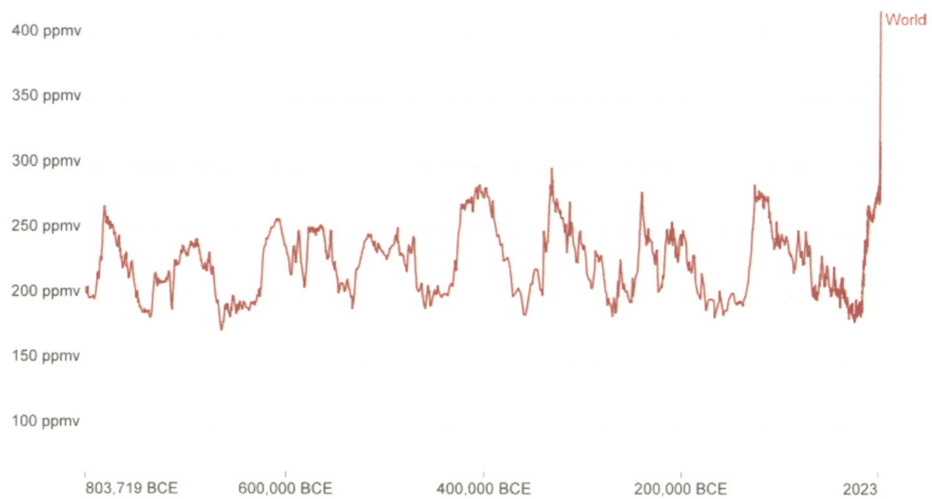

Figure 2.
Variation of atmospheric CO_2 concentration (Quoted: https://ourworldindata.org/).

impetus for economic growth but also promotes the transformation and upgrading of the economic structure.

The importance of carbon-neutral research is more prominent in the real estate industry due to its own characteristics of high energy consumption, high pollution emission, and rough construction method. The traditional mode of property development generates large amounts of carbon emissions throughout its life cycle, especially during the production, transport, construction, and operation phases of building materials. The cement production process generates about 4.1 billion tonnes of carbon dioxide annually, accounting for about 8% of total global emissions [6]. Therefore, examining carbon-neutral policies will help provide the real estate industry with cutting-edge low-carbon building materials, smart construction techniques, and the use of renewable energy, which in turn will significantly reduce carbon emissions across the property market. The integration of carbon-neutral concepts can not only improve the social responsibility of enterprises but also enhance their brand competitiveness. In today's increasingly competitive market environment, green building and energy-efficient building are gradually seen as the future development trend in the real estate sector. Therefore, an in-depth exploration of the role and significance of carbon neutrality in the real estate industry is indispensable for promoting the decarbonisation of the industry and enhancing the market competitiveness of enterprises.

2. Analysis of the current situation of property development in the context of carbon neutrality

2.1 Current status of carbon emissions from the property sector

As the global warming problem becomes more and more serious, the carbon emission problem has gradually received great attention from governments and all walks of life. The real estate industry, as a major energy consumer, is at the forefront of carbon emissions amongst all industries. According to statistics, about 40% of global carbon emissions are related to the energy consumption of construction equipment, and in China, the proportion is as high as 45% [7]. According to the National Bureau of Statistics, by the end of 2020, China's real estate sector has already exceeded 8 billion square metres of floor space, equivalent to a quarter of the world's, and continues to grow. This makes the issue of carbon emissions from the real estate industry particularly salient and in urgent need of in-depth analysis and assessment.

Looking at the structure of energy consumption in the real estate industry, the heating, ventilation, and air conditioning (HVAC) systems of buildings are the main source of energy consumption, accounting for 60–70% of the total energy consumption [8]. In addition, the energy consumption of lighting, home appliances, and other amenities should not be overlooked. According to China's National Development and Reform Commission (NDRC), energy consumption in China's construction industry has been growing at an average annual rate of 9% over the past decade, with electricity consumption accounting for 55% of the total. This shows the huge demand for energy in the real estate industry and signifies its importance in reducing emissions.

An in-depth look at the entire lifecycle of a property gives us a more holistic view of its carbon emissions. The whole life cycle of a building is divided into four key stages: design, construction, operation, and demolition, with the operation stage accounting for the largest amount of energy consumption and carbon emissions. According to scientific research, when a building is in operation, its carbon emissions account for 80% of its total life cycle carbon emissions. For example, an average house with an expected lifespan of 50 years will emit about 200 tonnes of carbon during the construction phase, but more than 800 tonnes during the operation phase. Such figures suggest that resources should be focused on achieving energy efficiency and reducing carbon emissions during the operational phase of the property sector [9].

From a building materials perspective, the process of manufacturing building materials also results in significant carbon emissions. In the case of steel and cement, for example, the carbon emissions generated during the manufacturing process of steel account for 25% of global industrial carbon emissions, whilst the carbon emissions used to manufacture cement are about 8% [10]. As the world's largest construction market, China's production of cement and steel reached 2.3 billion tonnes and 960 million tonnes, respectively, in 2019, and the total carbon emissions from their manufacturing process were about 2.9 billion tonnes, accounting for about 30% of the country's total carbon emissions. Therefore, when exploring the current state of carbon emissions in the real estate industry, how to select and utilise materials becomes particularly critical.

In examining the current state of carbon emissions in the real estate industry, we cannot ignore the positive impact of green and low-carbon buildings. Green buildings adopt low-energy and low-pollution technologies in design, material selection, construction, and operation management, effectively reducing the carbon footprint of buildings. At this stage, the promotion of green buildings in China is increasing year by year. By the end of 2020, there will be a cumulative total of 24,000 green building label projects with a floor area of 2.1 billion square metres, accounting for 15% of the newly built floor area. The popularisation and promotion of green buildings will provide an important technical support and cooperation platform for carbon emission reduction in the real estate industry.

Finally, policy orientation and technological innovation are also key factors affecting the current status of real estate carbon emissions. In recent years, the Chinese government has introduced a series of policies and standards, such as the Green Building Evaluation Standards and the Building Energy Efficiency Regulations, aiming to promote the development of the real estate industry in the direction of low-carbon, energy-saving, and environmental protection. At the same time, the application of technological innovation in the field of building energy efficiency has become increasingly widespread, such as solar photovoltaic technology, ground source heat pump technology, and smart home systems, all of which have achieved carbon emission reduction effects to varying degrees. Some studies have pointed out that the more mature building energy-saving technologies can achieve carbon emission reduction of 20–30% for individual buildings [11]. Therefore, the promotion of policy orientation and technological innovation provides strong support for carbon emission reduction in the real estate industry.

The carbon emission status of the real estate industry is jointly influenced by many factors, including energy consumption, choice of building materials, green building, policy orientation, and technological innovation. In order to achieve the goal of carbon neutrality, it is necessary to deeply analyse and study the current situation of carbon emissions in the real estate industry and formulate corresponding carbon reduction policies and measures in a targeted manner. By strengthening policy support, increasing technological innovation, and promoting green buildings, it is expected to further reduce the carbon emissions of the real estate industry and make positive contributions to achieving carbon neutrality.

2.2 Impact of carbon-neutral policies on the real estate sector

In recent years, the issue of climate change has become increasingly prominent globally, and countries have implemented a series of carbon-neutral policies to achieve carbon reduction targets [12]. According to the International Energy Agency, the energy consumption of the construction industry accounts for nearly 40% of the total global energy consumption, and the real estate industry accounts for the majority of it. Against this background, the real estate industry, as an area of high energy consumption and high carbon emissions, has received widespread attention. In response to carbon-neutral policies, real estate companies have begun to re-examine their business operation models and project development strategies, with the aim of reducing carbon emissions and improving energy utilisation. Therefore, with the implementation of carbon-neutral policies, real estate companies are adopting a series of energy-saving and emission-reduction measures in the development, design, and management stages.

For the real estate industry, the carbon-neutral policy has brought about significant impacts in the following areas.

The carbon-neutral policy has promoted the transformation and upgrading of the real estate sector. Real estate enterprises have re-planned and redesigned their projects in accordance with carbon-neutral targets, strengthened guidance and training on energy-saving and environmental protection awareness, and promoted the application of green building and renewable energy technologies. For example, real estate developers can reduce energy consumption by choosing more environmentally friendly and energy-efficient building materials, adopting energy-efficient systems and equipment, and optimising building structure and ventilation design. At the same time, they can further reduce their reliance on traditional energy sources by introducing renewable energy sources such as solar photovoltaic power generation and wind power generation. These measures not only reduce carbon emissions and environmental pollution but also conserve resources and improve corporate image and market competitiveness.

Secondly, the carbon-neutral policy has promoted the development of green finance. Against the backdrop of carbon neutrality, environmental protection and energy conservation and emission reduction have become one of the important criteria for financial institutions to assess real estate projects. More and more financial institutions have launched financial products to support green buildings and low-carbon projects, such as green bonds and green loans, in order to provide financial support and preferential interest rates to guide real estate enterprises to transition to low-carbon development. Real estate companies also need to proactively adapt to the requirements of green finance and improve the sustainability and return on investment of their projects in order to obtain more financial support.

Thirdly, carbon-neutral policies promote technological and management innovation in the real estate sector. In order to meet the needs of low-carbon and environmental protection, real estate companies should continue to introduce intelligent, automated, and information technology to ensure the optimal allocation and efficient use of energy. For example, with the help of modern technologies such as the Internet of things, big data, and artificial intelligence, they can monitor and control the energy consumption of buildings in real time, providing intelligent means for energy management and adjustment.

The carbon neutral policy has had a far-reaching impact on the real estate industry, promoting the development of real estate enterprises in energy-saving and emission reduction, green building, green finance, and technological innovation. Real estate enterprises should pay close attention to the trend of carbon-neutral policy, understand the policy requirements in depth, and actively explore and practise green and low-carbon development paths, in order to respond to the challenges of climate change and sustainable development and achieve sustainable and high-quality development.

3. Exploring carbon-neutral pathways for future full-cycle property development

3.1 Study on the full cycle of property development under the carbon-neutral target

Against the backdrop of the global move towards carbon neutrality, the real estate industry, as a high-energy consumption and high-emission sector, is facing the pressure and challenge of achieving carbon emission reduction and green sustainable development. Carbon neutrality in the whole cycle of real estate involves a systematic project of planning, design, construction, operation, and management as well as demolition [13], in which the focus is placed on the control of carbon emissions and the application of green technologies in each link.

At the planning and design stage, priority should be given to low-carbon and energy-saving planning and design concepts to achieve the goal of carbon neutrality. Therefore, the planning and design stage should reduce energy consumption and carbon emissions through measures such as rational site selection, optimal layout, and material selection. The development vision and objectives of a low-carbon city can be clarified through urban master plans and control plans, and corresponding land-use plans as well as supporting plans for transport, energy, water resources, and other infrastructures should be formulated. Emphasis should be placed on the greening of building design, such as improving the energy efficiency and resource utilisation efficiency of buildings. In the design process, the design concept of eco-city can be invoked to build a sustainable ecosystem by focusing on the planning of natural elements such as green space and water bodies. The demand for artificial energy can be significantly reduced by optimising the orientation, shape, structure, and material of the building and enhancing its natural lighting and ventilation performance. Some studies have shown that building energy consumption can be reduced by 50–70% by adopting passive house design techniques [14]. The application of building information modelling (BIM) technology during design can significantly improve the accuracy and efficiency of building design, thereby reducing material waste and construction errors.

During the construction phase, the promotion of green building materials and green construction techniques is an important initiative to achieve carbon emission reduction. Data show that carbon dioxide emissions generated during concrete production account for about 8% of global human carbon emissions. Therefore, building materials with superior performance, low energy consumption for production and recyclability, such as high-strength steel reinforcement, fly ash cement, and recycled concrete, should be selected. These materials can not only meet the functional requirements of buildings but also reduce carbon emissions during the production and transport of building materials. During the construction process, green construction technology is promoted to reduce resource consumption and environmental pollution. For example, the use of advanced prefabricated building technology can not only improve construction efficiency but also significantly reduce waste emissions and noise pollution during the construction process. At the same time, green building technologies such as solar photovoltaic systems and integrated intelligent lighting systems can also be used to achieve efficient use of building energy.

In the operation and management phase, the introduction of intelligent management systems is the key to reducing carbon emissions in buildings. According to statistics, energy consumption in the use phase of buildings accounts for about 80% of total energy consumption [15]. Therefore, IoT, big data, and artificial intelligence technologies should be utilised to monitor and regulate building energy consumption in real time. For example, through intelligent heating, ventilation, and air-conditioning (HVAC) systems, dynamic regulation of indoor environmental parameters can be achieved to maximise energy utilisation and reduce carbon emissions. Promoting the use of renewable energy, such as solar, geothermal, and wind energy, can significantly reduce carbon emissions during the operational phase of a building.

During the demolition phase of a building, construction waste and hazardous substances should be handled scientifically in order to achieve the goal of carbon neutrality. According to research data, waste concrete generated during the demolition of buildings accounts for more than 50% of all construction waste [16]. Therefore, green construction guidelines for demolition works should be implemented, and recycling of waste materials should be promoted. Through scientific demolition management and waste resource utilisation, the reduction, resourcing, and harmless treatment of construction waste can be achieved.

Overall, to achieve the goal of carbon neutrality in the full cycle of real estate development, it is necessary to comprehensively use a variety of means to enhance the energy utilisation rate and carbon emission control capacity at all stages. Through scientific planning, green design, environmentally friendly construction, efficient operation, and resource utilisation, the goal of carbon neutrality can be achieved in a comprehensive manner to promote the sustainable development of the real estate industry. Relevant studies have shown that strengthening carbon neutral management in every aspect, from macroeconomic policy formulation to specific technology implementation, can reduce carbon emissions from the real estate industry by more than 40% [17].

3.2 Study on carbon-neutral full-cycle development of real estate with financial security

Against the backdrop of carbon neutrality, the real estate sector faces enormous challenges and opportunities. Amongst them, financial security is crucial. Achieving the goal of carbon neutrality requires a large amount of capital investment. According to the Green Finance Development Report [18] released by the People's Bank of China, it is expected that by 2030, to achieve the goal of carbon peaking, the scale of green investment required in China will reach RMB 2.7 trillion to RMB 3.5 trillion per year. This portion of the funds will mainly be used in areas such as clean energy, green buildings, and low-carbon transport. In the process of property development, developers need to raise funds in various ways such as issuing green bonds and setting up green funds. Taking green bonds as an example, China's green bond issuance scale reached 286.3 billion yuan in 2020, accounting for nearly 19% of the global total, making it one of the largest green bond markets in the world.

Government policy guidance and incentives are also an important way to raise funds. The Chinese Government has in recent years actively introduced a series of policies and regulations to encourage green buildings and low-carbon communities, such as the Green Building Action Programme and the Regulations on Energy Conservation in Civil Buildings. These policies and regulations have incentivised enterprises to increase green investment through financial subsidies, tax incentives, preferential loan interest rates, and other forms. Local governments can also set up special funds to support the development and operation of green projects, such as the Shenzhen Green Building Special Fund and the Beijing Green Building Demonstration Project Subsidy Fund.

Financial security is a core element of carbon-neutral full-cycle development of real estate. Close cooperation between the government and enterprises is needed to jointly explore and promote green and low-carbon pathways. Only in this way can we truly achieve the goal of carbon neutrality and promote the sustainable development of the real estate industry.

4. Future directions for the full cycle of real estate development

4.1 Application of carbon-neutral technologies in future property development

With the growing problem of global climate change, carbon neutrality is becoming particularly important in the future of property development. In the current context of carbon neutrality, it is necessary for the real estate industry, as one of the important areas of carbon emission, to take corresponding measures to promote the sustainable development of the real estate industry. Green building design is one of the important means to achieve carbon neutrality. By improving energy efficiency, utilising renewable energy, and reducing waste and carbon emissions, green buildings can reduce carbon emissions during the operation of buildings. For example, The Shard in London, UK, has significantly reduced its energy consumption and carbon emissions through the use of a highly efficient glass curtain wall system, rainwater harvesting, energy reuse systems, and a highly efficient HVAC system. According to the operational data of The Shard in London, it consumes 30% less energy and emits 20% less carbon than conventional buildings [19].

The application of renewable energy is also an important way to achieve carbon neutrality. The widespread application of renewable energy sources such as solar, wind, geothermal, and biomass is an inevitable trend in future property development. For example, by installing solar photovoltaic panels and wind turbines, the Win Morrison Building (Morrison Campus Center) at the University of California, San Francisco, USA, has achieved an average annual reduction of about 120 tonnes of carbon dioxide emissions. This not only effectively reduces the building's carbon emissions but also provides the building with a sustainable energy supply.

Intelligent building management systems also play an integral role in achieving carbon neutrality. With the application of the Internet of things, artificial intelligence, and big data, intelligent building management systems can monitor and optimise a building's energy consumption in real time and improve energy efficiency. For example, IBM's global "Green Hoffman Initiative" has reduced energy consumption at its Hoffman Estates by 25% through an intelligent building management system, reducing carbon dioxide emissions by approximately 15,000 tonnes per year.

Carbon capture and storage (CCS) technology is also attracting attention in real estate development [20]. CCS technology reduces carbon emissions by capturing carbon dioxide produced during the use of buildings and storing or utilising it. For example, in the Urban Nodule Project launched by Japanese construction company Shimizu Corporation, carbon neutrality is achieved by installing carbon dioxide capture devices in high-rise buildings and using the captured carbon dioxide for the production of biofuels from plants and algae.

With regard to the selection and use of carbon materials, the development and application of low-carbon building materials are also particularly important in achieving carbon neutrality. For example, the application of ultra-high performance concrete (UHPC) and recycled concrete can significantly reduce carbon emissions during building construction. According to statistics, the carbon emissions of UHPC are about 25% lower than those of conventional concrete, whilst the carbon emissions of recycled concrete are only 50% of those of conventional concrete [21].

As can be seen from the above, the application of carbon-neutral technologies in future property development is diverse and extensive, including green building design, the use of renewable energy, intelligent building management systems, carbon capture and storage technologies, and the application of low-carbon building materials. Through the comprehensive application of these technological means, carbon emissions in the process of real estate development can be substantially reduced to achieve the goal of sustainable development. In the future, with the continuous progress of technology and policy support, carbon-neutral technology will be more widely and deeply applied in the real estate industry, making positive contributions to the mitigation of global climate change.

4.2 Trends of intelligent, ecological, and sustainable development of real estate

Against the backdrop of carbon neutrality, the real estate industry is facing unprecedented challenges and opportunities. In order to cope with global environmental issues, especially to mitigate climate change, the future development direction of real estate is bound to shift towards intelligence, ecology, and sustainability. This shift is not only a technological advancement but also a profound change in the development model of the entire industry.

The development trend of real estate intelligence is mainly reflected in the extensive application of emerging technologies such as Building Information Modelling (BIM), Internet of Things (IoT), Big Data, and artificial intelligence (AI) in real estate development [22]. Through BIM technology, accurate 3D modelling can be achieved at the building design stage, significantly improving the synergy and efficiency of design and construction, shortening the development cycle, and reducing construction costs. According to data from the China Building Information Modelling (BIM) Development White Paper, the application of BIM technology can reduce the budget overrun rate of a project by 20% and the rework rate in the design phase by 50%. Through the application of IoT technology, all kinds of facilities and equipment inside the building can achieve intelligent and automated management and improve energy efficiency and user comfort. For example, the application of smart home systems can automatically adjust the indoor temperature, humidity, and brightness according to the user's behavioural habits, further achieving energy-savings.

The trend of ecological development of real estate is mainly reflected in the full implementation of green buildings and the wide application of environmentally friendly materials. A green building is a building that maximises resource conservation, protects the environment and reduces pollution throughout its life cycle, and provides people with healthy, suitable, and efficient spaces for use. According to the International Green Building Trends Report, as of 2020, the global green building market has exceeded US$230 billion and is expected to reach US$280 billion by 2023. In China, according to statistics, by the end of 2021, the country's green building area has reached 2.5 billion square metres, accounting for more than 60% of the new building area. The development of green buildings can not only effectively reduce energy consumption and carbon dioxide emissions during the operation of buildings but also improve the service life and comfort of buildings to meet people's pursuit of a high quality of life.

The trend of sustainable development in real estate emphasises the integrated consideration of economic, social, and environmental benefits in the development process and the sustainable management of buildings throughout their life cycle. The core concept of sustainable development is to meet the needs of the present generation without compromising the ability of future generations to meet their needs. Under the framework of sustainable development, real estate companies need to take measures in all aspects of project planning, design, construction, and operation to improve the efficiency of resource utilisation, reduce environmental loads, and enhance social values. For example, through the promotion and application of energy-saving and emission-reduction technologies and the use of green building materials, they can promote the energy transition of real estate and the sustainable reduction of carbon emissions.

The trend of intelligent, ecological, and sustainable development of real estate requires the industry to use advanced scientific and technological means, adopt green materials and energy-saving technologies, and carry out sustainable management of the whole life cycle. These measures will not only achieve the goal of carbon neutrality but also create a safer, healthier, and more comfortable living and working environment for human beings. In the face of the future, the real estate industry needs to continue to innovate and explore greener, smarter, and more sustainable development paths in order to meet the increasingly severe environmental challenges and to achieve the coordination and development of society, the economy, and the environment.

4.3 Policies and concepts for future full-cycle property development

As the global challenge of combating climate change becomes more and more severe, the goal of carbon neutrality has become a common mission for governments and various industries. Against this backdrop, the real estate industry, as one of the major areas of carbon emissions, is bound to follow this trend in its full-cycle development model and to promote its sustainable development through policy guidance and conceptual updating, amongst other things.

At present, governments have introduced a series of policies to promote the real estate sector to achieve the goal of carbon neutrality. For example, in China, the Carbon Neutral Action Plan specifies the goal of achieving carbon neutrality by 2060.

At the same time, governments at all levels are actively formulating and implementing policies such as green building standards, energy efficiency labelling systems, and green financial policies to provide policy support for carbon emission reduction in the whole cycle of real estate development. Specifically, the promotion of green building standards will effectively improve the energy efficiency of buildings, whilst the implementation of green financial policies guides enterprises to invest in green projects and reduce carbon dioxide emissions through financial instruments. In the future, the government should also further improve the carbon trading market to incentivise real estate enterprises to actively participate in carbon trading and to promote the achievement of carbon neutrality.

At present, the real estate industry is gradually establishing a new development concept centred on green, low-carbon, and sustainable development. This new concept requires real estate development enterprises to comprehensively implement low-carbon technologies and green management modes in all aspects of the project's life cycle, from planning and design to selection of building materials and construction management to operation and maintenance. It advocates a green building culture, raises the awareness and acceptance of owners and residents of green buildings, and guides them to more actively realise a low-carbon and environmentally friendly lifestyle.

Lastly, full-cycle real estate development aiming at carbon neutrality requires not only the joint efforts of the Government and enterprises but also the extensive participation and support of the public. The government can enhance the public's awareness of environmental protection and energy conservation through publicity and education and encourage the public to choose green living patterns. At the same time, enterprises should also promote low-carbon products and services through green marketing to guide consumers to choose environmentally friendly housing. Strengthen cooperation between industries, universities, and research institutes, train and introduce professionals in the field of green building, and promote the innovation and development of green building technology. Only through the joint efforts of the government, enterprises, and the public to form a positive interaction can we achieve the goal of carbon neutrality in the full cycle of future property development and realise the sustainable development of the construction industry.

5. Conclusion

The goal of carbon neutrality is not simply energy-saving and environmental protection, but a profound change that touches the fundamentals of the global economy and energy system. In the near future, low-carbon sustainable development will be the way forward for all enterprises, especially real estate enterprises. Although the real estate industry continues to invest and innovate in carbon reduction, it still faces challenges on the road to carbon neutrality, and all sectors of society need to work closely together to promote the real estate industry to achieve the goal of carbon neutrality. Real estate enterprises need to pay more attention to the symbiosis and co-prosperity between buildings and the environment, set appropriate carbon reduction targets, and participate in the construction of an environmental governance system "led by the government, with enterprises as the main body, and with the participation of social organisations and the public" through a series of methods such as green design, green construction, and green operation so as to fully realise the sustainable development of the enterprise and the industry development of the enterprise and the industry.

Author details

Rui Liu[1,2*] and Yu Chen[2]

1 School of Economics and Management, Nanjing University of Aeronautics and Astronautics, Nanjing, China

2 Nanjing Chixia Construction Co., Ltd., Nanjing, China

*Address all correspondence to: 16221966@qq.com

References

[1] Pretis F, Schwarz M, Tang K, Haustein K, Allen MR. Uncertain impacts on economic growth when stabilising global temperatures at 1.5°C or 2°C warming. Philosophical Transactions of the Royal Society A. Mathematical, Physical and Engineering Sciences. 2018;**376**(2119):20160460

[2] Meinshausen M, Lewis J, McGlade C, Gütschow J, Nicholls Z, Burdon R, et al. Realisation of Paris agreement pledges may limit warming just below 2°C. Nature. 2022;**604**(7905):304-309

[3] Zhou Y. Worldwide carbon neutrality transition? Energy efficiency, renewable, carbon trading and advanced energy policies. Energy Reviews. 2023;**2**(2):100026

[4] Sodiq A, Abdullatif Y, Aissa B, Ostovar A, Nassar N, El-Naas M, et al. A review on progress made in direct air capture of CO2. Environmental Technology & Innovation. 2023;**29**:102991

[5] Gil GO, Chowdhury JI, Balta-Ozkan N, Hu Y, Varga L, Hart P. Optimising renewable energy integration in new housing developments with low carbon technologies. Renewable Energy. 2021;**169**:527-540

[6] Mohamad N, Muthusamy K, Embong R, Kusbiantoro A, Hashim MH. Environmental impact of cement production and solutions: A review. Materials Today: Proceedings. 2022;**48**:741-746

[7] Niu Y. Study on the prediction of carbon emissions from residential buildings in Beijing under the background of "Dual Carbon" [dissertation]. Beijing Architecture University; 2023

[8] Kuo B. Post-use assessment of green buildings in small and medium-sized terminals in northwest China [master's thesis]. Xi'an University of Architecture and Technology; 2024

[9] Wei B. Carbon emission measurement and control strategies for public building projects, master of science [dissertation]. Fujian University of Engineering; 2023

[10] Ma Y. A review of carbon footprint research on the whole life cycle of buildings. Green Building. 2024;(03):124-130

[11] Economidou M, Todeschi V, Bertoldi P, D'Agostino D, Zangheri P, Castellazzi L. Review of 50 years of EU energy efficiency policies for buildings. Energy and Buildings. 2020;**225**:110322

[12] Runa A, Zhang Z, Zhang H. Carbon emission peak and carbon neutrality under the new target and vision. In: 2021 International Conference on Advanced Electrical Equipment and Reliable Operation (AEERO). IEEE; 2021. pp. 1-5

[13] Wei J, Li J, Zhao J, Wang X. Hot topics and trends in zero-energy building research-A bibliometrical analysis based on CiteSpace. Buildings. 2023;**13**(2):479

[14] Li FW, Li YB. An overview of domestic and international research on passive house development. Urban Architecture. 2021;(21):137-139

[15] Feng G, Cui H, Chang S, Huang KL, Wang XR. Analysis of carbon emissions and factors affecting near-zero energy buildings. Advances in Climate Change Research. 2022;(02):205-214

[16] Lang H. Research on influencing factors of construction waste reduction

based on the whole life cycle master [dissertation]. Chongqing Institute of Science and Technology; 2022

[17] Shi Q-W. Research on carbon peak simulation and emission reduction responsibility sharing in China's building sector [PhD dissertation]. Chongqing University; 2021

[18] Summary of the Report on Local Green Finance Development in China. Journal of Central University of Finance and Economics. 2021;(11):2

[19] Aboulnaga MM. High-rise buildings in context of sustainability; urban metaphors of greater Cairo, Egypt: A case study on sustainability and strategic environmental assessment. In: Sustainable High Rise Buildings in Urban Zones: Advantages, Challenges, and Global Case Studies. Sustainable High Rise Buildings in Urban Zones; 2017. pp. 163-217

[20] Du L. Development and utilisation of carbon capture and storage technology to reduce greenhouse gas emissions. Leather Making and Environmental Technology. 2023;(20):144-146. DOI: 10.20025/j.cnki. CN10-1679.2023-20-49

[21] Luan F. Research on the development of assembly building technology under the goal of "double carbon". Engineering and Construction. 2024;(01):14-15 +46

[22] Wang J. Research on the key influencing factors constraining the application of big data in the construction field and countermeasures to explore master [dissertation]. Chongqing Jiaotong University; 2023. Available from: https://link.cnki.net/doi/10.27671/d.cnki.gcjtc.2023.000192

Chapter 9

Sustainable Low-Income Housing: Analyzing Green Building Initiatives within the LIHTC Program across U.S. States

Anaid Yerena

Abstract

This study delves into the sustainability initiatives within the Low-Income Housing Tax Credit (LIHTC) program across U.S. states. By reviewing state Qualified Allocation Plans (QAPs), the research assesses the emphasis placed on green building standards within the LIHTC program over time. The research questions this work seeks to answer are: (1) To what extent are green building standards a priority in the LIHTC program across states? (2) How has this priority changed from 2015 to 2023? The data come from QAPs and states' Housing Finance Agency (HFAs) website. Through content analysis methods, I explore how green building standards contribute to enhancing the sustainability of low-income housing projects, shedding light on the program's role in promoting sustainable housing solutions.

Keywords: green building standards, LIHTC program, low-income housing, Qualified Allocation Plans (QAPs), sustainability, United States

1. Introduction

As housing costs continue to rise and income inequality pervades the United States' society, programs that support the construction of housing for low-income households have become essential. Beyond the development of affordable housing, some programs are designed to serve multiple social goals. The Low-Income Housing Tax Credit (LIHTC) program is one of the few federal housing programs that addresses affordable housing needs through the creation of new subsidized housing while considering multiple other goals [1].

For nearly 30 years, the LIHTC program has been providing capital to developers to produce low-income housing. Today, LIHTC is the federal government's primary supply-side housing program [2–5]. As part of the program's design, states are responsible for distributing the tax credits to developers on a competitive basis. As mandated by the U.S. Department of Housing and Urban Development, state finance agencies adopt a policy called the Qualified Allocation Plan (QAP) to guide their funding decisions. The QAP varies by state and depends on the priorities each state sets. States, for example, may identify green or sustainable development as an important goal and use the QAP scoring system to incentivize developers to propose energy/resource-efficient LIHTC projects. The purpose of this study is to analyze the current policies in QAPs concerning energy efficiency or green building standards for LIHTC programs across every state in the country. The research questions this chapter addresses are: to what extent is green building a priority in the LIHTC program? And how has this priority changed over time in the period between 2015 and 2023? The data comes from the states' QAPs, which were coded using content analysis.

The U.S. Department of Housing and Urban Development has interpreted the federal housing law, which states that rent shall be no more than 30% of a household's income for the term rent to include both shelter and the costs for reasonable amounts of utilities. This way of defining housing affordability, in turn, includes operation costs. In the LIHTC program, there is no federal mandate for developers to build energy-efficient units that would reduce tenants' utility costs; instead, whether the units reduce utility costs (through water and energy conservation) or are built to overcome climate disasters [6] is decided at the state level.

2. The low-income housing tax credit program

The LIHTC is a crucial tool for affordable rental housing production and preservation in many cities across the United States [7]. Private investors receive federal income tax credits to incentivize equity investments in affordable rental housing. Since its inception in 1986, the LIHTC has facilitated the production of nearly 3 million affordable housing units by providing federal income tax credits to private investors as incentives for equity investments in affordable rental housing [8]. The allocation of tax credits is governed by federal regulations and managed by state housing finance agencies, which prioritize projects based on annual criteria. The successful execution of LIHTC deals involves a collaboration between federal, state, and local government agencies, equity investors, legal experts, and project developers. This collaborative approach often includes leveraging additional housing funds to support the financing of these projects.

From 1987 to 2015, LIHTC facilitated the creation of 45,905 projects and nearly 3 million housing units [8]. LIHTC encourages private investors to fund affordable housing by offering federal income tax credits, which helps reduce reliance on more expensive financing methods like bank loans. The program supports the construction of new properties, acquisition and renovation of existing buildings, and refinancing of previously funded affordable housing. Despite its success and efficiency compared to other federal rental housing programs, LIHTC remains complex and less understood due to its intricate financing structures and diverse partner involvement, including private investors, government entities, and specialists. State agencies primarily monitor affordability requirements, while HUD gathers basic data on LIHTC projects.

LIHTC mandates that rents remain affordable, defined as housing costs (including utilities) not exceeding 30% of household income, for at least 30 years. Initially, at least 20% of units had to be affordable for households earning 50% or below the area median income (AMI) or 40% of units for those earning 60% or below AMI. The 2018 Consolidated Appropriations Act introduced more flexibility by allowing households earning up to 80% of AMI, provided the average income of all assisted households remains at or below 60% of AMI. This change aims to subsidize lower rents for extremely low-income households and offers stability for residents whose incomes rise slightly over time. Compliance requires regular certification of tenant income and housing quality inspections, with non-compliance risking the loss of tax benefits for investors. Some states enforce longer affordability periods beyond the federal minimum.

LIHTC provides two types of tax credits: the 9% credit and the 4% credit, both of which support the construction, rehabilitation, or refinancing of rental properties meeting income eligibility and affordability standards [3]. The funding for 9% credits is allotted annually by the Internal Revenue Service (IRS) to states, who then competitively allocate these tax credits through their state housing finance agencies. The allocation criteria and process are guided by each state's Qualified Allocation Plan (QAP), which is updated yearly to reflect the state's housing priorities and scoring criteria. States with larger populations receive more credits, and the Consolidated Appropriations Act of 2018 increased the availability of these credits, ensuring that through 2021, all states and territories receive at least $3,105,000 or $2.70 per capita in 9% credits, whichever is greater [9]. In contrast, the 4% credits are available to projects that receive at least 50% of their funding through tax-exempt bond financing. These projects do not require state allocation, and there is no annual cap on the total 4% credits available. Both types of credits are claimed by investors over 10 years and must comply with affordability requirements for at least 15 years.

2.1 Qualified Allocation Plans (QAPs)

A state's Qualified Allocation Plan (QAP) sets forth the guidelines that determine how projects will be scored, ranked, and, in turn, receive LIHTC awards. QAPs can impose minimum requirements for applications, like extending the affordability period to 50 years, and reserve tax credits for projects with specific attributes, such as preservation efforts, resident services, or those that support residents' health [10, 11]. They also influence project locations by setting opportunity standards (e.g., low-poverty rates, access to public amenities), access to public transit, or prioritizing high-poverty neighborhoods [12, 13]. Additionally, QAPs offer scoring incentives for meeting criteria like sustainable green building standards or disaster-related provisions for preparedness or recovery [14]. Due to the high competition for the 9% credits, the preferences outlined in QAPs significantly shape the nature and location of LIHTC-funded housing projects [15].

3. Sustainability in affordable housing

According to the 2020 Census, 80.0% of the U.S. population lives in urbanized areas or urban clusters. This marks a slight increase from the 79% reported in the 2000 Census. This data highlights the continuing trend of urbanization in the United States, where most of the population resides in densely populated urban areas, reflecting the ongoing importance of urban regions in shaping social, economic, and environmental policies. Housing in these environments is intricately linked to spatial organization, transportation systems, environmental concerns, social inequalities, and regional economies, making it indispensable to urban functioning and outcomes [16]. The complex interconnections within urban settings mean that actions and policies in one domain can have far-reaching effects across multiple domains. Housing not only shapes the physical, social, and economic structures of regions but also influences individual well-being, social dynamics, and economic conditions through industries like construction and lending. This interconnectedness underscores why housing is central to urban sustainability, necessitating a comprehensive understanding of its role within urbanized regions to foster sustainable development and well-being for current and future urban populations.

Moreover, sustainability is critically important in low-income housing for its multifaceted benefits across environmental, economic, and social dimensions. Environmentally, sustainable practices in low-income housing reduce carbon footprints through energy-efficient designs, renewable energy integration, water conservation strategies, and responsible waste management [17]. These measures not only mitigate climate change impacts but also conserve natural resources. Economically, sustainable housing lowers utility costs for residents, particularly beneficial for low-income households that spend a significant portion of their income on energy and water expenses. Improved building efficiency and durability also reduce maintenance and operational costs over time, ensuring long-term affordability. Moreover, sustainable housing contributes to better health outcomes by enhancing indoor air quality, providing natural lighting, and minimizing exposure to pollutants and allergens. Socially, it fosters community resilience and cohesion by engaging residents in sustainable practices and promoting equitable access to safe and healthy housing options. Overall, integrating sustainability into low-income housing initiatives supports environmental stewardship, economic stability, health improvement, and social equity in communities.

3.1 Challenges and benefits of incorporating sustainable building practices into affordable housing

Implementing sustainable building practices in affordable housing presents a range of challenges and benefits. On the challenging side, costs remain a primary concern. Initial investments in sustainable technologies and materials can be high, potentially exceeding typical budgets for affordable housing projects. Overcoming these cost barriers often requires securing additional funding sources or subsidies to make green initiatives financially viable for developers. Moreover, the technical expertise needed for designing and constructing green buildings poses another hurdle. Ensuring that architects, engineers, and contractors are well-versed in sustainable practices can be challenging, especially in regions where such expertise is limited or underdeveloped.

Perceptions around affordability also present a challenge. There is a common misconception that integrating green standards increases the overall cost of affordable housing. Educating stakeholders about the long-term cost savings associated with energy-efficient designs and materials is crucial to dispelling these misconceptions. Additionally, navigating regulatory frameworks and local building codes that may not fully support or incentivize green building practices can be complex. Advocating for policy changes and regulatory updates is essential to create an environment conducive to sustainable, affordable housing development.

Despite these challenges, implementing green standards offers numerous benefits. First and foremost is the potential for energy efficiency and cost savings. Green buildings typically consume less energy, resulting in lower utility bills for residents. Over time, these savings can offset initial investment costs, making housing more affordable in the long run. Green standards also promote improved indoor air quality and overall health and well-being among residents by using sustainable building materials and designs that enhance ventilation and natural lighting.

From an environmental perspective, green standards contribute to resource conservation, waste reduction, and lower carbon footprints. This aligns with broader sustainability goals and reduces the ecological impact of housing developments. Furthermore, green buildings tend to be more durable and resilient to climate impacts, requiring less maintenance and extending the lifespan of affordable housing units. Beyond these tangible benefits, green housing initiatives can foster community pride and engagement, create local job opportunities in green industries, and increase property values over time, benefiting both residents and the surrounding neighborhoods.

Successfully addressing the challenges while maximizing the benefits of green standards in affordable housing requires collaboration among developers, government agencies, financial institutions, and local communities. By working together, stakeholders can overcome financial barriers, enhance technical capabilities, advocate for supportive policies, and ultimately create healthier, more sustainable housing solutions for low-income populations.

4. Sustainable or green building standards

Green housing incorporates various elements designed to enhance sustainability, reduce environmental impact, and promote healthier living environments. A key component is energy efficiency, achieved through proper insulation, airtight construction, high-performance windows, and energy-efficient appliances. Integrating renewable energy systems like solar panels and wind turbines further minimizes energy consumption. Water efficiency and conservation are also critical, with low-flow fixtures, efficient irrigation systems, and rainwater harvesting reducing water usage.

Material conservation and resource efficiency are emphasized by using recycled, reclaimed, or sustainably sourced materials and implementing construction waste management plans to reduce, reuse, and recycle materials. Durable materials are chosen for their longevity, reducing the need for frequent replacements. Indoor environmental quality is improved through adequate ventilation systems, low-VOC materials, and maximizing natural lighting, which reduces the need for artificial lighting and enhances occupant well-being.

Site planning and design play a significant role in green housing. Sustainable site selection minimizes environmental impact, and green landscaping with native plants and xeriscaping reduces water and maintenance requirements. Transportation accessibility is considered to ensure easy access to public transportation, bike paths, and pedestrian-friendly infrastructure. Climate resilience is another crucial aspect, with measures for flood and stormwater management, such as permeable surfaces, green roofs, and rain gardens, as well as reflective roofing materials and tree planting, to mitigate urban heat island effects. Buildings are designed to withstand extreme weather events and adapt to changing climate conditions.

Health and well-being are prioritized through biophilic design, which integrates natural elements and greenery to enhance mental and physical health, and noise reduction strategies to minimize noise pollution. Innovation and performance tracking are incorporated with smart home technologies that monitor and optimize energy and water usage alongside performance metrics to ensure sustainability goals are met. By integrating these elements, green housing significantly contributes to environmental sustainability, reduces operating costs, and provides healthier living conditions for occupants.

In the United States, some states and local jurisdictions have developed their own green building standards tailored to regional priorities and requirements. These systems address specific environmental concerns and promote sustainable construction practices. Significant milestones and the establishment of key national frameworks, such as LEED and Energy Star, have marked the development of green building standards.

4.1 Leadership in Energy and Environmental Design (LEED) building standards

Leadership in Energy and Environmental Design (LEED) is an independent, third-party verification rating system created in the early 1990s. It aims to standardize and oversee building environmental performance, focusing on sustainable site development, water savings, energy efficiency, materials selection, and indoor environmental quality [18]. LEED's first pilot version was launched in 1998, followed by significant updates and expansions in subsequent years, including the transition of the USGBC (United States Green Building Council) to nonprofit status in 2003. LEED building standards are widely applied to housing construction to promote sustainability and enhance environmental performance [19]. LEED certification for homes focuses on several key areas to ensure a comprehensive approach to sustainable building.

Firstly, sustainable site development ensures that the site is used in a way that minimizes environmental impact. This includes encouraging sustainable landscaping and promoting access to transportation and amenities. By selecting appropriate sites and managing their development sustainably, LEED standards help reduce the overall ecological footprint of housing projects.

Water efficiency is another critical focus area. LEED standards incorporate strategies to reduce water usage through efficient plumbing fixtures, landscaping that requires less irrigation, and innovative water reuse technologies. These measures help conserve water, a vital resource, and reduce the long-term costs of water consumption in residential buildings.

Energy-efficient design, construction, and appliances emphasize energy efficiency. Proper insulation, high-performance windows, and energy-efficient HVAC systems are integral components of LEED-certified homes. These features collectively reduce overall energy consumption, lowering homeowners' utility bills and decreasing greenhouse gas emissions.

Materials and resources are also a significant aspect of LEED standards. The focus is on using sustainable building materials with low environmental impacts, such as recycled content, locally sourced materials, and sustainably harvested wood. Additionally, construction waste management and recycling are encouraged to minimize the environmental footprint of building projects.

Indoor environmental quality is crucial for ensuring healthy living spaces. LEED standards promote using non-toxic building materials, effective ventilation systems, and strategies to reduce indoor pollutants. These measures contribute to healthier indoor air quality, enhancing the well-being of residents.

Lastly, LEED standards encourage innovation and design. This involves creative and innovative approaches to sustainable building practices and technologies that go beyond existing standards. LEED certification for homes involves a rigorous verification process to ensure that the building meets these comprehensive criteria.

Overall, LEED certification helps create housing that is not only energy-efficient and environmentally friendly but also healthier and more comfortable for residents. This comprehensive approach ensures that sustainable building practices are integrated into every aspect of residential construction.

4.1.1 LEED certification levels

LEED certification levels recognize varying degrees of sustainability achievement in buildings. At the entry-level, certification indicates that a building meets basic sustainability requirements by earning a minimum number of points across LEED categories. Moving up to Silver demonstrates a higher level of environmental performance, requiring more points across categories such as energy efficiency, water conservation, and materials selection. Gold certification signifies significant sustainability efforts, achieved through accumulating even more points across LEED criteria. Buildings with Gold certification excel in areas like energy savings, indoor air quality, and site design. Finally, Platinum represents the highest level of LEED certification, indicating exceptional commitment to sustainability with substantial points across all categories. Platinum buildings are recognized leaders in energy efficiency, resource conservation, and occupant well-being. Each level reflects a building's overall performance and dedication to sustainable practices, offering a clear benchmark for environmental stewardship in construction and design.

4.2 EPA's Energy Star program

Established to certify energy-efficient products and buildings, Energy Star reduces energy consumption and greenhouse gas emissions, emphasizing energy performance and recognizing high-performing buildings and appliances [20]. These systems have evolved to incorporate industry insights, research, and growing environmental awareness, reflecting their dynamic nature and ongoing commitment to sustainability. The Energy Star program is a voluntary labeling initiative designed to promote energy efficiency. Under this program, the U.S. Environmental Protection Agency (EPA) establishes stringent energy efficiency criteria for various products. Manufacturers have the option to display the Energy Star logo on products that meet these specifications, which helps consumers and businesses identify energy-efficient choices when making purchasing decisions. By opting for Energy Star-certified products, consumers can benefit from lower electricity bills due to reduced energy consumption. Moreover, these products play a crucial role in environmental protection by contributing to fewer harmful emissions from power plants. Energy Star thus encourages both economic savings and sustainable practices, making it a pivotal initiative in promoting energy efficiency across various sectors. Although not a green building rating system, Energy Star is pivotal in promoting energy efficiency.

4.3 Enterprise Green Communities (EGC)

Enterprise Green Communities (EGC) provides a rigorous framework tailored for affordable housing, focusing on energy efficiency, water conservation, resource management, lot development, maintenance practices, and indoor air quality [21]. It aims to create healthy and cost-effective homes. This comprehensive green building framework is specifically designed to address the unique needs of affordable housing. EGC sets forth rigorous standards aimed at creating healthy, efficient, and sustainable living environments for low-income families. The EGC framework encompasses various critical aspects of sustainable building, including energy efficiency, water conservation, resource management, and indoor environmental quality. For instance, EGC standards mandate using energy-efficient appliances and lighting to reduce energy consumption and lower utility costs for residents. Additionally, water efficiency measures such as low-flow fixtures and drought-resistant landscaping help conserve water and reduce expenses.

The resource efficiency component emphasizes using recycled and locally sourced materials, minimizing the environmental footprint of construction projects. In terms of lot development, EGC encourages the integration of green spaces and community gardens, which not only enhance the esthetic value of the housing development but also provide residents with opportunities for recreation and food production. To ensure long-term sustainability, EGC includes guidelines for the operation and maintenance of buildings, promoting practices such as regular HVAC maintenance and the use of non-toxic cleaning products to improve indoor air quality.

For example, EGC standards include the requirement for Energy Star-rated windows and appliances, which significantly reduce energy usage and costs. Another example is the installation of water-saving fixtures, like low-flow toilets and faucets, which can reduce water usage by up to 30%. Furthermore, EGC promotes using advanced insulation techniques and air sealing to improve the building envelope, enhancing thermal comfort and energy efficiency. These standards collectively contribute to creating resilient, cost-effective, and environmentally friendly housing solutions that benefit both residents and the broader community.

4.4 National Green Building Standards (NGBS)

The National Green Building Standards (NGBS) offer a rigorous and flexible framework for constructing green residential buildings, including homes and apartments [21]. Approved by the American National Standards Institute (ANSI), NGBS provides a credible and affordable means to verify and promote sustainable construction practices. The standards cover a broad range of areas, such as energy efficiency, water efficiency, resource conservation, lot development, and indoor environmental quality.

In terms of water efficiency, NGBS includes standards for installing low-flow plumbing fixtures, rainwater harvesting systems, and drought-resistant landscaping. These measures help conserve water and reduce utility costs. Resource efficiency standards encourage using sustainable building materials, such as recycled content products and locally sourced materials, which minimize the environmental impact of construction and support the local economy. Additionally, NGBS emphasizes the importance of indoor environmental quality by setting guidelines for proper ventilation, using low-VOC (volatile organic compound) paints and finishes, and mold prevention strategies, ensuring healthier living environments for occupants.

For example, NGBS mandates that buildings achieve specific energy performance levels, often requiring the use of high-efficiency HVAC systems, Energy Star-rated appliances, and advanced insulation methods to reduce energy consumption and enhance thermal comfort. Other NGBS standards include the requirement for high-performance windows and doors to improve building envelope efficiency and the implementation of site design practices that preserve existing vegetation and manage stormwater effectively.

4.4.1 NGBS certification levels

Homes can achieve NGBS certification at different levels—Bronze, Silver, Gold, or Emerald—based on the number of points earned in these categories, reflecting their overall commitment to sustainability. The NGBS standards collectively contribute to creating residential buildings that are not only environmentally responsible but also cost-effective and comfortable for residents.

4.5 Passive House (Phius) standards

Passive House (Phius) standards focus on creating buildings that prioritize energy conservation and quality construction, achieving significant reductions in energy consumption while maintaining high levels of comfort [22, 23]. The Phius CORE certification emphasizes passive and active conservation strategies to create buildings that use minimal energy for heating and cooling. For instance, Phius standards require extremely high insulation levels and airtightness, ensuring that buildings maintain a consistent indoor temperature with minimal energy input. Windows are triple-glazed and strategically oriented to maximize solar gain in the winter and minimize overheating in the summer, reducing reliance on artificial heating and cooling systems.

An example of Phius standards in action is the requirement for mechanical ventilation with heat recovery (MVHR) systems. These systems provide a continuous supply of fresh air while recovering heat from exhaust air, enhancing indoor air quality and energy efficiency. Phius also mandates the elimination of thermal bridges, which are points in the building envelope where heat can escape, ensuring that the building remains thermally efficient throughout.

Furthermore, Phius ZERO certification takes these principles further by setting a net-zero energy target, meaning the building produces as much energy as it consumes over a year. This involves incorporating renewable energy systems such as solar panels and ensuring no fossil-fueled combustion on-site, thus drastically reducing the building's carbon footprint. These stringent standards result in buildings that are not only incredibly energy-efficient but also contribute to long-term sustainability and occupant comfort, making them exemplary models of modern, environmentally responsible architecture.

4.6 The International Living Future institute's Living Building Challenge (LBC)

The International Living Future Institute's Living Building Challenge (LBC) sets a high bar for sustainable building practices, aiming for buildings that are regenerative and have a positive impact on both human and natural systems [24, 25]. The LBC focuses on seven performance areas, known as Petals: Place, Water, Energy, Health + Happiness, Materials, Equity, and Beauty. Each Petal includes specific imperatives that projects must meet to achieve certification. For example, under the Water Petal, a building must achieve net-positive water, meaning it captures and treats all its water needs on-site. This often involves systems for rainwater harvesting, greywater recycling, and composting toilets.

In the Energy Petal, buildings must achieve net-positive energy by producing more energy than they consume, typically through renewable energy sources like solar or wind power [26]. This requires integrating energy-efficient design strategies and technologies, such as high-performance insulation, passive solar heating, and energy-efficient appliances.

The Materials Petal demands that buildings use non-toxic, sustainably sourced materials and avoid products on the LBC's "Red List" of hazardous chemicals. An example of this is using locally sourced, reclaimed wood and low-VOC (volatile organic compounds) paints and finishes to ensure a healthy indoor environment.

The Health + Happiness Petal promotes physical and mental well-being through design elements that maximize natural light, improve indoor air quality, and provide access to nature. For instance, biophilic design principles might be applied, incorporating natural elements like indoor plants and views of greenery.

The Equity Petal ensures that buildings support social justice and community well-being, requiring universal access to public spaces and promoting fair labor practices. Lastly, the Beauty Petal encourages the creation of spaces that inspire and delight, emphasizing esthetics, design innovation, and connection to the local culture.

Overall, the LBC standards push the boundaries of sustainability by integrating comprehensive environmental, social, and economic goals, creating buildings that not only minimize harm but actively contribute to the regeneration of their surroundings.

4.7 Department of Energy's (DOE's) Zero Energy Ready Homes (ZERH) Program

The Department of Energy's (DOE's) Zero Energy Ready Homes (ZERH) Program sets rigorous standards for homes to be energy-efficient, comfortable, and environmentally friendly [27, 28]. ZERH was developed by the Department of Energy and mandates rigorous energy efficiency and indoor air quality standards. ZERH homes aim for self-sufficiency and resource efficiency and are certified by third-party verifiers. These homes are designed to operate at high-performance levels and are so energy-efficient that a renewable energy system can offset most or all their annual energy consumption. To achieve this, ZERH requires compliance with Energy Star for Homes criteria, ensuring superior insulation, air sealing, and high-efficiency windows and appliances. For example, homes must have advanced insulation that exceeds standard building codes, often involving spray foam or rigid foam insulation to minimize heat loss.

Another key standard is the integration of high-efficiency HVAC systems that improve indoor air quality and reduce energy consumption. This might include heat recovery ventilators (HRVs) or energy recovery ventilators (ERVs) that provide fresh air while recovering the energy from exhaust air and air-source or ground-source heat pumps that provide efficient heating and cooling.

ZERH also mandates water efficiency, typically requiring low-flow fixtures, energy-efficient water heaters, and drought-resistant landscaping to reduce water usage. For instance, homes may include tankless water heaters that only heat water on demand, thus saving energy compared to traditional storage water heaters.

Indoor air quality is another critical aspect addressed by the EPA's Indoor airPLUS program, which is part of ZERH. This program ensures that homes are built with materials that minimize indoor pollutants, include radon-resistant construction techniques, and have effective ventilation systems to reduce indoor air contaminants.

Lastly, ZERH homes are constructed with solar readiness in mind, meaning they have the necessary structural and electrical provisions to easily add solar panels in the future. This includes proper roof orientation and space for solar panels, as well as pre-installed conduits for wiring.

Overall, the ZERH standards ensure that homes not only meet high energy efficiency and sustainability benchmarks but are also healthier and more comfortable for occupants, providing long-term cost savings and environmental benefits. These standards collectively advance sustainable building practices, offering diverse approaches to enhancing environmental performance, occupant health, and operational efficiency in housing. Each standard addresses specific aspects of sustainability, reflecting a commitment to reducing environmental impact and improving living conditions across diverse communities.

The green building standards—Leadership in Energy Efficient Design (LEED), Energy Star, Enterprise Green Communities (EGC), National Green Building Standards (NGBS), Passive House (Phius), Living Building Challenge (LBC), and DOE's Zero Energy Ready Homes (ZERH)—all share a common commitment to enhancing sustainability, energy efficiency, and indoor environmental quality in buildings. They aim to reduce energy consumption, improve resource efficiency, and minimize environmental impacts, ultimately fostering healthier and more sustainable living environments. These standards also promote using renewable energy sources and sustainable building materials.

However, these standards differ in their specific focus, certification processes, and levels of stringency. LEED provides a broad framework applicable to various building types and emphasizes various sustainability aspects, including site selection and materials. ENERGY STAR focuses primarily on energy efficiency and is widely recognized for its consumer-friendly labeling of appliances and building performance. EGC targets affordable housing, balancing cost-effectiveness with sustainability criteria. NGBS offers a flexible, tiered certification system that allows builders to achieve different levels of green building practices, from Bronze to Emerald. Passive House (Phius) standards are highly stringent, concentrating on passive design strategies to achieve ultra-low energy consumption. The LBC sets the highest standard by requiring buildings to be regenerative, net-zero, or net-positive in energy and water. ZERH aims to achieve near-zero energy consumption by integrating advanced energy efficiency measures and solar readiness, ensuring comprehensive performance through additional certifications like ENERGY STAR and Indoor airPLUS.

In essence, while all these standards strive toward the common goal of sustainable and efficient building practices, they cater to different aspects of green building and offer varying degrees of flexibility, focus, and comprehensiveness in their criteria.

5. Methods

5.1 Unit of analysis

The unit of analysis for this project is the state. States decide what their funding priorities beyond HUD's guidelines will be for allocating tax credits. The data were collected for each of the fifty U.S. States.

5.2 Data collection procedures

The data for all fifty states were gathered through a thorough reading of each state's Qualified Allocation Plan (QAP). Each QAP was obtained from the state's Housing Finance Agency's website. While reviewing these documents, each of the two coders highlighted the specific policy related to green building standards. The coder also documented the (potential) point value the QAP assigns to each of these policies, any specific certification programs the document refers to comply with energy-efficient design standards, and the total possible points a proposed project can obtain.

5.3 Measures

To assess the priority level, the state gives energy efficient design, I added all the potential points a project can accumulate related to these policies. The sum of energy-efficient point values was then divided by the total possible points; the result was a percentage. This percentage is the priority level at which the state assigns energy-efficient design as a part of LIHTC projects. In cases where the state ascribes no point value for energy efficient design, the state was assigned a value of zero. On the other hand, a state that requires the project to obtain a certification in order to be considered for the tax credits was assigned a value of one.

I completed this process for the data in 2015 and 2023. Intermediate data for 2017 was collected from a report by the National Housing Trust (National Housing Trust, 2024). The reports were then content-coded for specific mentions of sustainable or green building standards and organized into themes.

To assess the reliability of the coding, I calculated Kripendorff's alpha of a subset of the documents (10% of the QAPs). Two independent coders reviewed the documents and compared the results. The resulting Krippendorff's alpha coefficient was 0.90 ($p = 0.05$) considered a sign of very high consistency between coders.

6. Findings

The findings of this study provide a comprehensive analysis of state policies regarding the integration of green and sustainable building standards within the Low-Income Housing Tax Credit (LIHTC) program, reflecting the evolving priorities of state finance agencies. For example, in 2015, the states of Missouri, North Carolina, and Vermont, for example, required developers to propose LIHTC projects that are energy efficient, which can be demonstrated by obtaining a specific certification. Most other states, such as California and Washington, place the decision in the hands of each developer, who can decide whether the "reward" (specific point allotment) for building energy-efficient units is worth their while. Finally, four states, Maine, Minnesota, New Hampshire, and Texas, make no mention of energy conservation as part of the requirements to be considered in evaluating the proposal or as a potential way for a proposed project to accrue points. Colorado's 2015 Qualified Allocation Plan (QAP) awards points for projects that generate 5% of their energy on-site or provide renewable generation for 75% of their annual common area electricity needs.

As of 2023, Pennsylvania incentivizes new construction projects that meet green building standards by awarding points for achieving EGC 2020, LEED v4 Silver (Building Design and Construction), and NGBS Silver certifications. While Ohio requires that new construction projects must meet the certification criteria of either the Enterprise Green Communities (EGC), Leadership in Energy and Environmental Design (LEED), or National Green Building Standard (NGBS).

Between 2015 and 2023, there has been a significant trend toward increasing sustainable building standards for LIHTC-funded projects. Over this period, the number of states requiring sustainable building practices in all new LIHTC construction grew from 3 to 29. This shift underscores a widespread recognition of the benefits of energy-efficient and environmentally responsible housing. Additionally, the number of states offering incentives for incorporating green building standards also rose, from 19 to 29. These findings highlight a strong movement toward sustainability in affordable housing, driven by diverse state-level policies that encourage or mandate green building practices. This trend not only addresses housing affordability but also promotes environmental stewardship and energy efficiency.

7. Conclusions

The findings in this study reveal a strong movement toward sustainability in affordable housing, driven by diverse state-level policies that encourage or mandate green building practices. This trend not only addresses housing affordability but also promotes environmental stewardship and energy efficiency. Analyzing these policies provides a nuanced understanding of how states can leverage the Low-Income Housing Tax Credit (LIHTC) program to achieve sustainability goals.

Recent scholarship supports these findings, showing that while state housing finance agencies have increasingly adopted green building standards, the rate of adoption has decelerated, and some states have even removed crucial criteria such as on-site renewable energy generation [29]. These observations suggest that adopting green building standards is heavily influenced by government motivation, financial resources, and external factors. The present study strongly advocates for LIHTC agencies to rigorously enforce compliance with green building rating systems and to periodically update these criteria based on planned changes and life-cycle analyses. Such measures can ensure sustained progress in integrating sustainability into affordable housing projects, ultimately enhancing both environmental and economic outcomes.

A similar methodology was employed in another study [10], which explored the impact of the physical environment on health within affordable housing through the LIHTC program. This research assessed the extent to which various housing quality provisions related to health are embedded in the Qualified Allocation Plans (QAPs). The study found that the most frequently required healthy housing provisions address housing quality, while the most incentivized ones pertain to proximity to neighborhood services and amenities. Few states bundle high-priority provisions relevant to asthma, respiratory health, and toxic exposures, which are major health concerns for vulnerable children. The top motivators for incorporating healthy housing provisions in the LIHTC process were championship and initiation by agency staff and learning of similar practices in other states. This study, along with the present one, underscores the importance of state-level decision-making in shaping the LIHTC program's focus on either green building standards or health-related housing quality provisions.

Owens and Smith [1] examined how QAP incentives impact the development of affordable housing in higher-resource areas. Their study analyzed the location of proposed and funded LIHTC developments and found that the probability of incentivized LIHTC projects being proposed in higher-resource areas increased while the probability of locating developments ineligible for incentives in higher-resource areas declined. Both Owens and Smith's and the present study highlight the importance of state-level decision-making and incentives in shaping the priorities and outcomes of the LIHTC program. Together, these studies provide a deeper understanding of how various policies and incentives can enhance the effectiveness of the LIHTC program in addressing both environmental sustainability and social equity in affordable housing.

In conclusion, the LIHTC program has become an indispensable tool in addressing the affordable housing crisis in the United States. By analyzing the QAPs across various states, I have gained insights into the diverse priorities and policies that shape the implementation of energy-efficient and green building standards in LIHTC projects. This research highlights the significant role that state-level decision-making plays in promoting sustainability within the framework of affordable housing. It underscores the complexity and variability of state policies, which are influenced by many political, economic, environmental, and social factors. The analysis of these policies provides a nuanced understanding of the ways in which states can leverage the LIHTC program to not only address housing affordability but also to achieve broader social and environmental goals. The conclusions drawn from this study offer valuable implications for policymakers, developers, and stakeholders seeking to enhance the sustainability and effectiveness of affordable housing initiatives nationwide.

Acknowledgements

I would like to express my deepest gratitude to Dr. Victoria Basolo for her invaluable contributions to this research. Her expertise, guidance, and support were instrumental in shaping the direction and outcomes of this study. Dr. Basolo's insightful feedback and encouragement throughout the research process have been greatly appreciated, and her commitment to advancing knowledge in the field of affordable housing and urban sustainability has been a source of inspiration. I also thank my loving partner, Eric B. Jordie, for his support and encouragement.

The author acknowledges the use of Grammarly for the language polishing of the manuscript.

Conflict of interest

The author declares no conflict of interest.

Author details

Anaid Yerena
University of Washington, Tacoma, WA, United States of America

*Address all correspondence to: yerena@uw.edu

References

[1] Owens A, Smith RB. Producing affordable housing in higher-opportunity neighborhoods: Incentives in California's LIHTC program. Journal of Urban Affairs. 2023:1-29 [Published online]

[2] Schwartz AF. Housing Policy in the United States. New York, NY: Routledge Books; 2021

[3] Kuai Y. A missed opportunity? The 4% low-income housing tax credit program. Housing Policy Debate. 2024;**34**(3):372-395

[4] Oluku U, Cheng S. The low-income housing tax credit program: A multicity rent savings analysis. Housing Policy Debate. 2024;**34**(3):326-352

[5] Wijburg G. Tax-incentivized housing production and the affordability crisis: International lessons from the low-income housing tax credit program in the United States. Housing Studies. 2024;**39**(7):1632-1657

[6] Shamsuddin S, Leib G. Weather or not: Tracking hurricanes and changes to low-income housing tax credit program plans. Housing Policy Debate. 2022;**32**(1):128-151

[7] Schwartz A, McClure K. The geography of absence: Cities, towns, and suburbs with No LIHTC housing. Housing Policy Debate. 2023;**33**(4):963-982

[8] Scally CP, Gold A, DuBois N. The Low-Income Housing Tax Credit. Washington, DC: The Urban Institute; 2018. Available from: https://www.urban.org/sites/default/files/publication/98758/lithc_how_it_works_and_who_it_serves_final_2.pdf

[9] Royce ER. TARGET Act (Targeted Rewards for the Global Eradication of Human Trafficking). 2018. Available from: https://www.govinfo.gov/content/pkg/PLAW-115publ141/pdf/PLAW-115publ141.pdf

[10] Ahrentzen S, Dearborn L, Momen-Heravi A, Sadoughi A. Shaping a healthier LIHTC housing stock: Examining the role of states' qualified allocation plans. Housing Policy Debate. 2023;**33**(5):1206-1227

[11] Mansour A, Bentley R, Baker E, Li A, Martino E, Clair A, et al. Housing and health: An updated glossary. Journal of Epidemiology and Community Health. 2022;**76**(9):833-838

[12] Adkins A, Sanderford A, Pivo G. How location efficient is LIHTC? Measuring and explaining state-level achievement. Housing Policy Debate. 2017;**27**(3):335-355

[13] Dawkins C. The spatial pattern of low-income housing tax credit properties: Implications for fair housing and poverty deconcentration policies. Journal of the American Planning Association. 2013;**79**(3):222-234

[14] Mehta A, Brennan M, Steil J. Affordable housing, disasters, and social equity: LIHTC as a tool for preparedness and recovery. Journal of the American Planning Association. 2020;**86**(1):75-88

[15] Ellen IG, Horn K, Kuai Y, Pazuniak R, Williams MD. Effect of QAP Incentives on the Location of LIHTC Properties. Washington, DC: U.S. Department of Housing and Urban Development; 2015

[16] Basolo V. Housing and Urban Sustainability in the United States. East Lansing, MI: Michigan State University Press; 2013

[17] Wells W, Vermeer K. Blueprint for Greening Affordable Housing. Washington, DC: Island Press; 2020

[18] USGBC. USGBC | U.S. Green Building Council. 2023. Available from: https://www.usgbc.org/

[19] Garde A. Sustainable by design?: Insights from US LEED-ND pilot projects. Journal of the American Planning Association. 2009;**75**(4):424-440

[20] Hecht AD, Sanders Iii WH. How EPA research, policies, and programs can advance urban sustainability. Sustainability: Science, Practice and Policy. 2007;**3**(2):37-47

[21] Losoya J, Walker J, Fuller A, Seefeldt J. Ensuring One Water Works for All: Opportunities for Realizing Water Reuse in Affordable Housing. Austin, TX: National Wildlife Federation; 2022

[22] Whitmyre GK, Pandian MD. Probabilistic assessment of the potential indoor air impacts of vent-free gas heating appliances in energy-efficient homes in the United States. Journal of the Air and Waste Management Association. 2018;**68**(6):616-625

[23] Nadel S. Programs to Promote Zero-Energy and Zero-Carbon New Homes and Buildings. 2024

[24] Basińska M, Koczyk H. Analysis of the possibilities to achieve the low energy residential buildings standards. Technological and Economic Development of Economy. 2016;**22**(6):830-849

[25] Keck L. How to implement high-performance design: High-performance design drives owners and designers to integrated thinking to gain a holistic perspective on how to maximize building performance outcomes. Consulting Specifying Engineer. 2020;**57**(7):10-17

[26] Von Hausen MA. Dynamic Urban Design: A Handbook for Creating Sustainable Communities Worldwide. Bloomington, IN: iUniverse; 2013

[27] Albrecht CF, Iulo LD. Healthful, sustainable design in community-engaged architecture education. In: Educating the Sustainability Leaders of the Future. Springer Nature Switzerland: Cham; 2023. pp. 431-451

[28] Satre Meloy A, Webster B, Toffoli L, Campbell M, Badger L, Donnovan A, et al. Accelerating Residential Building Decarbonization: Market Guidance to Scale Zero-Carbon-Aligned Buildings. Washington, DC: Advanced Building Construction Collaborative; 2024

[29] Yeganeh A, McCoy AP, Reichard G, Schenk T, Hankey S. Green building and policy innovation in the US Low-Income Housing Tax Credit programme. Building Research and Information. 2021;**49**(5):543-560

Chapter 10

Sustainable Makerspace Pedagogy in Architecture Education

Suneela Ahmed

Abstract

Design studios are an essential and significant part of the curriculum of Bachelor of Architecture degree. Typically design studios are conducted in a specific realm, and the results are gauged as per charted outcomes, which are often pre-determined. There are some studies that research different methods of teaching design studio as part of the architecture degree. Inquiry by design is one of them. The pedagogy of "makerspace" within this paradigm was recently implemented as a test in the design studio of third-year architecture students at Indus Valley School of Art and Architecture, Karachi. The implementation of this pedagogy resulted in various interesting outcomes that were connected to sustainable design practices in responsive design, which are presented in this paper.

Keywords: system design tools, mountain studio, interdisciplinary, reflective pedagogy, design studio

1. Introduction

The notion of making the makerspace a part of the architectural pedagogy surfaced in the late eighties, as a response to surface-driven and glossy architecture, that did not respond to contextual and social realities of a place [1, 2]. This type of architecture also lacked depth of research and functional understanding and was very superficial. The idea of the makerspace was, as the name suggests, entrenched in making. It included equipment and tools that facilitated the making aspect. Tools enabling working in different mediums like fabric, cardboard, metal, and wood would be part of the makerspace. A makerspace, by making the various tools available to work in different materials, would train architects to think creatively and objectively, while relying on certain observation and analytical skills. A makerspace also brought together a group of makers, where collaboration, problem-solving, skill development, and innovation became the highlights of the space.

Based on these ideas, various studio exercises were initiated within the Bachelor of Architecture program at Indus Valley School of Art and Architecture, in Karachi, Pakistan. The first studio assignment lasted over four weeks, and third-year students were a part of the design and hands-on making of a flood-responsive model house. The materials used for the construction of this model house ranged from sun-dried blocks, to concrete blocks, to bamboo screens and bamboo poles, and to polyvinyl chloride (PVC) sheet. The foundation was made of concrete blocks, and the roof was made from PVC sheets. The choice of materials was based on an in-depth research of the context, structural necessities in the event of a flood, and other requirements like structure, cost, adaptability, and availability of the materials. The students were part of the various stages of design and construction, ranging from calculating quantities of material, purchasing the materials from various local markets, budgeting the cost, preparing the bill of quantities, and identifying various vendors.

During the process of design and construction of the flood-responsive model house, the students took on the roles of a project estimator, a contractor, a mason, a project manager, quantity surveyor, and above all, a maker. The process enhanced their comprehension of proportions, space, esthetics, scale, and material joinery. Another outcome achieved via this project was the development of an understanding within the students about technical joinery details, structural systems, and how various materials can be put together in actual construction. Furthermore, the students learned how to work as a team as they were required to plan daily activities, delegate tasks, and schedule productive hours. The students also developed problem problem-solving attitude as at times they had to think of alternative strategies when faced with challenges during the project. For instance, at a couple of points the joinery details of materials had to be rethought as they were not working as they had been designed on paper. Thus, a "can-do mentality" [3] was developed, which is an essential ingredient in design thinking and helps in overcoming any mental and design blocks.

As part of these studio projects, the design thinking process was utilized to pinpoint issues, generate ideas for fixes, create prototypes, and test various concepts. To find problems, generate ideas for solutions, create prototypes, and test concepts, the design thinking approach was employed. The second studio was designed and built in Pakistan's northern mountains. The students were exposed to a new environment and had the opportunity to engage with a village community in Gilgit Baltistan's upper Hunza Valley through the Mountain Studio. In this specific studio, the students discovered the value of using contextually and locally relevant materials. The site's natural geology and terrain, as well as the existing Poplar trees on the site, dictated this structure's design. Furthermore, the students got exposed to the idea of responding to natural terrains and topography, as they were required to take into consideration an existing water channel on the site. Thus, the students gained knowledge about site gradients and contours and resultant watersheds for identifying the location of the structure, which was a storage shed.

The students' exposure multidisciplinary learning was further cemented by the Mountain Studio, where they were again able to incorporate the perspectives of surveyors, builders, and project managers. This aspect encouraged innovation and teamwork. They also practiced reflection as they jotted down their ideas, observations, and design concepts as the Mountain Studio stressed the importance of critical thinking and self-evaluation in the profession of architecture.

Yet another studio that took place in the winters as a follow-up was named "City Studio". This six-week studio was devised as an experiment, and the students were encouraged to learn from their failures. Strategies were planned to illustrate and translate concepts into architectural compositions, make connections between appropriate functional relationships and user requirements, and formulate and organize project background research and design strategies. A water purification system was designed and constructed here as part of a system design approach where various components of the process were designed and constructed by the third-year students.

When the students returned to their typical studio the next semester, they were seen using what they had learned in these three studios to debate and think about materials, joinery details, scale and proportions, and details of wall sections, form, and elevations, in a way that prioritized problem-solving.

2. Literature review

Design studio is an intrinsic part of any architectural curriculum. The aspects addressed within this studio range from understanding of space, scale, proportions, esthetics, materials, textures, and certain construction principles. Different architectural schools prescribe different thought processes, and based on the focus of the studio, the approach of design changes from space, to technical details, and to larger social understanding to esthetic qualities and proportions. According to Salama and Wilkinson [4, 5], the four different kinds of the pedagogical approaches to architecture teaching can be classified into: "academic, craft, technological, and sociological". Each of these approaches emphasizes a different aspect of the built form, which becomes the driving force of the design studio, and is governed and linked to the larger vision and design philosophy of the teaching institution. While the academic approach emphasizes the traditional principles of design, linking it up to design and compositional theories, the craft training views architecture as a discipline in which hands-on experience is of supreme importance, is to be achieved by direct exposure to building sites, or can be achieved as an apprentice to a master crafts person. Thirdly, in schools where technological aspects of design are considered of supreme importance, beauty and esthetics may take a back seat and be preceded by pragmatics and scientific principles.

Lastly, in design studios being backed with "sociological" approaches, the understanding and requirements of the society take precedence, and studios are oriented towards an urban scale addressing social aspects with emphasis on contextual connections in terms of society, economy, and geography [4, 5]. Whichever school of thought is followed by an academic institution, the studio is always the main forum for assimilation, acquisition, and processing of knowledge gained through various inputs. Studio assignments are designed to develop a sense of inquiry, enhance critical thinking, and analyse the stimulus being received from the environment in various forms. The student is taught to process the information in order to not only make sense of the various stimuli, but also to conceptualize and create a meaningful built environment, which connects back to the original cues. Over the years, a number of tools have been developed to create this connection, which ranges from heavy reliance on digital technologies [6], working on live sites, hands-on material exploration, and engagement with community [4, 5, 7–13]. In developing an understanding of design and its outcome, the drawing board may take a back seat, if these alternative tools of design are used. The emphasis then is on using these alternative tools to achieve similar results of developing critical thinking, sense of inquiry, and taking cues from the environment, but there is greater emphasis on consciously enhancing and using all five senses.

The alternate pedagogy revolved around the art of self-education [14], project-based work, rapid iteration of design solutions, formal and informal critiques, use of precedent, and thinking about the whole and the creative use of constraints [15].

Reliant on the craft and technological models, the following three studios were designed and executed, and the results and learnings are presented here.

3. The flood responsive studio

A four-week flood-responsive housing design and build exercise was undertaken within the Department. The idea behind this project was to conceptualize a low-cost house that was an appropriate response and was resilient to natural disasters associated with climate change, particularly flooding. The third-year students (a group of 28 students), along with their tutors and mentors, completed the exercise of developing the agreed design. In developing the design, details were worked around what materials are to be procured, the quantity of the materials, and construction joinery and details. In the process, the students gained knowledge about how to develop contextually relevant design both in terms of physicality as well as sociocultural aspects of rural areas of Sindh. The students also gained an understanding of methods of developing design in terms of responding to local vernacular materials and customs, climate, material joinery, working with a mixed material palette, procurement of materials, role and workings of labour on site, and various other construction details and practices. The materials explored were bamboo, sun-dried brick, concrete block, bamboo screens [16], and recyclable PVC sheets.

The idea was to construct a model house which was sensitive to the idea of affordable housing, was a practical solution, integrated within the context, and was owned by the locals in terms of construction and habitability [16]. When presenting solutions, the students were encouraged to think innovatively about methods and ideas that can help combat or embrace the water that inundates the locality in the scenario of flooding [16]. As flooding can cause structures to float, shift, submerge, or block, thus it was emphasized that the solutions can be challenging and out of the box as long as the function of the architecture is retained. “The house was to accommodate a family of six and was to be thought of as a unit within a larger village settlement having communal toilets and at times communal kitchens too” [16].

As a requirement of the project, it was mandatory for the students to acquaint themselves with the living conditions and norms in villages in Sindh and elsewhere in Pakistan. Local building methods were to be researched upon and understood to encourage designs for construction and reconstruction using local vernacular methods. Usage of local materials and techniques also ensured cost-effective solutions. Designing successful modules took advantage of local materials and techniques while also preserving the context. Students were also encouraged to be sensitive to the physiological and psychological requirements of the locals and to design while respecting the community structure [16].

Thus, the design ideas revolved around vernacular housing solutions that respectfully conversed with the context of rural Sindh. Therefore, a study was initiated by the students about the lifestyle and daily routines of the locals. Eventually, the designs were based on the idea of permanence and outdoor living, as well as the flood resistance solutions. Studying the communities helped the students understand the lifestyle choices and desires of the people in Gharo, Sindh (**Figures 1** and 2).

During the design phase, the students attempted to respond to the flood conditions by catering for the lifestyle of rural Sindh. There was a ground plus one module designed that could be connected to other modules through ladders. The design had nine bamboo columns reaching to the top. To meet the requirements for strength and load, four columns were joined together on the ground floor and extended vertically. The overall design consisted of nine bamboo columns which were supported by

DOI: http://dx.doi.org/10.5772/intechopen.1006314

Figure 1.
Study models developed in the third-year studio.

Figure 2.
Various studies to develop contextual understanding during third-year studio.

repurposed oil canisters. The bamboo was prevented from absorbing moisture from the ground by treating it with a sealant. This also strengthened the structural capacity of the bamboo columns (**Figure 3**) [16].

Figure 3.
Final prototype house designed and constructed by third-year students.

Figure 4.
Structure during construction.

Locally sourced building materials that were readily available on-site were used for the material palette. The materials were either indigenous or readily available at the location. The materials of choice were bamboo, clay bricks, and limestone. Cement gave the required strength to the foundation and plinth by strengthening the bamboo structure and mud brick wall. The primary bamboo structure was supported by a mud brick wall. These two building materials helped to ensure structural integrity. Providing a cost-effective solution was the goal of the design, and bamboo proved to be a reliable and straightforward material, while mud bricks bolstered the structure (**Figure 4**) [16].

The main strength of the design stemmed from the simplicity of form leading to an easy construction process. A comfortable solution was facilitated by the material palette, which was based on contextual familiarity (mud bricks and bamboo). The versatility and durability of the structure ensured sustenance during storms and embraced the inhabitants [16].

4. The mountain studio

In this era of rapid changes in construction industry and related architectural practices, unpredictable weather patterns, diminishing resources, and growing population trends, the practice of architecture calls for eco-friendly passive solutions and designs around the globe. The physical distance between Hunza and Karachi is 2000 km, with Karachi being at the southern tip of Pakistan and sitting right next to the Arabian Sea, and Hunza being the northern edge of the country housing the Karakoram and Hindu Kush Mountain ranges. The two cities are connected and dependent via the country's numerous vital but diminishing resources, such as the rivers, agricultural land, and energy networks. Thus, the objective of this studio was to develop a larger geographical understanding in the students about these related geographical and environmental patterns, learn from physical and culturally sensitive design interventions in the context of ethnically homogenous and relatively smaller towns of Hunza, and explore possibilities of replicating them in the multicultural and global city of Karachi in the south (where the school of architecture is located). Both Karachi and Hunza face the brunt of the climate change in the form of urban flooding and heat island impacts; however, the small-scale and place-based lifestyles of the Hunza Valley have many learnings to offer.

The focus of this studio was to harness a dialog between practice and academia and move beyond, by questioning what role does architecture play in climatic challenges? Is a shift in the process of making space able to counter the draining resources possible? What kinds of materials and construction methods, inspired by nature, allow us to alter our lifestyles? Can our practices in the cities be aligned with the cultural values of the mountain landscape?

The Mountain Studio was a rigorous and demanding exercise, where the tone of possibilities was a crucial and vital part of the studio. The students explored how the land, rivers, and mountains and all inseparably connected to the idea of a place via history, culture, materiality, and methods of construction. Simultaneously, the students examined the symbiotic relationship between spatial language, morphology, topography, structure, and material and techniques of construction.

The interconnection between spatial realities and the influence on the tectonic character of the place was focused upon in the studio. The interest in architecture was the reaction to the specifics of the context, both physical and cultural, and the aim was to include social value by offering conscientiousness, Diligence, and sensitivity.

The success of this studio depended on vigorous engagement of the students. As part of the Mountain Studio, the students focused on the following aspects:

1. Craftsmanship—The students were not only expected to craft their skills, but to craft their minds as well. The daily work required an appropriate level of refinement to reflect the degree to which the project itself was developed. Sketches and diagrams were a vital and constant part of the studio.

2. Documentation of work—The students were required to provide documentation of their work at the end of the studio. Documentation included drawings, mapping, site studies, writings, poetry, and thoughts.

3. Fundamental design skills—The students were required to show the ability to effectively use architectural principles in an intuitive and iterative process of drawing, critiquing, and making.

4. Critical thinking skills—The students were expected to demonstrate the capability to establish connections between various ideas and comprehend the implications of these concepts based on examining multiple political, cultural, environmental, theoretical, and economic contexts.

5. Design thinking skills—The students were expected to showcase the aptitude to ask precise and clear questions, use ideas to understand information, reflect upon varied viewpoints, arrive at well-thought-out conclusions, and test alternate results against pertinent standards and criteria.

6. Collaboration skills—The students were required to show the ability to work together on group exercises in a positive and respectful way that encourages optimism, respect, sharing, engagement, and innovation.

7. Communication skills—The students were required to show the ability to listen, speak, read, and write effectively and show the capability to use adequate representational skills, ranging from use of conventional digital and graphic technology.

Figure 5.
The Mountain Studio in progress in Gulmit, Hunza.

8. Making skills—All students were required to show the ability to use various tools and machines and make decisions around methods for construction with respect to materials and outcomes to be achieved.

It is important that as a part of the first Mountain Studio team, the students considered the terrain, landscape, and geographical features as sacred and fragile and respected the traditions, way of life, and community structure of Hunza Valley. The need was to tread carefully and sensitively in a place that is in a state of rapid transitions and societal transformation. The primary objective of the studio was to develop the ability to see and make architectural space that were sensitive to human needs and requirements in terms of use, scale, and materiality—always through critical thinking (**Figure 5**).

5. The City studio

As a follow-up to the above two studios, this six week's workshop was devised as an experiment, and the students were encouraged to learn from their failures. Strategies were planned to illustrate and translate concepts into architectural compositions, make connections between appropriate functional relationships and user requirements, and formulate and organize project background research and design strategies.

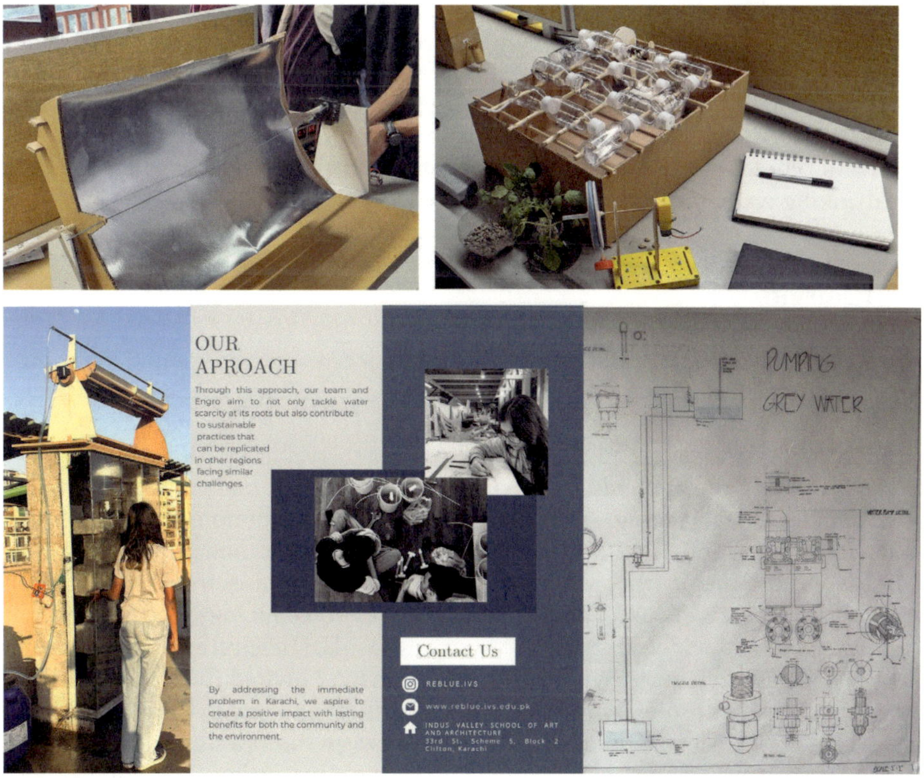

Figure 6.
Outcomes and process of the City Studio.

In order to obtain distilled water at the end of the process, a system was designed and built by the students and the mentor where various processes were included, ranging from chemical cleansing to bioremediation and to aqua phonics. The design of a solar trough, a solar still, and a solar dish, along with an automation and sensor-based management system was also part of the entire system design.

The chemical cleansing system worked around the idea of pumping the gray water up the alum chamber, using an agitator to chemically treat the water, and finally mixing the water in a flocculator. Eventually the water entered the septic tank and went through sedimentation. The water moved through the barrels acting as baffles and cleansing on its way. The second step of biomediation had water pass through water plants which were placed vertically, one on top of the other, and acted as filters that took away impurities from the water. The next step involved filling water with nitrates and inhabited by catfish, to convert nitrates into nitrogen and purify the water further. This process is known as aqua phonics and was designed and built by the students to plug into the overall system of water purification. Next, the solar still was designed and constructed. A solar parabolic through was used here to capture solar energy. This was monitored by pressure and temperature sensors, with the automation and sensor designed and constructed by the students themselves. The solar still received heated water from the through. As the water entered the still the temperature of the water rose, and it transitioned from liquid to gas state, while separating all impurities in the process. Eventually the steam was condensed back into liquid form, and distilled water was collected. This studio created a strong link between spatial design and understanding of systems, as students of architecture designed and executed various parts of the system design based on their spatial and functional requirements (**Figure 6**). They worked with materials, tools, and machines and were also introduced to the use and experimentation of the 3D printer. The roof of the University was the physical site, and the larger premise was the availability of many such rooftops in the urban scape of Karachi. These can become potential sites for installing similar systems to obtain pure drinking water for the city.

6. Conclusion

The pedagogy models of self-education and project-based work, rapid iteration of design solutions, formal and informal critiques [14], use of precedent and thinking about the whole, and the creative use of constraints [15] based on multi-disciplinary methodology were the most appropriate models for these studios. Within these larger research methods, the methods operationalized relied on the makers' studio, experimentation, learning by doing, and incorporation of expert opinions and collective sessions with peers and tutors which followed cue from other models [4].

Some of the strategies that helped during the studio exercises from the part of the tutors were to stay focused, accept the challenges of technology, and use it to maximum potential, video recordings of progress by the students every week, peer reviews, and feedback, juries where external jurors were invited from various local contexts and various professional backgrounds, critical and creative thinking about mitigation measures, constant communication between tutors and students, shortlisting and convergence of scattered ideas towards one or two options, development of limited options in accordance with the student capacity and available resources/time, and lastly consolidation of ideas into selected options.

Acknowledgements

This is to acknowledge Architecture Design and Research Lab (ADRL) at Indus Valley School, Mountain Getaway, Architect Ebne Sajjad, Architect Mehmooda Maqsood respectively for conducting the Mountain and City Studios and to the students and faculty members (Architects Ira Kazi, Mujahid Sadiq, Yasir Rai, and Shahid Khan) of the Department of Architecture, Indus Valley School of Art and Architecture, Karachi, Pakistan, who were a part of these studios at various stages.

All the images used in the chapter are owned by the department and are the author's own reproduction. I have permission from the persons visible in the pictures to be included in this research.

Conflict of interest

The authors declare no conflict of interest.

Author details

Suneela Ahmed
Indus Valley School of Art and Architecture, Karachi, Pakistan

*Address all correspondence to: suneela.ahmed@ivs.edu.pk

References

[1] Chau H-W, Newton C, Yang S, Geng S. Emerging architectural makerspaces: Analysis and design strategies. Nordic Journal of Architectural Research. 2020;**32**(2):37-62

[2] Hackel M, Gaube A, Lampe MS. Intercultural Hands on Projects: Experiences in Architectural Education. Wismar: SHS Web of Conferences; 2002

[3] Jiao J, Tang P. Application of bamboo in a design–build course: Lianhuadang farm project. Frontiers of Architectural Research. 2019;**8**(4):549-563

[4] Salama AM, Wilkinson N. Design Studio Pedagogy: Horizons for the Future. Gateshead, UK: The Urban International Press; 2007

[5] Salama A. Seeking Responsive Forms of Pedagogy in Architectural Education [Online]. 2022. Available from: https://strathprints.strath.ac.uk/49929/1/Seeking_new_forms_of_pedagogy_Salama_Field_5_1_Salama_11_.pdf

[6] Kvan T. Teaching architecture, learning architecture: Technology in support of design learning. In: Proceedings of CAADRIA. Department of Architecture, University of Hong Kong; 2000

[7] Alalouch C. A Pedagogical approach to integrate parametric. Archnet-IJAR. 2018;**12**(2):162-181

[8] Dutton TA. Design and studio pedagogy. Journal of Architectural Education. 1987;**41**(1):16-25

[9] Jenson MK. Educating the 21st century architect: Complexity, innovation, interdisciplinary methods, and research in design. In: Design Studio Pedagogy: Horizons for the Future. Gateshead, UK: The Urban International Press; 2007. pp. 47-62

[10] Fernando N. Decision making in design studios: Old dilemmas-new strategies. In: Design Studio Pedagogy: Horizons for the Future. Gateshead, UK: The Urban International Press; 2007. pp. 143-152

[11] Dunin-Woyseth H. Book review: Design studio pedagogy: Horizons for the future. Archnet-IJAR, International Journal of Architectural Research. 2007;**1**(3):194-206

[12] Groats L, Wang D. Architectural Research Methods. New Jersey: John Wiley & Sons; 2013

[13] Maturana B, Salama AM, McInneny A. Architecture, urbanism and health in a post-pandemic virtual world. Archnet-IJAR. 2021;**15**(1):1-9

[14] McClintock R, McClintock J. Architecture and pedagogy. The Journal of Aesthetic Education. 1968;**2**(4):59-77

[15] Kuhn S. Learning from the architecture studio: Implications for project-based pedagogy. International Journal of Engineering Education. 2001;**17**(4 & 5):349-352

[16] Ahmed S. Flood Responsive House Design, Archi TImes. 2023. Available from: https://www.architimesonline.com/site/flood-responsive-house-design/

Chapter 11

Sustainable and Resilient Housing in Tropical Climates: Best Practices for Construction and Energy Security

Miguel Chen Austin, Thasnee Solano, Olga Yuil Valdés, Hatvany Gómez Concepción, Dafni Mora and Yazmín Mack-Vergara

Abstract

Ensuring that ecosystem services are effectively maintained and integrated while also proactively adapting to the challenges posed by climate change is essential for developing sustainable and resilient housing in both new and existing settlements. The tropics cover roughly 40% of the world and are home to about 40% of the global population. By the late 2030s or 2040s, 50% of the world's population is projected to live in the tropics. This research project investigates the intersection of sustainable construction and energy security for achieving sustainable and resilient housing in tropical climates. Examining (1) resilience strategies against climate change impacts and (2) case studies led to highlighting best practices regarding construction and energy security aspects. Finally, a reference framework is provided for architects, policymakers, and stakeholders involved in tropical housing development.

Keywords: energy security, resilient housing, sustainability, tropical climate, best practices

1. Introduction

Emphasizing the urgency of sustainable and resilient housing is crucial, given the significant time spent inside buildings and this sector's escalating impact on greenhouse gases and climate change [1, 2].

Climate change effects have led to global migration [3], with people leaving rural or suburban areas that are no longer habitable and relying on basic human services. These migrations, often initiated by the people themselves and sometimes supported by governments, present a complex challenge. To achieve successful relocations, numerous factors need to be considered, and understanding the peoples' ways of life is crucial. The task of achieving a sustainable and resilient relocation is daunting, especially given the limited time frame and the need for available land for new settlements or the adaptation of existing settlements.

On the other hand, climate change effects have caused the enforcement of strategies to develop new settlements that are able to adapt to future climate change scenarios. This endeavor to design and build new settlements increases the anthropogenic metabolism of new land resources for ecosystem services. Thus, it becomes imperative to focus on both new and existing settlements, where the latter may produce a higher negative environmental impact. As most existing settlements did not consider sustainability practices or climate resiliency, infrastructure losses and waste (due to floods, storms, earthquakes, among others) will increase land, atmosphere, and ocean pollution [4].

Many initiatives around the world have been launched aiming to achieve sustainable and resilient housing, such as the United Nations Sustainable Development Goals (SDGs) [5], the Paris Agreement [6], Global Alliance for Building and Construction (GlobalABC) [7], International Green Construction Code (IgCC) [8], Leadership in Energy and Environmental Design (LEED) [9], the World Green Building Council [10], and the World Resources Institute [11]. Although many countries have compromised for at least one such initiative, achieving sustainable and resilient housing is a highly context-related task [4], and thus, starting by understanding the context-related ecosystem services first may help establish equitable social well-being.

Therefore, this chapter addresses this crucial issue by exploring the intersection of sustainable construction, energy security, and best practices for tropical climates in shaping the future of habitation. Tropic latitude countries may face diverse challenges compared to Northern and Southern latitude countries, withstanding significantly different approaches for sustainable housing [12]. Tropical latitude countries generally experience dry and rainy seasons with slight temperature variations and highly variable precipitation regimes [13]. However, in each country, significant temperature variations are found between regions with different altitudes and forestry amounts [14].

Thus, through case studies analysis, this chapter aims to identify actionable best practices for architects, policymakers, and stakeholders involved in tropical housing development. The project seeks to contribute valuable insights into creating environmentally sustainable and resilient dwellings in tropical regions like Panama and beyond by addressing these critical aspects.

2. Addressing resiliency in tropical housing: construction and energy

Fulfilling ecosystem services while adapting to climate changes is key to achieving sustainable and resilient housing in new or existing settlements. While sustainability aims to meet the needs of the present without compromising future generations [15], resilience focuses on the power to act on urban systems to maintain continuity or to adapt amid shocks and stresses [16].

On the one hand, resilience can be understood as a process. When a community faces a disaster, the robustness, redundancy, and rapidity of the resources and how they are managed are fundamental to overcoming the catastrophe and functioning in an altered environment, considering both the material and psychological aspects as individuals and as a collectivity. On the other hand, in the practical world, resilience is also about lessons and practices that can be implemented after a disaster. It is about providing communities with the tools to prevent, adapt, and bounce back from disasters. Among those lessons and practices are [1] never thinking communities can be immune to disasters, [2] learning from old experiences, [3] thinking and planning for possible hazards, and [4] implementing green technology and practices in the design of infrastructures as in developing a design [17].

Particularly, tropical housing is characterized by high cooling demands due to abundant solar radiation, high temperatures (in most regions), high humidity, and high flood risk due to high heavy rain frequency throughout the year [18]. Housing near coastal zones is under high corrosion risk due to wind and sea salt. Besides, tropical housing can benefit from lower temperatures at high altitudes, reducing cooling demand, although there is a high risk of humidity problems.

The tropics, which cover approximately 40% of the world, are home to about 40% of the global population (**Figure 1**). With a higher-than-average population growth in sub-Saharan Africa, it is projected that by the late 2030s or 2040s, 50% of the world's population will reside in the tropics. This rapid shift in population underscores the urgent need for sustainable and resilient housing solutions in tropical regions [19]. In turn, climate change effects are currently increasing the intensity of such housing characteristics, challenging the new and existing housing in terms of construction approaches, energy security, cooling technologies, urban planning, and resource management.

2.1 Construction for sustainable and resilient housing

By the year 2050, 70% of the population is estimated to live in cities, leading to significant land occupation for uncontrolled urban expansion without green infrastructure to address challenges such as the increased urban heat island effect (UHI). This problem, combined with the effects of climate change, increases the vulnerability of cities to severe storms, intensified heat, and rising sea levels compared to surrounding rural regions [20, 21]. Therefore, it is crucial to develop resilient urban environments capable of adapting to these challenges. Interdisciplinary collaboration is key, as recent studies on urban planning multidisciplinary methods underline the need to create knowledge and expertise on sustainable urban planning for tropical regions. Climate analysis methodologies are rarely implemented in local climate plans fostered by tropical cities [22]. Research suggests that urban development strategies have been mostly examined in temperate zones and, therefore, must be tailored to local climates, socio-economic contexts, and cultural landscapes—in tropical regions—to be accurate

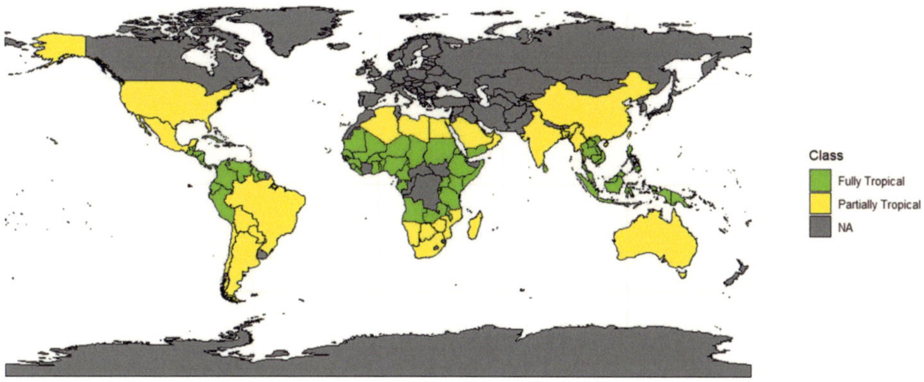

Figure 1.
Worldwide map highlighting countries considered tropical according to the world population review [19].

and effective. It is imperative to target challenges tropical cities face, such as high population density, economic disparities, and vulnerability to climate change [23].

Sustainable urban planning is essential for promoting long-term social, economic, and environmental well-being. This is crucial for bridging the urban gap and ensuring that cities are inclusive, sustainable, economically dynamic, culturally rich, and safe for everyone [24]. Planning systematically helps to anticipate future problems and decide in advance the best solutions, maximizing land use and optimizing urban resources while restoring ecosystem services.

2.1.1 Urban planning for tropical regions

There are different approaches to planning for facing climate change, such as adaptation, mitigation, and emergency response plans. There are planning schemes at diverse scales in ranges from regional to local plans. On the other hand, the planning will depend on the type of hazards the community faces according to its location and geographical characteristics [22–24]. Nevertheless, whatever the scale or focus of the plan, there are essential principles that should not be overlooked when planning to overcome climate change challenges in tropical regions. Those have been summarized and categorized as shown below and in **Figure 2**.

a. Comprehensive data collection, mapping, and collaboration: Accurate data on weather and climate, such as temperature, precipitation, wind patterns, and sea levels, is essential to plan, understand climate trends, and predict future scenarios. Accurate risk maps and thorough assessments of flood-prone areas susceptible to sea level rise, storm surges, or heavy rainfall, among others, help plan and implement effective defense strategies. Sharing data and best practices with multiple governmental agencies, research institutions, and international organizations improves monitoring capacity and response to Climate Change Hazards [22, 25].

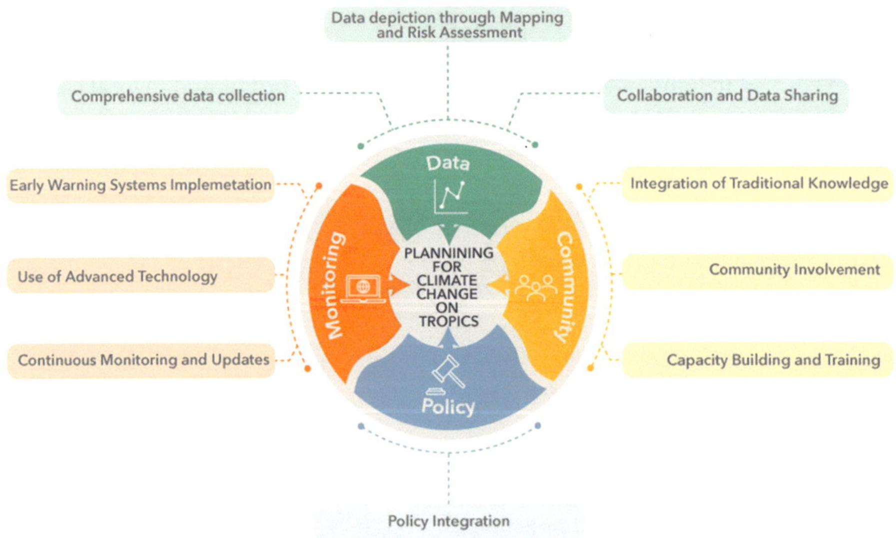

Figure 2.
Elements that constitute the key principles of sustainable urban planning for tropical regions facing climate change.

b. Monitoring updates and robust early warning systems through advanced technology: New remote sensing, Geographic Information Systems (GIS), and climate modeling software enable precise monitoring through real-time data, updated climate models, and risk assessments. These facilitate wise decision-making and planning—ensuring that climate plans remain relevant and responsive to changing conditions—and that Early Warning Systems, integrated into local communication networks, provide timely alerts to vulnerable populations [22, 25–27].

c. Community participation, integration of traditional knowledge with scientific data, and capacity building: Local knowledge often provides valuable insights into weather patterns and environmental changes. Building capacity through engagement and training of community members, stakeholders, and local authorities to observe, report weather changes, operate monitoring equipment, and execute data analysis and interpretation empowers the community to participate in disaster preparedness activities and ensures that monitoring systems are socially inclusive, locally relevant, and sustainable. Local knowledge can enhance the accuracy and relevance of monitoring on-edge efforts [22, 28–31].

d. Policy integration: Fusing weather and climate monitoring into broader policy frameworks ensures that monitoring data informs urban planning, infrastructure development, and emergency response strategies. Policies must regulate the use of monitoring data in decision-making processes [22, 32, 33].

2.1.2 Protection against strong winds

In many tropical areas, the dry season is followed by strong winds associated with the onset of the wet season. These strong winds pose a danger to poorly located and protected buildings. Protection against strong winds must begin with the proper location. For small buildings, mature trees and buildings within 6 to 12 diameters can substantially reduce wind forces. For larger buildings, such protection may not be available, and wind protection must be provided at the site location. A rule of thumb that can be used is that the annual percentage of time there are no winds greater than 24 km/h is 89% for an open site, 93% for a suburban site with 30% tree coverage (approximately 1.62 hectares of 2YO forest in 0.4 km radius of a building), and 97% for an urban site with 50% building coverage (approximately 92% of the area built 6.1 m high) [34–36].

2.1.3 Adaptation to rising temperatures

Tropical regions have been projected to experience continuous urbanization over the past decades. One of the most common issues experienced by cities worldwide and in the humid and warm climates of the tropics is the urban heat island (UHI) effect. The UHI occurs when urban areas have higher temperatures than rural areas due to urbanization. It is primarily caused by reduced vegetation cover, open burning, wildfires, changes in land use, building environment configuration and its construction materials, deforestation, and emissions from industrial activities and traffic, all contributing to increased temperatures. It is predicted that some cities within tropical regions could become up to 4°C warmer by 2050 due to these factors, and some are already showing these trends [37, 38]. Cooling strategies, particularly passive ones, could mitigate heat island and indoor discomfort problems, reduce the energy for mechanical cooling, and promote walking rather than motorized forms of movement [39, 40].

At the same time, these cities are affected by rising temperatures associated with climate change [41]. In these circumstances, an increase in the frequency and severity of heat waves and tropical nights is predicted by the end of the twenty-first century. Extreme heat events are projected to increase significantly, especially in urban areas. Shore and building facets, solar and anthropogenic heating, increasing human population density, and growing energy demand, especially during the pre-monsoon months, boost the land- and sea-breeze circulations, which ultimately result in the stagnation of air masses over larger cities [42].

Adequate urban morphology and open space, the use of planning elements to direct air movement, effective sunshades, and local evaporative and other cooling techniques are important passive cooling strategies [43]. Shading, applying reflective and high-emissivity materials on roofs and pavements, optimizing the sky view factors, and an urban energy balance about form factor, plant density, and site coverage could cool the heat island [44].

The climatic environment throughout the year at the street level should be studied, monitored, and counteracted by a combined optimized street width and height-to-width ratio, optimal vegetation placement for localized UHI reduction, pedestrian air conditioning, and the shading of the facades [45, 46].

2.1.4 Adaptation to floods and rising sea levels

Currently, nearly 500 million people live near coastal areas, making them vulnerable settlements at risk of flooding due to extreme precipitation events or rising sea levels [47]. Some of the most severe urban, local, and regional problems are associated with flooding of rivers and ponds. The problems created by floods can only be solved if the spatial allocation of different functions and land uses are modified so that they are compatible with the recurrence of flooding and the increased flows expected to be caused by global warming. To solve flood problems, it is usually necessary to use land that is now classified as low value for agriculture or nature conservation. This requires recognition of the value of the services performed by green areas to society and the environment.

However, this can be achieved by (a) increasing the demand for residential and work-related services that are utilized under the present conditions of natural presence and protection without isolated towers and (b) achieving their recognition in income estimations and value-added beauty awards to the parcels that shelter them [39, 48]. Using the areas that are prone to flooding and do not have economic value can reduce flood risks for other areas that are relative at lower elevations. These include parks, buffer areas, detention basins, and natural flood retention areas (**Figure 3**). Although the occurrence of floods has decreased the economic value of the unoccupied areas, these recreational and sports grounds can be integrated with the water body and demonstrate ecological problems for residents [49].

Numerous towns and cities in tropical zones already experience regular flooding due to heavy rainfall events in catchment areas, leading to rapid inflow into rivers and the obstruction of water channels by accumulated solids (illicit connections, discharges of solid waste, illegal occupation of waterways, etc.). Therefore, due to the serious shortage of space in urban areas, building on flood-prone lands cannot be avoided. However, sustainable and controlled urban planning in these regions can mitigate this risk by maintaining or creating watercourses that regulate fluid drainage and by planning for the xeric use of the urban area [50–52].

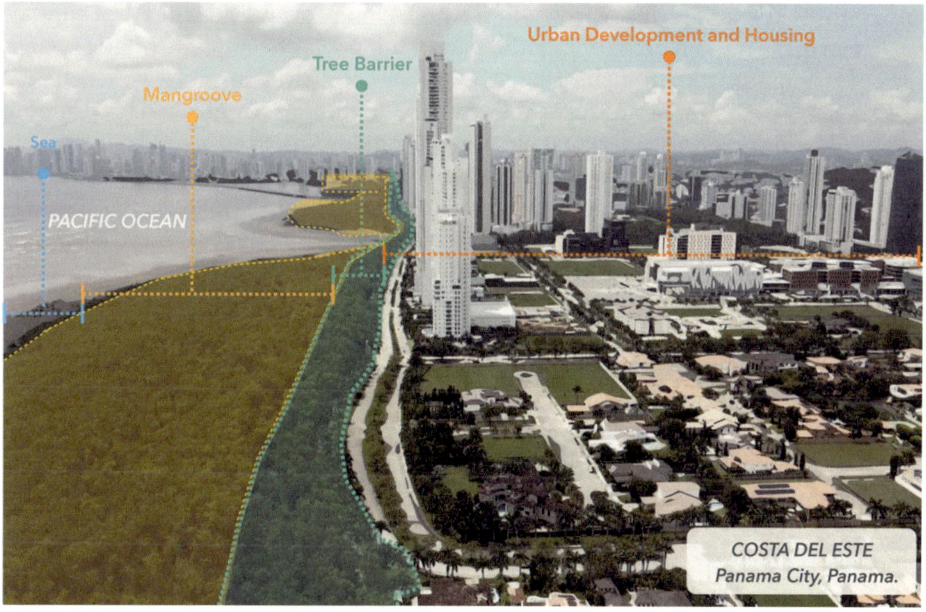

Figure 3.
Coastal urban settlement in Costa del Este, Panama City, Panama.

2.2 Energy security for resilient housing

Achieving sustainable and resilient housing requires relying on ecosystem services such as clean or low-carbon energy sources and their harnessing. Securing the ability to harness such energy sources contributes to energy security. **Figure 4** represents key aspects of energy security, such as renewable energy integration, diversification of energy sources, grid resilience and decentralization [53], energy efficiency, climate resilience, and community engagement and empowerment.

Each of these aspects of housing in tropical climates is challenged. Regarding renewable energy integration, solar energy is abundant and can be harnessed effectively through photovoltaic systems or solar water systems. However, the high

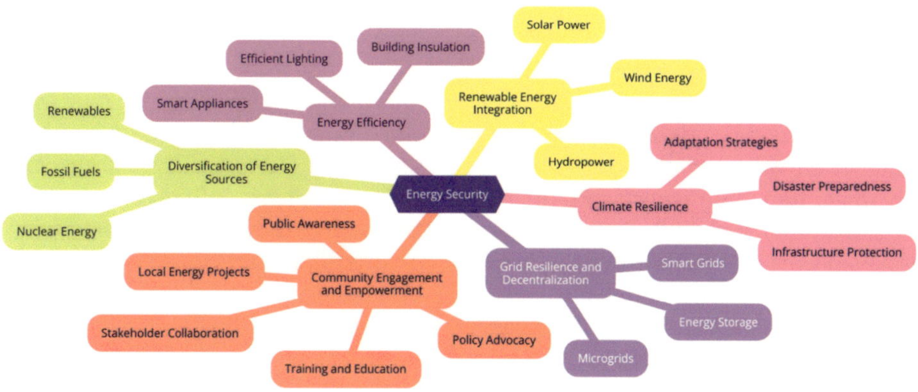

Figure 4.
Mind mapping the aspects of energy security to focus on in the achievement of sustainable and resilient housing.

temperatures during the dry season and the high cloud coverage during the rainy season can make it difficult to obtain the maximum power from photovoltaic systems. This calls for innovation in such systems. The affordability of such systems is changing over time, becoming more accessible even in developing countries. This promotes decentralization through individual self-energy generation. Such an individual generation opportunity will need to be supported by a resilient electrical grid capable of maintaining high performance against any climatic event [54]. Such energy decentralization, relying on solar energy, among residential and commercial sectors may increase the risk of instability of the distribution network. Nonetheless, several non-conventional strategies can be found to harness solar energy for cooling, such as daytime radiative passive cooling [55], and electricity generation, such as bidirectional reflectance photovoltaics [56], purposes at a residential level.

Despite the well-known recommendations from bioclimatic design approaches for tropical climates, multifamily and office buildings are very tall (**Figure 3**). Many of the office buildings can be found to have fully glazed facades. The cooling demand reaches tremendously high levels. This is a call for increasing the compliance of construction standards to increase energy efficiency in buildings. For multifamily buildings and house construction and development, deficiencies can be found when considering indoor air or indoor environment quality.

To cope with the cooling demand to reach comfort levels, which vary greatly, the use of air conditioning split units may have been overrated because of the accumulation risk of carbon dioxide (CO_2) and volatile organic components (VOC) concentrations [57] and this may be a result of their affordability and accessibility. As the split unit only recirculates indoor air in a closed room, CO_2 and VOC concentration levels can rise beyond the healthy limits [58, 59]. This negatively impacts the indoor air quality and, thus, the occupants' health.

3. Best practices: strategies for tropical housing

3.1 Standardization and regulation

Many existing standards were established before the importance of sustainability and resilience was fully understood, resulting in a framework that may not adequately support modern construction and needs. Traditional standards often prioritize cost-efficiency and speed over environmental considerations, making it challenging to implement innovative solutions that reduce ecological footprints, enhance energy efficiency, and improve overall sustainability. Additionally, these standards may lack provisions for resilience against climate change impacts, such as extreme weather events.

In recent decades, there has been a growing incorporation of sustainability and resilience into international construction standards [60]. However, few countries have fully integrated these aspects into their construction standards. While construction standards provide crucial guidance for safety and quality, they can sometimes delay the advancement of sustainable and resilient construction practices [61]. The main cause is that updating and revising construction standards can be slow and bureaucratic, hindering the adoption of new technologies and methods that could benefit the environment and society.

Construction standards are useful for establishing the design criteria, materials, construction methods, and operational guidelines for buildings and other structures (**Figure 5**). Primarily, these standards focus on safety aspects such as job

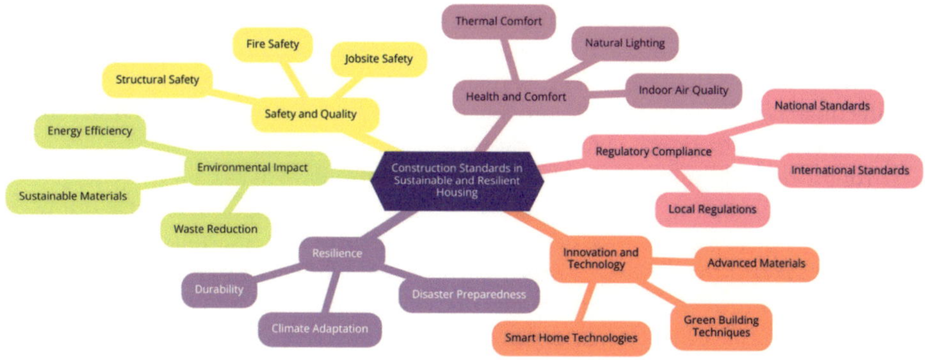

Figure 5.
Mind mapping of the role of construction standards in the sustainable and resilient housing goal.

site, structural, and fire safety, as well as basic services like water, sanitation, and energy. They also cover dimensions for functionality, quality of construction, energy efficiency during operation, accessibility, indoor air quality, and urban planning. Construction standards exist at both national and international levels, with national standards often mandated by law. However, construction standards must be tailored to local conditions, such as topography, geology, and climate [62], especially considering climate change.

Building efficiency codes set minimum energy and resource efficiency standards for construction and design. These codes are most effective within a policy framework that includes mandatory regulations, financial programs, and incentives [63]. For instance, the Panamanian Regulation for Sustainable Building [64] aims to establish minimum standards to encourage the development of high-performance buildings, thereby reducing the environmental impact of the building. Despite its mandatory nature, this regulation still faces implementation challenges.

3.2 Certifications

Other construction standards are voluntarily led by corporations that are compromised with sustainability, offering advisory services and recognition throughout the achievement of sustainable housing. Such is the case of the Leadership in Energy and Environmental Design (LEED) for Homes, Multifamily Low-rise, and Multifamily Midrise, created by the U.S. Green Building Council (USGBC) which serves as a voluntary framework and tool for planning, constructing, operating, and maintaining green buildings. Its objectives include optimizing natural resource use, promoting restorative practices, minimizing the environmental footprint of the building industry, and enhancing indoor environmental quality for occupants.

A minimum requirement that a LEED for Homes project must meet is to be defined as a "dwelling" according to all applicable codes, including, but not limited to, the International Residential Code stipulation that a dwelling unit must include "permanent provisions for living, sleeping, eating, cooking, and sanitation" [65]. Applicability is for single-family dwellings (homes), multifamily residential buildings with 1 to 3 stories (multifamily low-rise), and multifamily residential buildings with four or more occupiable stories above grade (multifamily midrise) [65].

Similarly, the Excellence in Design for Greater Efficiencies (EDGE) for Homes and Apartments is a sustainable building certification system with the aim of making buildings more efficient. EDGE was created by the International Finance Corporation, a member of the World Bank Group, and is a quick and easy-to-use online calculator for massive market transformation. This calculator allows developers and construction companies to quickly identify the most effective ways to reduce energy, water, and resources in building materials [66].

Through key energy-saving and cost-saving solutions such as energy-efficient lighting, thermal glazing, and water-conserving bathroom fixtures, and therefore lower water and electricity bills, developers can meet the expectations of buyers and residents who want to achieve returns on investment while living in comfortable and enjoyable spaces [66]. A comparison summary between LEED and EDGE is shown in **Table 1**. The costs of the necessary professionals and third parties (e.g., Green Rater, Energy Rater, Auditor) must be considered in the investment and development of projects with green housing certification to maintain the economic balance and sustainability.

Table 2 shows the number of units registered and certified under the LEED and EDGE green building programs in fully and partially tropical countries. The data hint that the sustainable and resilient housing movement and market is still weak but slowly gaining ground.

Moreover, based on 25 interviewed and surveyed LEED certified green projects, covering a built-up area of 830,000 m^2, of which 4% are within the residential use,

	LEED certification	EDGE certification
Verification	LEED Green Rater, Energy Rater.	EDGE Auditor.
Certification	Green Business Certification Inc. (GBCI).	Green Business Certification Inc. (GBCI).
Certification levels	Certified, Silver, Gold, Platinum.	Certified, Advanced, Zero Carbon.
Strategies	Location and Transportation, Sustainable Sites, Water Efficiency, Energy and Atmosphere, Materials and Resources, Indoor Environmental Quality, Innovation and Regional Priority.	Energy, Water, and Materials.
Costs	Registration and certification fees, LEED Green Rater, Energy Rater, Consultancy.	Registration and certification fees, EDGE Auditor, Consultancy.

Table 1.
Comparison between LEED and EDGE certification.

Registered and certified LEED projects (2016) [40]	Registered and certified EDGE projects (2022–2024) [67]
Homes, Multifamily Lowrise, and Multifamily Midrise (25): Chile (15), México (4), Colombia (2), Perú (2), Brazil (1), Costa Rica (1).	Apartments (308): Perú (146), Colombia (83), Brazil (32), México (21), Guatemala (10), Ecuador (9), Chile (4), Argentina (3), Panamá (2).
	Homes (116): México (42), Brazil (13), Colombia (13), Ecuador (13), El Salvador (13), Argentina (12), Guatemala (3), Panamá (3), Costa Rica (2), Dominican Republic (1), Chile (1).

Table 2.
Registered and certified LEED and EDGE projects in fully and partially tropical countries.

it was found that the main sustainability strategies to be implemented are mainly decided during the early planning phase [68]. **Table 3** presents the main strategies and practices implemented among the aforementioned study.

Furthermore, the selection of green certifications will depend on specific project details, sustainability goals, and unique local conditions and opportunities presented by tropical climates in terms of sustainable and resilient housing design and construction.

The Green Mark Scheme from Singapore withstands, among others, because it combines various legal and regulatory mechanisms to undergird this green certification regime and various financial incentive schemes, risk-sharing programs, and rewards, which have made the government's commitment credible, lowering entry barriers for new participants [69].

Certain green building certifications are increasingly prioritizing both embodied and operational carbon emissions. This shift includes incorporating mandates for life cycle assessments of building materials and applying purchase of carbon offsets [70].

Carbon trading mechanisms can facilitate emission reductions by incentivizing cleaner practices and the implementation of carbon offset projects. Studies explore the global transition to carbon neutrality, focusing on energy efficiency, renewable energy, carbon trading, and the need for advanced and integrated energy policies [71]. On the other hand, there is also a critical analysis of carbon offsetting and its impact on maintaining consumption patterns. While carbon offset projects can contribute to reducing greenhouse gas emissions, their effectiveness raises concerns because they also risk legitimizing and perpetuating high-consumption lifestyles in developed countries. Carbon offset projects allow individuals or organizations to

Strategy	Practice
Lighting	• LED lighting technology. • Access to natural light through translucent elements in the envelope, such as windows and skylights. • Daylight sensors and dimming of electric lighting. • Occupancy sensors, and additional controls such as timers, daylight control, or scene control.
Ventilation and Air conditioning	• Mechanical ventilation systems have on/off monitoring and temperature control. • Have minimum efficiency reporting value (MERV) or high efficiency particulate air (HEPA) filters. • Use natural ventilation strategies as the primary method to ensure minimum or supplemental indoor air quality. • Natural conditioning of the spaces in their totality or partially in common areas.
Plumbing	• Implement water-saving and low-flush toilets and low-consumption faucets. • Collection of rainwater for non-potable use, such as flushing toilets and irrigation. • Greywater treatment for non-potable uses.
Landscaping	• Implementation of native or adapted vegetation that does not require irrigation after planting.
Energy Sources	• Implementation of renewable energy systems based on solar energy, with solar panels or solar collectors for water heating.

Table 3.
Main strategies and practices implemented in certified green projects.

compensate for their carbon emissions by investing in emission reduction projects elsewhere, this may serve as a distraction and divert attention from the need for fundamental changes in consumption and production patterns, therefore, systemic solutions to climate change [53].

3.3 Strategies and technologies

3.3.1 Facing heavy rains

Open spaces and building facades designed for natural ventilation become a challenge during the rainy seasons of tropical areas, therefore, a balance between ventilation and rain protection is needed when designing and building housing.

Research on wind-driven rain barriers in tropical regions revealed that tiled overhangs are the most effective elements to protect building facades designed for natural ventilation from heavy rains. This strategy was followed by a full horizontal louvered facade configuration among other three strategies that proved less effective. Thus, considering the design and implementation of tiled overhangs for facades with large openings is key in adapting to heavy rain patterns facing climate change in the tropics. This allows maintaining the openings necessary for breeze and ventilation while performing as a buffer for heavy rains and winds [72, 73].

Flooding is an increasing problem in many cities facing urban growth and climate change. A great part of flood prevention should be based on urban planning and flood risk mapping to avoid housing development in flood-prone areas. Therefore, it is worth emphasizing in another aspect of flood prevention and preparedness as is water infiltration for rainy tropical regions [74–77].

The impervious surfaces hamper natural rainwater infiltration into the soil, boosting runoff to lower areas and causing flooding. Green roofs have been popularized in temperate climates for the thermal comfort they provide, nonetheless, studies brought to light the importance of implementing these strategies to adapt to heavy rains and prevent flooding events. The implementation of green roofs demonstrated an average reduction of the runoff of 60.3% for the 9 months of the study [73]. More studies on local plant use should be made to adapt green roofs to different tropical environments.

Concrete-based constructions have important advantages against floods. For example, concrete structures such as Xblocs are being used to mitigate the effects of sea level rise (**Figure 6a**) [81], permeable concretes are being used to allow infiltration and in some cases, the harvesting of rainwater in the face of the intense rainfall in tropical areas [82] and linear parks using permeable concrete are presented as a drainage strategy against flooding in urban areas.

3.3.2 Facing heat increase

In tropical climates, high temperatures combined with high humidity led to uncomfortable thermal sensations, making thermal comfort one of the main challenges, while reducing energy consumption and its associated impacts to improve the occupants' quality of life during the building's operation stage [83].

At the building level, this effect can be mitigated by implementing passive strategies and vernacular architecture approaches, such as using sustainable materials, reflective paints (cool roofs) [84], green roofs, and even installing solar panels on the

Figure 6.
Resilient construction strategies for tropical climates: (a) Shore protection of the Cinta Costera Boulevard in Panama City using concrete XBlocs [78], (b) Concrete retaining wall against landslides in Panama City, (c) Hurricane-resistant housing (monolithic dome) made of reinforced concrete [79], (d) Concrete houses within the former Albrook Air Force Station in Panama [80] which have been in use for more than 90 years.

roofs of buildings [85]. It is necessary to consider the orientation of the building and natural cross-ventilation [86, 87]. In addition, it should have windows and an envelope that improves thermal insulation.

At the city level, adequate urban planning that includes open and green spaces to cool cities through air circulation, shade, and evapotranspiration is necessary [88]. Permeable concrete pavements that reduce the concentration of heat due to their porosity and reflective paints are recommended [89].

3.3.3 Facing landslides and strong winds

Focusing on site selection, land use planning, slope stabilization, and building design can address landslides and strong wind effects. Historical data and land use planning help avoid construction on steep slopes and areas with a history of landslides [90, 91]. Zoning regulations that restrict building in high-risk areas can be implemented.

Using slope stabilization approaches, such as retaining walls (**Figure 6b**), terracing, and vegetation, is effective in creating landslide safety. However, it is equally important to employ geotechnical engineering techniques, such as soil nailing, rock bolting, and geosynthetics.

Design buildings with strong foundations and structural reinforcements to withstand ground movement. Use flexible building materials that can absorb and dissipate energy from landslides. Concrete walls and foundations retaining works against possible landslides [92]. Regarding building materials, a review of resilient and sustainable housing models found that they had in common the use of concrete as a building material [93]. The robustness of concrete allows the construction of structures resistant to extreme weather events such as storms and hurricanes (**Figure 6c**) with a lifespan of at least 50 years that could be further expanded with appropriate maintenance (**Figure 6d**).

3.3.4 Construction technologies

Emerging construction methods and tools address current limitations and deliver additional benefits in constructing sustainable and resilient housing. Many construction companies build primarily with conventional techniques without utilizing the various technological advances, such as 3D printing with concrete, whose strength is in printing and utilizing the exact number of materials required.

Although concrete is a cement-based material with high CO_2 emissions, the industry is constantly researching and implementing technologies to reduce these emissions [94]. For instance, 3D printing can potentially reduce embodied carbon, which is the total carbon dioxide (CO_2) associated with material manufacturing, transportation, and construction methods used throughout the life cycle of a building. The carbon footprint is reduced by using alternative products in the concrete mix, decreasing the number of materials needed on site, reducing timber formwork, and minimizing the volume of workers commuting to the site.

To lower the environmental impact of housing and infrastructure, it is recommended to design and build with less concrete through topological optimization and 3D printing [95], optimize concrete mix to use less cement, and use fillers and supplementary cementitious materials in cement to reduce clinker [96]. A limitation in some regions may be construction standards that require a minimum of clinker, although the possibility of using up to 70% less clinker while maintaining the strength and durability of concrete has been demonstrated [97].

In terms of resilience, one of the challenges of 3D printing is that more research is needed, as flammable structural and architectural materials or products (roofs, windows, doors) used in conjunction with concrete must be studied to ensure the resilience of the entire system in the event of a site-specific catastrophe or extreme weather event. Other constraints are related to variable weather conditions and topography of the construction site, gaps in regulations and building codes not updated to include specific guidelines for 3D printing, and the impact of reduced employment in the construction industry [98, 99].

Proper insulation is crucial in the housing envelope, particularly in tropical climates. It plays a key role in avoiding heat gain and reducing cooling demands in air-conditioned buildings. For these buildings, high insulating capacity or low U-value should be installed on the hotter side of the dwelling. Conversely, it is advisable to avoid applying thermal insulation on the walls for non-air-conditioned buildings, as this could trap heat. Roofs exposed to direct solar radiation and excessive heat gain should be protected by appropriate insulation [100].

One of the best practices in building sustainable and resilient houses is to conduct a comprehensive and robust site assessment. This assessment should identify possible alternatives and disaster mitigation measures, ensuring that the construction is well-prepared for potential challenges.

3.4 Proposed reference framework toward sustainable and resilient tropical housing

By implementing the best practices from the previous section, cities in tropical regions can significantly improve their resilience to climate change and the management of risks associated with extreme weather events. The principles of sustainable urban planning in tropical climates can be a guideline for developing sustainable and resilient housing. **Figure 7** shows how each principle relates to others and to urban planning, as well as how all these factors of planning influence local people.

Depending on the housing, whether existing or new, the starting principle to apply the framework varies. For instance, for new housing data to come first, historical and forecast data of the new settlement zone must be analyzed to uncover possible risks of extreme weather events. Such data serves to establish contextualized regulations and policies in defining the benchmark for construction, i.e., suitable materials, best-performance building envelope, and construction technologies. Similarly, it helps define a benchmark to achieve energy security (availability, decentralization, among others) for energy generation technologies, the best energy-saving strategies given.

However, to ensure the sustainability of such resilient housing, monitoring the weather variables and forecasting is crucial to generate alerts, allowing the community to be prepared and improve the development of future resilient urban planning.

Moreover, for existing housing, achieving resilience should start by analyzing data on weather and climate events and the existing community. This data-driven approach allows for a comprehensive assessment of the risks and the effectiveness of current mitigation and adaptation measures, helping to make informed decisions about the future of our communities. In such a scenario, the community resilience concept can be implemented, where health, safety, and environmental improvement of communities are the primary goals. It establishes the basis for a wider comprehension of communities and their behavior when facing an unexpected traumatic situation, as well as what to do regarding planning and design to be ready for a disaster, before, during, and after it [17, 101].

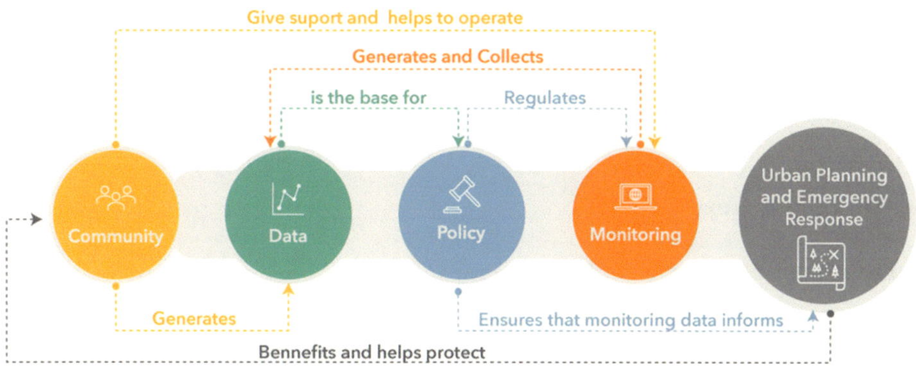

Figure 7.
Links between key principles of sustainable urban planning for tropical regions facing climate change.

4. Final remarks

The implications of sustainable construction and energy security related to sustainable and resilient housing, specifically in tropical climates, were addressed in this study. Tropical regions face significant climate change challenges such as strong winds, rising temperatures, heavy rain, floods, rising sea levels, and landslides. Developing resilient and sustainable housing is essential to mitigate these challenges.

Addressing such requires tailored solutions that consider local environmental and economic conditions. Uniform approaches are ineffective given the diverse challenges even within the same country. Emphasizing localized strategies ensures that housing solutions are more effective in meeting the particular needs of each area, leading to more sustainability and resilience.

Such local and tailored strategies are best framed as part of data-informed construction standards and regulations. Updating construction standards and certifications is crucial for promoting sustainability and resilience. The lack of regulations for tropical climates demands timely community efforts.

Incorporating new technologies and adapting to local needs can ensure that buildings are not just better equipped, but significantly improved to withstand climate change impacts. This proactive approach not only enhances building performance but also aligns with global sustainability goals, fostering a more resilient and optimistic built environment.

Finally, the principles of sustainable urban planning, combined with innovative design and construction practices, sustainable harvest energy technologies, and responsible energy consumption, can significantly improve the safety and comfort of inhabitants in these vulnerable regions, increasing their chances of resilience.

Acknowledgements

The authors would like to thank the Research Group Energy and Comfort in Bioclimatic Buildings (ECEB, https://eceb.utp.ac.pa/) part of the Faculty of Mechanical Engineering (https://fim.utp.ac.pa/) and the Research Group Sustainable Construction UTP, both within the Universidad Tecnológica de Panamá (https://utp.ac.pa/) for their collaboration. Likewise, special thanks to the Faculty of Architecture and Design (FADUP) from the University of Panama for their collaboration.

This research was funded by Sistema Nacional de Investigación (SNI, https://sni.senacyt.gob.pa/) and a Panamanian Institution Secretaría Nacional de Ciencia, Tecnología e Innovación (SENACYT, https://www.senacyt.gob.pa/) under the projects FIED22-09 and FIED24-11.

Conflict of interest

The authors declare no conflict of interest.

Author details

Miguel Chen Austin[1], Thasnee Solano[1], Olga Yuil Valdés[2],
Hatvany Gómez Concepción[3], Dafni Mora[1] and Yazmín Mack-Vergara[4*]

1 Research Group Energy and Comfort in Bioclimatic Buildings (ECEB), Faculty of Mechanical Engineering, Technological University of Panama, Panama City, Panama

2 CBRE Panama, Plaza Credicorp Bank, Ciudad de Panamá, Panama

3 Facultad de Arquitectura y Diseño, Universidad de Panamá, Panama City, Panama

4 Research Group Sustainable Construction UTP, Experimental Center for Engineering, Technological University of Panama, Panama City, Panama

*Address all correspondence to: yazmin.mack@utp.ac.pa

References

[1] Monteiro A, Ankrah J, Madureira H, Pacheco MO. Climate risk mitigation and adaptation concerns in urban areas: A systematic review of the impact of IPCC assessment reports. Climate. 2022;**10**:115

[2] World Meteorological Organization. United in Science 2023-Sustainable Development Edition [Internet]. Geneva, Switzerland: World Meteorological Organization. Available from: https://library.wmo.int/idurl/4/68235

[3] Hoffmann R, Dimitrova A, Muttarak R, Crespo Cuaresma J, Peisker J. A meta-analysis of country-level studies on environmental change and migration [Internet]. Nature Climate Change. 2020;**10**(10):904-912. Available from: https://www.nature.com/articles/s41558-020-0898-6 [Accessed: July 29, 2024]

[4] Maquiling KSM, De La Sala S, Rabé P. Urban resilience in the aftermath of tropical storm washi in the Philippines: The role of autonomous household responses. Environment and Planning B: Urban Analytics and City Science 2021;48(5):1025-1041. Available from: https://journals.sagepub.com/doi/10.1177/2399808321998693 [Accessed: June 30, 2024]

[5] United Nations. Transforming our world: The 2030 agenda for sustainable development. In: Resolution Adopted by the General Assembly on 25 September 2015. New York, NY, USA: United Nations; 2015

[6] United Nations. Paris Agreement. London: United Nations; 2015

[7] United Nations. The Global Alliance for Buildings and Construction [Internet]. Paris, France: United Nations. Available from: https://globalabc.org/; [Accessed: July 2, 2024]

[8] International Code Council I, Ashrae. International Green Construction Code (IgCC). Massachusetts, Washington DC: International Code Council I, Ashrae; 2021

[9] US Green Building Council Inc. LEED [Internet]. Available from: https://www.usgbc.org/leed [Accessed: July 2, 2024]

[10] World Green Building Council [Internet]. Available from: https://worldgbc.org/ [Accessed: July 2, 2024]

[11] World Resources Institute. REHOUSE (Resilient, Equitable Housing, Opportunities and Urban Services) [Internet]. Available from: https://www.wri.org/initiatives/rehouse-resilient-equitable-housing-opportunities-and-urban-services [Accessed: July 17, 2024]

[12] Aste N, Butera FM, Adhikari RS, Leonforte F. Sustainable building design for tropical climates. In: Research for Development [Internet]. Cham: Springer; 2020. pp. 37-46. Available from: https://link.springer.com/chapter/10.1007/978-3-030-33323-2_4 [Accessed: June 30, 2024]

[13] Hartshorn GS. Tropical forest ecosystems. In: Encyclopedia of Biodiversity. 2nd ed. Cambridge, Massachusetts, United States. 2013. pp. 269-276. DOI: 10.1016/B978-0-12-384719-5.00146-5

[14] Ismaeel A, Tai APK, Santos EG, Maraia H, Aalto I, Altman J, et al. Patterns of tropical forest understory temperatures. Nature Communications [Internet]. 2024;**15**(1):1-10. Available from: https://www.nature.com/articles/s41467-024-44734-0 [Accessed: June 30, 2024]

[15] Keeble BR. The Brundtland report: 'Our common future. Medicine and War [Internet]. 1988;**4**(1):17-25. Available from: https://www.tandfonline.com/doi/abs/10.1080/07488008808408783 [Accessed: June 21, 2024]

[16] Meerow S, Newell JP, Stults M. Defining urban resilience: A review. Landscape and Urban Planning. 2016;**147**:38-49

[17] Schwab JC. The vision of a resilient community. In: Planning for Post-Disaster Recovery: Next Generation [Internet]. Chicago, IL: APA Planning Advisory Service; 2014. pp. 16-21. Available from: https://www.planning.org/publications/report/9026899/

[18] Pan J, Tang J, Caniza M, Heraud JM, Koay E, Lee HK, et al. Correlating indoor and outdoor temperature and humidity in a sample of buildings in tropical climates. Indoor Air. 2021;**31**(6):2281-2295

[19] World Population Review. Tropical Countries [Internet]. 2024. Available from: https://worldpopulationreview.com/country-rankings/tropical-countries [Accessed: July 2, 2024]

[20] Jamei E, Ossen DR, Seyedmahmoudian M, Sandanayake M, Stojcevski A, Horan B. Urban design parameters for heat mitigation in tropics. Renewable and Sustainable Energy Reviews [Internet]. 2020;**134**:110362. Available from: https://www.sciencedirect.com/science/article/pii/S136403212030650X

[21] Solano T, Bernal A, Mora D, Chen AM. How bio-inspired solutions have influenced the built environment design in hot and humid climates. Frontiers in Built Environment. 2023;**9**:1267757

[22] Tiepolo M, Ponte E, Cristofori E. Planning to Cope with Tropical and Subtropical Climate Change [Internet]. Warsaw, Poland: De Gruyter Open Poland; 2016. pp. 1-380. DOI: 10.1515/9783110480795. Available from: http://creativecommons.org/licenses/by-nc-nd/3.0/

[23] Chaiechi T. Sustainable tropical cities: A scoping review of multidisciplinary methods for urban planning. eTropic. 2020;**19**(2):25-51

[24] Planning Sustainable Cities Un-Habitat Practices and Perspectives [Internet]. Available from: www.unhabitat.orgacknowledgements

[25] Adedeji O, Olusola A, Babamaaji R, Adelabu S. An assessment of flood event along lower Niger using Sentinel-1 imagery. Environmental Monitoring and Assessment. 2021;**193**(12):858

[26] Dutrieux LP. Multidimensional remote sensing based mapping of tropical forests and their dynamics [thesis]. Netherlands: Laboratory of Geo-Information Science and Remote Sensing, Wageningen University; 2016. 147 p

[27] Starzec M, Kordana-Obuch S, Słyś D. Assessment of the feasibility of implementing a flash flood early warning system in a small catchment area. Sustainability. 2023;**15**(10):8316. Available from: https://www.mdpi.com/2071-1050/15/10/8316/htm [Accessed: May 10, 2024]

[28] Abenir MAD, Manzanero LIO, Bollettino V. Community-based leadership in disaster resilience: The case of small island community in Hagonoy, Bulacan, Philippines. International Journal of Disaster Risk Reduction [Internet]. 2022;**71**:102797. Available from: https://www.sciencedirect.com/science/article/pii/S2212420922000164

[29] Singh P, Tabe T, Martin T. The role of women in community resilience to climate change: A case study of an indigenous Fijian community. Women's Studies International Forum [Internet]. 2022;**90**:102550. Available from: https://www.sciencedirect.com/science/article/pii/S0277539521001138

[30] Tobin GA. Sustainability and community resilience: The holy grail of hazards planning? Global Environmental Change Part B: Environmental Hazards [Internet]. 1999;**1**(1):13-25. Available from: https://www.sciencedirect.com/science/article/pii/S1464286799000029

[31] Curtis AG, Allred SB, Murphy Esq RC, Roberts BA. Exploring the role of social capital in community flood resiliency in Binghamton, NY. International Journal of Disaster Risk Reduction [Internet]. 2024;**104**:104322. Available from: https://www.sciencedirect.com/science/article/pii/S2212420924000840

[32] Hurlimann A, Moosavi S, Browne GR. Urban planning policy must do more to integrate climate change adaptation and mitigation actions. Land Use Policy. 2021;**101**:105188

[33] Wanner MST. Change in policy regimes for disaster risk reduction in Fiji and Nepal. International Journal of Disaster Risk Reduction [Internet]. 2022;**77**:103030. Available from: https://www.sciencedirect.com/science/article/pii/S2212420922002497

[34] Wang Z, Li Y, Song J, Wang K, Xie J, Chan PW, et al. Modelling and optimizing tree planning for urban climate in a subtropical high-density city. Urban Climate. 2022;**43**:101141

[35] Zheng S, Guldmann JM, Liu Z, Zhao L, Wang J, Pan X, et al. Predicting the influence of subtropical trees on urban wind through wind tunnel tests and numerical simulations. Sustainable Cities and Society. 2020;**57**:102116

[36] Jackson TD, Sethi S, Dellwik E, Angelou N, Bunce A, Van Emmerik T, et al. The motion of trees in the wind: A data synthesis. Biogeosciences. 2021;**18**(13):4059-4072

[37] Gomez H, Rojas I, Peren JI. Una Aproximación a Los Efectos Del Diseño Urbano En El Microclima Y Calidad De Espacios Urbanos De Una Ciudad Cálida-Húmeda: Panamá. SusBCity [Internet]. 2021;3(1):31-38. Available from: https://www.revistas.up.ac.pa/index.php/SusBCity/article/view/2009

[38] MacLachlan A, Biggs E, Roberts G, Boruff B. Sustainable city planning: A data-driven approach for mitigating urban heat. Frontiers in Built Environment. 2021;**6**:519599

[39] Dharmarathne G, Waduge AO, Bogahawaththa M, Rathnayake U, Meddage DPP. Adapting cities to the surge: A comprehensive review of climate-induced urban flooding. Results in Engineering. 2024;**22**:102123

[40] U.S. Green Building Council. LEED for Homes: International Market [Internet]. Washington, DC: U.S. Green Building Council; 2024. Available from: https://www.usgbc.org/articles/leed-homes-international-market-update-2016 [Accessed: July 3, 2024]

[41] Marcotullio PJ, Keßler C, Quintero Gonzalez R, Schmeltz M. Urban growth and heat in tropical climates. Frontiers in Ecology and Evolution. 2021;**9**:616626

[42] Perkins-Kirkpatrick SE, Lewis SC. Increasing trends in regional heatwaves. Nature Communications. 2020;**11**(1):3357

[43] Tan JKN, Belcher RN, Tan HTW, Menz S, Schroepfer T. The urban heat island mitigation potential of vegetation depends on local surface type and shade. Urban Forestry & Urban Greening [Internet]. 2021;**62**:127128. Available from: https://www.sciencedirect.com/science/article/pii/S1618866721001539

[44] Araque K, Palacios P, Mora D, Chen AM. Biomimicry-based strategies for urban heat island mitigation: A numerical case study under tropical climate. Biomimetics. 2021;**6**(3):48

[45] Hermawan H, Švajlenka J. Building envelope and the outdoor microclimate variable of vernacular houses: Analysis on the environmental elements in tropical coastal and mountain areas of Indonesia. Sustainability (Switzerland). 2022;**14**(3):1818

[46] Sari DP. A review of how building mitigates the urban heat island in Indonesia and tropical cities [internet]. Earth. 2021;**2**:653-666. Available from: https://www.mdpi.com/2673-4834/2/3/38/htm [Accessed: July 19, 2024]

[47] Latina Caribe A, Vera Jennifer Doherty Soledad Patiño Jeannette Sordi F. Infraestructuras verdes urbanas y espacio público en Diseño Ecológico: Estrategias Para La Ciudad Vulnerable. Washington, D.C. USA. 2020. DOI: 10.18235/0004388

[48] Pour SH, Wahab AKA, Shahid S, Asaduzzaman M, Dewan A. Low impact development techniques to mitigate the impacts of climate-change-induced urban floods: Current trends, issues and challenges. Sustainable Cities and Society. 2020;**62**:102373

[49] Van Long N, Cheng Y, Le TDN. Flood-resilient urban design based on the indigenous landscape in the city of Can Tho, Vietnam. Urban Ecosystems [Internet]. 2020;**23**(3):675-687. Available from: https://link.springer.com/article/10.1007/s11252-020-00941-3 [Accessed: July 19, 2024]

[50] Chitwatkulsiri D, Miyamoto H, Irvine KN, Pilailar S, Loc HH. Development and application of a real-time flood forecasting system (RTFlood system) in a tropical urban area: A case study of Ramkhamhaeng polder, Bangkok, Thailand. Water [Internet]. 2022;**14**(10):1641. Available from: https://www.mdpi.com/2073-4441/14/10/1641 [Accessed: July 18, 2024]

[51] Liu WC, Hsieh TH, Liu HM. Flood risk assessment in urban areas of southern Taiwan. Sustainability [Internet]. 2021;**13**(6):3180. Available from: https://www.mdpi.com/2071-1050/13/6/3180 [Accessed: July 18, 2024]

[52] Quesada-Román A, Villalobos Portilla E, Campos-Durán D. Hydrometeorological disasters in urban areas of Costa Rica, Central America. Environmental Hazards. 2020;**20**:264-278

[53] Lovell H, Bulkeley H, Liverman D. Carbon offsetting: Sustaining consumption? Environment and Planning A: Economy and Space [Internet]. 2009;**41**(10):2357-2379. DOI: 10.1068/a40345. Available from: https://journals.sagepub.com/doi/10.1068/a40345 [Accessed: June 30, 2024]

[54] Ruiz M, Inga E. Despliegue óptimo de redes ópticas para comunicaciones en redes eléctricas inteligentes. I+D Tecnológico. 2019;**15**(2):79-85

[55] Li X, Peoples J, Huang Z, Zhao Z, Qiu J, Ruan X. Full daytime sub-ambient radiative cooling in commercial-like paints with high figure of merit.

Cell Reports Physical Science. 2020;**1**(10):100221

[56] Lee KH, Song YH. Analysis of energy reduction and energy self-sufficiency improvement effects by applying a bidirectional reflectance PV array with integrated external shading at a school building. Buildings. 2023;**13**(12):2915

[57] Rasli NBI, Ramli NA, Ahmad MZ, Badroldin NAM, Ismail MR. Assessment of indoor air quality in an air-conditioning split units (ACSU) office building. Journal of Sustainable Science and Management. 2021;**16**(4):266-284

[58] Cedeño Quijada M, Solano T, Mora D, Chen AM. Evaluación del desempeño de sistemas de ventilación en salones de clase: Estudio numérico en edificios universitarios en Panamá. Revista Digital de Ciencia, Ingeniería y Tecnología-Novasinergia. 2022;**5**(1):100-127

[59] González J, Quijada MC, Serrano J, Solano T, Mora D, Chen AM. Indoor air quality assessment via experimentally calibrated dynamic simulation: A case study in an office building in Panama. In: Proceedings of the LACCEI International Multi-Conference for Engineering, Education and Technology. Boca Raton, Florida, USA: Latin American and Caribbean Consortium of Engineering Institutions; 2022

[60] Meacham BJ. Research paper: Sustainability and resiliency objectives in performance building regulations. Building Governance and Climate Change [Internet]. 2019;**44**:8-23. Available from: https://www.taylorfrancis.com/chapters/edit/10.4324/9781351184212-2/research-paper-sustainability-resiliency-objectives-performance-building-regulations-brian-meacham [Accessed: June 21, 2024]

[61] Phillips R, Troup L, Fannon D, Eckelman MJ. Do resilient and sustainable design strategies conflict in commercial buildings? A critical analysis of existing resilient building frameworks and their sustainability implications. Energy and Buildings. 2017;**146**:295-311

[62] Xu Y, Zhao S, Fan J. Urban planning construction land standard and its revision based on climate and topography in China. Journal of Geographical Sciences [Internet]. 2021;**31**(4):603-620. Available from: https://link.springer.com/article/10.1007/s11442-021-1861-9 [Accessed: June 21, 2024]

[63] Becqué R, Weyl D, Stewart E, Mackres E, Luting J, Shen X. Accelerating Building Decarbonization: Eight Attainable Policy Pathways to Net Zero Carbon Buildings for all [Internet]. United States of America: World Resources Institute; 2019. Available from: https://publications.wri.org/buildingefficiency/ [Accessed: July 17, 2024]

[64] Ministerio de Obras Públicas & Junta Técnica de Ingeniería y Arquitectura. Resolución JTIA N 02: Reglamento de Edificación Sostenible [Internet]. 2023. Available from: https://www.gacetaoficial.gob.pa/pdfTemp/29726/96966.pdf [Accessed: June 21, 2024]

[65] Reference Guide for Homes Design and Construction [Internet]. 2019. Available from: file:///C:/Users/ersua/Downloads/HOMES_2019_intro.pdf [Accessed: July 3, 2024]

[66] International Finance Corporation. EDGE Methodology Report. Switzerland: EDGE Certified Foundation; 2019

[67] International Finance Corporation. EDGE Projects [Internet]. Zug, Switzerland: International Finance

Corporation; 2024. Available from: https://app.edgebuildings.com/project-studies [Accessed: July 3, 2024]

[68] Consejo Colombiano de Construcción Sostenible. Caso de negocio de LEED en Lationoamérica. 2da edición ed. [Internet]. Bogotá D.C., Colombia: Consejo Colombiano de Construcción Sostenible; 2024. Available from: https://www.cccs.org.co/wp/wp-content/uploads/2024/03/Caso_Negocio_LEED_2024-1.pdf [Accessed: July 3, 2024]

[69] Han H. Governance for green urbanisation: Lessons from Singapore's green building certification scheme. Environment and Planning C: Politics and Space [Internet]. 2019;**37**(1):137-156. Available from: https://journals.sagepub.com/doi/10.1177/2399654418778596 [Accessed: July 19, 2024]

[70] World Green Building Council. Bringing Embodied Carbon Upfront [Internet]. Fox Court, London: World Green Building Council; 2019. Available from: www.worldgbc.org/embodied-carbon [Accessed: June 25, 2024]

[71] Zhou Y. Worldwide carbon neutrality transition? Energy efficiency, renewable, carbon trading and advanced energy policies. Energy Reviews. 2023;**2**(2):100026

[72] Lim CH, Alkhair M, Mirrahimi S, Salleh E, Sopian K. Optimization of wind driven rain barrier for tropical natural ventilated building. Research Journal of Applied Sciences, Engineering and Technology. 2016;**12**(1):69-85

[73] Laar M, Grimme FW. Thermal comfort and reduced flood risk through green roofs in the tropics. In: PLEA 2006 - 23rd International Conference on Passive and Low Energy Architecture. Conference Proceedings, Geneva, Switzerland. 2006

[74] Douglas I. The challenge of urban poverty for the use of green infrastructure on floodplains and wetlands to reduce flood impacts in intertropical Africa. Landscape and Urban Planning [Internet]. 2018;**180**:262-272. Available from: https://www.sciencedirect.com/science/article/pii/S0169204616302456

[75] Chan FKS, Griffiths JA, Higgitt D, Xu S, Zhu F, Tang YT, et al. "Sponge City" in China—A breakthrough of planning and flood risk management in the urban context. Land Use Policy [Internet]. 2018;**76**:772-778. Available from: https://www.sciencedirect.com/science/article/pii/S0264837717306130

[76] Islam MT, Meng Q. Spatial dynamic analysis and thematic mapping of vulnerable communities to urban floods. Cities [Internet]. 2024;**145**:104735. Available from: https://www.sciencedirect.com/science/article/pii/S0264275123005474

[77] Gu X, Liu X. Planning scale flood risk assessment and prediction in ultra-high density urban environments: The case of Hong Kong. Ecological Indicators [Internet]. 2024;**162**:112000. Available from: https://www.sciencedirect.com/science/article/pii/S1470160X24004576

[78] Cinta Costera, Panama. Xbloc [Internet]. Available from: https://www.xbloc.com/node/48 [Accessed: June 21, 2024]

[79] Gradin O. Hurricane Home - Flickr - Olaf [Internet]. 2005. Available from: https://commons.wikimedia.org/wiki/File:Hurricane_Home_-_Flickr_-_Olaf.jpg [Accessed: July 29, 2024]

[80] Office of Air Force History. Air Force Combat Units of World War II. Washington, DC: Office of Air Force History; 1983

[81] van den Berg I. Effect of irregularities in the under layer on the stability of Xbloc[Plus] concrete armour unit [Internet] [thesis]. Netherlands: Delft University of Technology; 2018. Available from: https://repository.tudelft.nl/islandora/object/uuid%3Addd9fc49-b3dd-4e7f-b0a9-91a9caf2ff92 [Accessed: June 21, 2024]

[82] Torres, Molina LE, Torres CD. Feasibility of using pervious concrete in tropical climate. LACCEI [Internet]. Boca Raton, Florida: Florida Atlantic University; 2023;**1**(8):33431. Available from: http://3.134.240.183/index.php/laccei/article/view/3120 [Accessed: June 21, 2024]

[83] Bondareva NS, Sheremet MA. Heat transfer performance in a concrete block containing a phase change material for thermal comfort in buildings. Energy and Buildings. 2022;**256**:111715

[84] Khorat S, Das D, Khatun R, Aziz SM, Anand P, Khan A, et al. Cool roof strategies for urban thermal resilience to extreme heatwaves in tropical cities. Energy and Buildings. 2024;**302**:113751

[85] Masson V, Bonhomme M, Salagnac JL, Briottet X, Lemonsu A. Solar panels reduce both global warming and urban heat island. Frontiers in Environmental Science. 2014;**2**(Jun):81306

[86] Nasrollahi N, Ghobadi P. Field measurement and numerical investigation of natural cross-ventilation in high-rise buildings; Thermal comfort analysis. Applied Thermal Engineering. 2022;**211**:118500

[87] Ahmed T, Kumar P, Mottet L. Natural ventilation in warm climates: The challenges of thermal comfort, heatwave resilience and indoor air quality. Renewable and Sustainable Energy Reviews. 2021;**138**:110669

[88] Bandurski K, Bandurska H, Kazimierczak-Grygiel E, Koczyk H. The green structure for outdoor places in dry, hot regions and seasons—Providing human thermal comfort in sustainable cities. Energies [Internet]. 2020;**13**:2755. Available from: https://www.mdpi.com/1996-1073/13/11/2755/htm [Accessed: June 21, 2024]

[89] Ferrari A, Kubilay A, Derome D, Carmeliet J. The use of permeable and reflective pavements as a potential strategy for urban heat island mitigation. Urban Climate. 2020;**31**:100534

[90] MAA K, Hussin H, Ramli N, MFA G. Landslide analysis approaches in tropical environment region for disaster risk reduction. IOP Conference Series: Earth and Environmental Science [Internet]. 2022;**1102**(1):012025. Available from: https://iopscience.iop.org/article/10.1088/1755-1315/1102/1/012025

[91] Dille A, Dewitte O, Handwerger AL, d'Oreye N, Derauw D, Ganza Bamulezi G, et al. Acceleration of a large deep-seated tropical landslide due to urbanization feedbacks. Nature Geoscience. 2022;**15**(12):1048-1055

[92] Amarasinghe MP, SAS K, Robert DJ, Zhou A, HAG J. Risk assessment and management of rainfall-induced landslides in tropical regions: A review [Internet]. Natural Hazards. 2023;**120**(3):2179-2231. Available from: https://link.springer.com/article/10.1007/s11069-023-06277-3 [Accessed: June 21, 2024]

[93] Ruíz MA, Mack-Vergara YL. Resilient and sustainable housing models against climate change: A review. Sustainability [Internet]. 2023;**15**:13544. Available from: https://www.mdpi.

com/2071-1050/15/18/13544/htm [Accessed: May 10, 2024]

[94] Rocha JHA, Toledo Filho RD, Cayo-Chileno NG. Sustainable alternatives to CO_2 reduction in the cement industry: A short review. Materials Today: Proceedings. 2022;**57**:436-439

[95] Yang W, Wang L, Ma G, Feng P. An integrated method of topological optimization and path design for 3D concrete printing. Engineering Structures. 2023;**291**:116435

[96] Snellings R, Suraneni P, Skibsted J. Future and emerging supplementary cementitious materials. Cement and Concrete Research. 2023;**171**:107199

[97] John VM, Damineli BL, Quattrone M, Pileggi RG. Fillers in cementitious materials — Experience, recent advances and future potential. Cement and Concrete Research. 2018;**114**:65-78

[98] Everett B, Soto J, Bakhshi P, Pourmokhtarian A. Exploring 3D printing potentials for sustainable, resilient, and affordable housing. In: Creative Construction e-Conference 2022 [Internet]. Budapest, Hungary: Budapest University of Technology and Economics National Technical Information Center and Library; 2022. pp. 296-306. DOI: 10.3311/CCC2022-038. [Accessed: June 30, 2024]

[99] Sakin M, Kiroglu YC. 3D printing of buildings: Construction of the sustainable houses of the future by BIM. Energy Procedia. 2017;**134**:702-711

[100] United Nations Environment Programme. Eco-Housing. Guidelines for Tropical Regions. Bangkok, Thailand: United Nations Environment Programme; 2006

[101] Norris FH, Stevens SP, Pfefferbaum B, Wyche KF, Pfefferbaum RL. Community resilience as a metaphor, theory, set of capacities, and strategy for disaster readiness. American Journal of Community Psychology. 2008;**41**(1-2):127-150

Chapter 12

Techniques and Procedures That Reduce Energy Consumption and Minimize the Negative Environmental Impact of a Sustainable House

Guillermo Yorel Noriega Aquise
and Fredy Nicolás Molina Rodríguez

Abstract

This chapter explores the integration of various techniques and processes to achieve a sustainable house, focusing on reducing energy consumption and minimizing negative environmental impacts. Solutions such as in-house wastewater treatment, solid waste management, and the capture of solar thermal and photovoltaic energy are presented. These methods allow to produce recycled water, biogas for cooking, and electrical energy for lighting, all within the framework of a bioclimatic urban house. The implementation of these technologies in urban family homes not only generates energy to meet basic needs but also reduces the negative impacts of wastewater and solid waste, contributing to the city's environmental sustainability.

Keywords: wastewater, family biogas, domestic energy, solid waste, sustainable housing, family home

1. Introduction

The city of Arequipa, like many cities in Latin America, faces significant challenges due to rapid population growth and migration from rural areas to urban centers. This phenomenon has driven accelerated urban expansion, leading to the occupation of land without access to basic services such as potable water and electricity. Since 1984, the urban area of Arequipa has grown from 67.8 to 275.8 km^2 in 2020, reflecting not only the city's physical expansion but also the increasing pressure on public resources and services. Currently, Arequipa is home to 1,009,132 inhabitants and has 301,570 dwellings, yet it faces an overload in its capacity to provide essential services [1].

The rapid urbanization process has resulted in a reduction of green areas and exacerbated uncertainty about the quality of life in newly inhabited areas. This unplanned growth has also led to an increase in solid waste generation, with a notable rise from 4097.32 tons per day in 2001 to 8202.09 tons per day in 2014, and a projection of 16,053.80 tons per day by 2034 [2]. Additionally, sewage service coverage only reaches 69.65% of the urban population, and wastewater treatment remains inadequate [3].

Solid waste management continues to be a challenge; in 2019, the amount of solid waste generated was 262,831.75 T/year, with a daily municipal generation of 720.09 T/day and a per capita municipal generation of 0.64 kg/person/day [2, 4]. Sanitation Service Providers (EPS) only cover 69.65% of the urban population with sewage systems, and wastewater treatment is insufficient [3].

The National Building Regulations (RNE) in Peru regulate urban development and basic services, but current regulations have not been sufficient to effectively address the growing demand for resources and services [5]. In this context, the search for sustainable solutions for housing design and construction becomes crucial. Sustainability in housing is not just a trend, but an urgent necessity to mitigate climate change, conserve natural resources, and ensure a livable future.

This study focuses on urban family housing, exploring alternatives to address the critical situation of basic services in Arequipa. An integrated model is proposed that incorporates advanced technologies for wastewater management, solid waste management, and the use of renewable energy in bioclimatic housing. These solutions aim not only to improve the energy efficiency of homes but also to reduce dependency on external energy sources and decrease the ecological footprint.

The importance of this study lies in its integrated approach, which considers the integration of wastewater treatment systems, solid waste management, and the capture of both thermal and photovoltaic solar energy. By optimizing these systems, the model seeks to enhance the comfort of inhabitants and provide a replicable and adaptable example for different climatic and socioeconomic conditions. This approach allows for a more self-sufficient and environmentally responsible home, addressing the challenges of sustainable development in a comprehensive manner.

The purpose of this work is two-fold: first, to design a family housing system that is self-sufficient and sustainable; second, to evaluate its technical and economic viability. This system has the potential to significantly reduce potable water and energy consumption, efficiently manage waste, and at the same time provide an elevated level of comfort for the inhabitants. In a world increasingly aware of the limitations of natural resources, this approach represents a promising path to achieving sustainable development goals.

Compared to previous studies, this work stands out for its integrated approach and its adaptability to the local context of Latin America. While many studies focus on specific aspects such as energy efficiency or the use of sustainable materials, this project addresses sustainability more broadly, proposing solutions that combine the best available practices and technologies, adapting them to the specific needs of Arequipa. This comprehensive and multidisciplinary analysis seeks to fill a gap in the current literature, providing a model of sustainable housing that is not only theoretically sound but also practical and adaptable.

This study offers an innovative and practical solution for the design of sustainable housing in urban environments, promoting energy efficiency, responsible resource management, and reduced environmental impact. The conclusions and recommendations presented have the potential to guide future developments in the field of sustainable housing, both in Arequipa and in other cities across Latin America, effectively addressing the challenges of urban growth and sustainability.

2. Integration of techniques and processes in sustainable bioclimatic housing

This chapter describes an integrated system for a bioclimatic and sustainable house that optimizes the use of natural resources and renewable energies, while reducing the environmental impact through efficient water and waste management. The design approach allows the house to be self-sufficient and adaptable to local climatic conditions.

2.1 General description of the integrated system

The system uses a combination of techniques to efficiently manage water and energy resources and improve thermal comfort. The strategies adopted focus on maximizing the use of both thermal and photovoltaic solar energy, as well as the treatment and recycling of wastewater. Below, a flowchart (**Figure 1**) illustrates the articulation, synchronization, and interrelation of the processes involved in creating a sustainable home.

2.2 Inputs and main processes

The essential inputs identified for the system are potable water, food, products, and solar energy. These inputs feed various main processes, which are described as follows:

- *Wastewater treatment:* The system distinguishes between greywater and blackwater from the source. Greywater is channeled into compact subsurface artificial

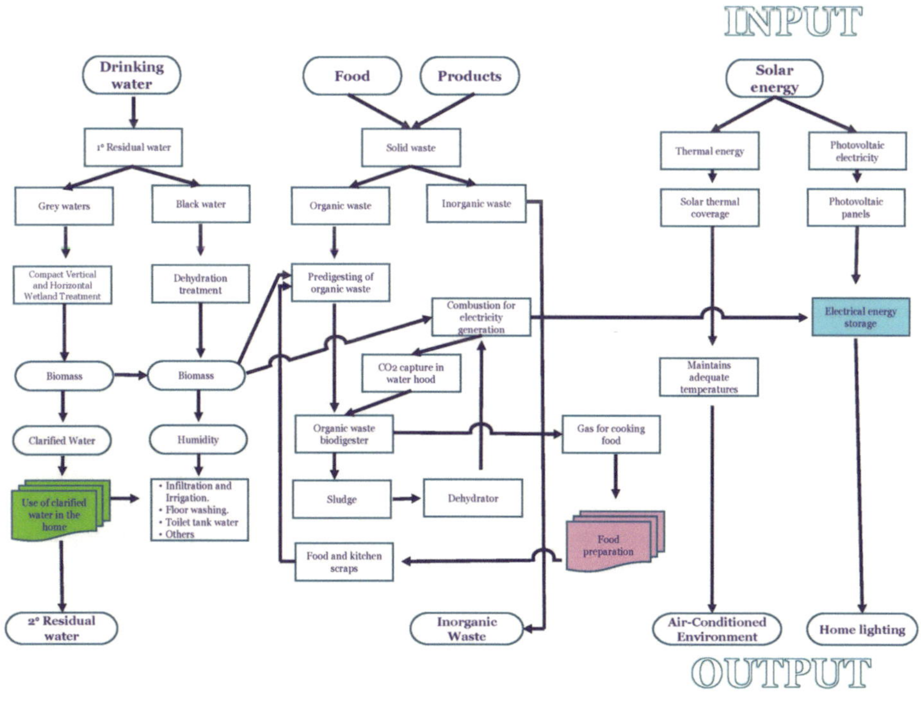

Figure 1.
Flowchart of the integrated sustainable housing system.

wetlands (horizontal and vertical), where it undergoes natural purification and clarification processes to be reused [6]. Blackwater is treated through solar dehydration, which significantly reduces the volume of waste, generating clarified water and dry biomass as by-products [7].

- *Solid waste management:* Solid waste is manually classified into organic and inorganic materials [8]. Organic waste undergoes pre-digestion and is later processed through anaerobic digestion in a compact biodigester, producing biogas and solid waste as by-products. Inorganic waste is responsibly managed to reduce its environmental impact [9].

- *Solar energy capture:* Both solar thermal and photovoltaic energy technologies are used. Solar thermal collectors and transparent covers increase ambient temperature and heat water for domestic use. Photovoltaic panels convert sunlight into electricity, which is stored in batteries and used for lighting and other electrical devices in the house [10].

2.3 Outputs and sustainability benefits

The system produces several outputs that contribute to the sustainability of the house:

- *Reused water:* The clarified water from the treatment of greywater and blackwater can be used for irrigation, washing, and toilet flushing, reducing the demand for potable water.

- *Biogas:* The biogas generated is primarily used for cooking. Any surplus can be used for electricity generation.

- *Renewable electricity:* The electricity generated by photovoltaic panels meets the lighting needs and other uses, reducing dependence on external energy sources.

2.4 Applied techniques and their impact

- *Compact artificial wetlands:* These systems use plants and microorganisms to naturally treat greywater, allowing its safe and efficient reuse.

- *Solar dehydration:* This technique harnesses solar energy to dry blackwater waste, producing dry biomass that can be used as fertilizer or for energy production.

- *Anaerobic digestion:* Anaerobic digestion in compact biodigesters allows the decomposition of organic waste in the absence of oxygen, generating biogas efficiently and sustainably.

- *CO_2 capture:* A system is implemented to capture the CO_2 generated during biomass combustion, using a circulating water hood. This not only reduces greenhouse gas emissions but also produces acidulated water that improves the efficiency of the biodigester.

- *Solar energy:* The combined use of thermal and photovoltaic solar energy maximizes the energy efficiency of the house, adapting to local climatic and economic conditions.

2.5 General benefits of the integrated approach

This integrated system promotes the self-sufficiency of the house in terms of water and energy, improves the thermal comfort of the inhabitants, and minimizes environmental impact by reducing dependence on external energy sources and improving waste management. Additionally, the adaptability of the design makes it suitable for various climatic and economic conditions, particularly in Latin America.

The implementation of these techniques and processes allows for more efficient use of natural resources, promoting a more ecological and environmentally conscious lifestyle. The integral approach not only enhances the sustainability of the house but also offers a replicable model for future sustainable constructions in different contexts.

3. Techniques and procedures to achieve a sustainable home

The integral approach to achieving sustainable homes involves the implementation of specific techniques and processes that optimize the use of natural resources and minimize environmental impact. A bioclimatic and sustainable house is defined by its ability to meet human needs for comfort and functionality while reducing dependence on conventional energy services [11]. Below, the necessary strategies to achieve these objectives are detailed.

3.1 Home design for energy optimization

The design of a sustainable house must consider architectural and construction elements that maximize the use of natural resources to regulate temperature and improve energy efficiency [12]. The basic principles include:

- *Home orientation:* Correct orientation takes advantage of sunlight during the winter and reduces thermal gain in summer, improving energy efficiency [13].

- *Solar utilization:* In temperate climates, the design should incorporate elements that optimize solar capture to cover a sizable portion of energy needs [14].

- *Thermal insulation:* The use of materials with good thermal insulation is crucial to maintain a stable indoor temperature, reducing the need for heating and cooling systems [15].

- *Natural ventilation:* The design should facilitate natural ventilation through strategically located windows and openings, improving indoor comfort without increasing energy consumption [16].

These techniques not only reduce energy consumption but also contribute to creating a healthier and more comfortable environment for inhabitants.

3.2 In-house wastewater treatment

In-house wastewater treatment is essential to reduce the demand for potable water and minimize environmental impact. This process includes the separation of greywater and blackwater for proper treatment [17], which includes:

- *Compact artificial wetlands:* Greywater is directed to horizontal and vertical subsurface artificial wetlands. These systems use microorganisms and plants, such as papyrus, to purify the water through natural biological processes [18, 19].

- *Solar dehydration of blackwater:* Blackwater undergoes a solar dehydration process that reduces the volume of waste and facilitates its management by converting it into dry biomass [20, 21].

In a temperate climate, with an average temperature of 22°C, wetland systems can operate with a treatment efficiency of 80–85%, depending on factors such as available space and specific treatment needs [22].

Procedure for greywater treatment:

- *Collection:* Capture of wastewater from showers, sinks, and washing machines.

- *Filtration:* Removal of large particles and impurities through sand and gravel filters.

- *Biological treatment:* Use of compact artificial wetlands that promote the natural purification of water.

- *Disinfection:* Application of methods such as chlorination, ozone, or ultraviolet light to eliminate remaining pathogens.

- *Reuse:* Use of treated water for irrigation, toilet flushing, and general cleaning.

These procedures ensure a significant reduction in potable water consumption and relieve the burden on public sewage systems.

3.3 Solid waste management

Efficient solid waste management is essential to minimize environmental impact and maximize resource recovery:

- *Source separation:* Initial classification of waste into organic, recyclable, and non-recyclable materials [23].

- *Pre-digestion chamber:* Organic waste is processed in a pre-digestion chamber before being introduced into an anaerobic digester, optimizing biogas production [24].

- *Anaerobic digestion:* A process that converts organic waste into biogas, used for cooking [25].

- *Recycling and final disposal*: Recyclable waste is sent to reprocessing centers, and non-recyclable waste is disposed of in controlled landfills.

Benefits of solid waste management:

- Reduction of waste sent to landfills.
- Production of biogas for domestic use.
- Generation of compost for gardening and urban agriculture.
- Conservation of natural resources through recycling.

3.4 Solar thermal and photovoltaic energy capture

Solar energy capture is crucial to reducing dependence on fossil fuels and lowering greenhouse gas emissions:

- *Solar thermal energy:* Installation of solar collectors on the roof and transparent covers to heat water and maintain a comfortable indoor temperature [26, 27].
- *Photovoltaic energy:* Installation of solar panels on the roof, inverters to convert direct current to alternating current, and batteries to store excess energy [10].

Procedure for solar energy capture:

- *Installation of collectors and panels:* Placement on the roof to maximize solar exposure.
- *Storage:* Use of tanks for hot water and batteries for electricity [28].
- *Distribution:* Connection to the domestic network to provide hot water and electricity.

Benefits of solar capture:

- Reduction of grid electricity consumption.
- Production of clean and renewable energy.
- Long-term savings on energy costs.

3.5 Renewable resource production

The integrated system also facilitates the production of renewable resources, such as recycled water and biogas:

3.5.1 Recycled water

Treated water is reused for non-potable applications, such as irrigation and toilet flushing, contributing to the conservation of water resources [29].

- *Collection and treatment:* Follow the procedure for in-house wastewater treatment.

- *Distribution:* Channel recycled water to irrigation systems and toilet cisterns.

Benefits:

- Reduction in potable water consumption.
- Conservation of water resources.

3.5.2 Biogas for cooking

Biogas, produced through the decomposition of organic waste in biodigesters, provides a renewable energy source for cooking:

- *Organic waste collection:* Collection of food scraps and biodegradable waste.
- *Digestion in biodigesters:* Controlled anaerobic decomposition process that generates biogas [9].

Benefits of biogas production:

- Reduction in dependence on fossil fuels.
- Efficient management of organic waste.
- Provision of a renewable energy source.

For optimal performance, it is essential to continuously monitor the digester temperature and adjust the heating or insulation system as necessary to maintain ideal conditions for biogas production. This will ensure efficient and stable biogas production, contributing to the energy sustainability of the bioclimatic home [30].

3.6 Implementation and maintenance

Proper implementation and maintenance of these techniques are essential to ensure long-term efficiency and sustainability:

- *Integrated design*: Consideration of sustainable technologies from the housing planning phase.
- *Professional installation*: Involvement of experts for the installation of solar and waste treatment systems.
- *Training*: Educating inhabitants on the use and maintenance of systems.
- *Regular maintenance*: Periodic inspections and preventive maintenance to ensure the optimal functioning of systems [31].

3.7 Application of techniques and procedures

The application of these integrated techniques and procedures not only ensures the sustainability of the home but also creates a more comfortable and environmentally friendly living environment. By adopting these strategies, homes can significantly reduce their environmental impact and improve the quality of life of their occupants.

4. Methodology

4.1 Sequential procedure

The development of sustainable family homes requires a meticulous approach that considers the infrastructure and integrated energy and resource management systems. Below, the methodological steps adopted for the design of these homes are described, based on the criteria of Peru's National Building Regulations (RNE).

4.1.1 Housing infrastructure design

For this study, six types of domestic family homes were defined, all classified as Type 3 according to the RNE. These homes are located on 160 m^2 land areas, with a constructed area on the first level ranging from 25.83 to 59.85 m^2, and a second level equal to or larger, in compliance with the RNE regulations. The design of these homes follows bioclimatic principles, taking advantage of solar energy capture for natural lighting and passive heating, aligning with the guidelines for sustainable housing design [13].

The implementation of sustainable homes through bioclimatic design and solar energy capture aligns with practices recommended in the literature. Duffie and Beckman [14], for example, emphasize the importance of maximizing solar energy capture to reduce dependence on external energy sources. This approach also involves the proper orientation of the home and the use of construction materials that provide good thermal insulation [15].

4.1.2 Piping connections for wastewater separation

The house design includes a piping system that separates greywater from blackwater, following the parameters specified in the RNE. This approach allows greywater to be treated through compact horizontal and vertical subsurface wetlands, which purify the water for reuse and generate biomass for biogas production [19]. The separation of the waters and their proper treatment not only reduces the load on sewage systems but also promotes sustainability by reusing treated water and generating useful biomass [19].

4.1.3 Design of the domestic biogas production process

A compact biodigester optimized for biogas production was designed, primarily for cooking purposes. This design considers the appropriate composition of organic waste, kitchen waste, and CO_2, and is optimized through a kinetic-stoichiometric analysis to maximize biogas production in the shortest possible time. Factors such as the available carbon composition, the granularity of the waste, and the effects of anaerobic pre-digestion are carefully analyzed to improve the process's efficiency [32].

4.1.4 Electricity generation from biomass and biogas

A process for electricity generation combining biomass combustion with biogas produced by the biodigester was established. This system is designed to be scalable to meet the home's energy needs. The electricity generated is stored in the photovoltaic solar system's batteries, ensuring a continuous supply even when biogas production exceeds the home's immediate needs [27].

The integration of photovoltaic solar energy with other electricity generation methods, such as biomass combustion, optimizes the energy efficiency of the house. Kalogirou [27] highlights that this combination can meet a sizable portion of energy needs, reducing both costs and greenhouse gas emissions.

4.1.5 Synchronization and articulation of processes

It is essential that all processes are efficiently synchronized and articulated to maximize sustainability and minimize costs. This involves continuously recording and monitoring the costs associated with each process to optimize resource management.

4.2 Comparative method

The comparative method is fundamental for evaluating the effectiveness of the different techniques and procedures implemented in the design of sustainable homes. This method, widely recognized in scientific research, allows empirical generalization and hypothesis verification by comparing results obtained under different conditions [33].

4.2.1 Application of the comparative method

- *Comparison of water treatment technologies*: Evaluation of the efficiency of different artificial wetland systems and dehydration methods for the treatment of greywater and blackwater.

- *Evaluation of biogas production systems*: Comparison of biogas production in diverse types of biodigesters and under various operating conditions.

- *Energy efficiency analysis*: Comparison of the performance of solar thermal and photovoltaic systems in generating energy for homes.

- *Environmental impact and sustainability*: Evaluation of the environmental impact of different processes and techniques and their contribution to the home's sustainability.

The use of the comparative method in this research provides a rigorous framework for evaluating the processes and techniques implemented, ensuring that the conclusions drawn are robust and based on empirical evidence [34].

4.2.2 Perspectives and benefits of the comparative method

The comparative method allows the development of best practice models that can be replicated in other sustainable housing projects. It also fosters innovation by identifying areas where current technologies can be improved and facilitates the adaptation of successful solutions to different climatic and socioeconomic contexts [35].

Evaluating environmental impact by comparing different systems and technologies allows the selection of the most sustainable options, minimizing the ecological footprint of homes. The results obtained can also inform the formulation of public policies and regulations in the field of sustainable construction, ensuring that they are evidence-based and promote efficient and sustainable practices [36].

In conclusion, the methodology adopted combines a bioclimatic design approach with a rigorous comparative analysis to evaluate and optimize the sustainability of homes. This comprehensive approach ensures not only resource efficiency but also the long-term viability of sustainable practices in home construction.

5. Application of techniques and procedures in housing

5.1 Study units

In this study, the unit of analysis is the infrastructure of domestic family housing, defined according to the parameters of the National Building Regulations (RNE) of Peru. For this research, Type 3 housing has been selected, designed for a plot area of 160 m^2 [5]. Below are the characteristics of the six homes analyzed, detailed in terms of dimensions, total built area over two levels, window, and door area, as well as the number of inhabitants, bedrooms, and bathrooms (**Table 1**):

These six study units reflect a variety of housing configurations designed to meet the needs of different family sizes in urban and developing urbanized areas. Each home is equipped with systems for wastewater recycling and solid waste management, aimed at generating energy for cooking and electric lighting. The selection of these homes responds to the need to evaluate their performance in terms of energy efficiency and sustainability within the urban context.

The designs of these homes are detailed in the following diagrams, illustrating the layouts of the first and second levels for each type of home (**Figures 2–7**):

Each of these designs considers aspects of energy efficiency and sustainability, seeking to maximize the comfort of the inhabitants while minimizing operational costs and environmental impact.

House	Length (m)	Width (m)	Total area Two levels (m^2)	Window and door area (m^2)	Inhabitants	Bedrooms	Bathrooms
VUF-01	8.48	4.56	77.34	8.22	4	2	1
VUF-02	8.48	6.38	108.20	13.5	6–8	4	2
VUF-03	9.4	6.38	119.94	9.9	6–8	4	2
VUF-04	8.93	7.68	137.16	9.9	6–8	4	2
VUF-05	9.18	8.48	155.69	9.9	8	4	2
VUF-06	10.58	8.48	179.44	12.1	8	4	2

Table 1.
Areas and codes of prefabricated family housing.

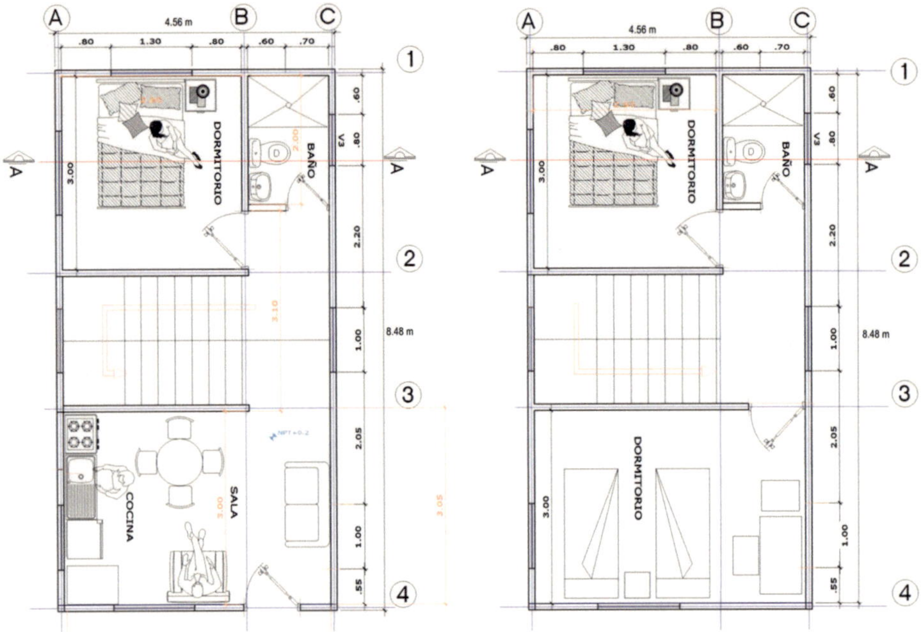

Figure 2.
VUF-01, first and second levels, 77.34 m².

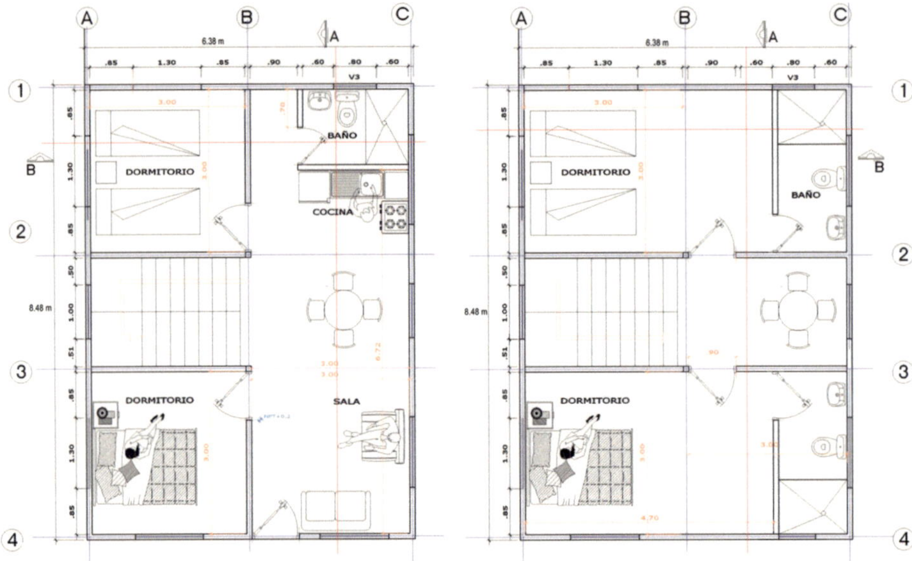

Figure 3.
VUF-02, first and second levels, 108.20 m².

5.2 Resource needs

To ensure the proper functioning of the family homes studied (VUF-01, VUF-02, VUF-03, VUF-04, VUF-05, and VUF-06), the resource needs in terms of potable water, wastewater disposal, gas consumption for cooking, and the amount of

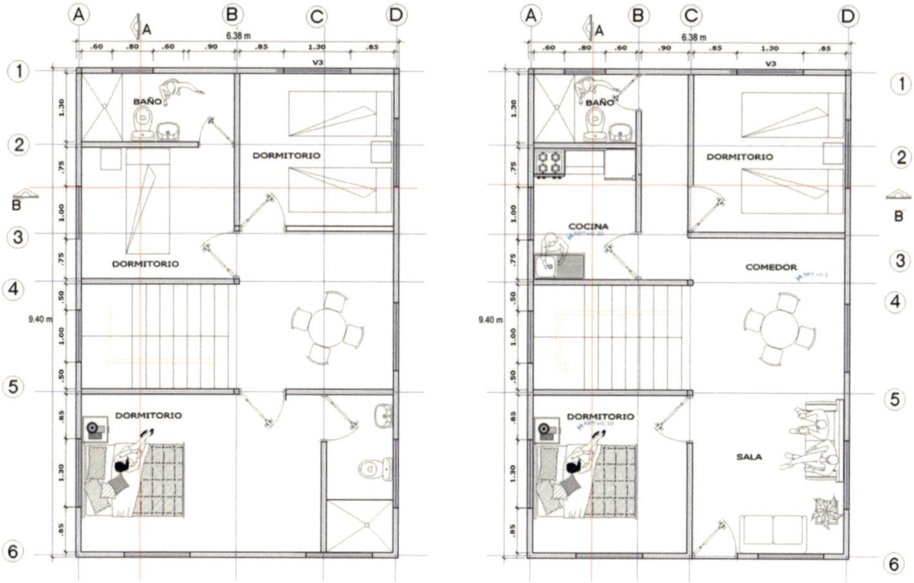

Figure 4.
VUF-03, first and second levels, 119.94 m².

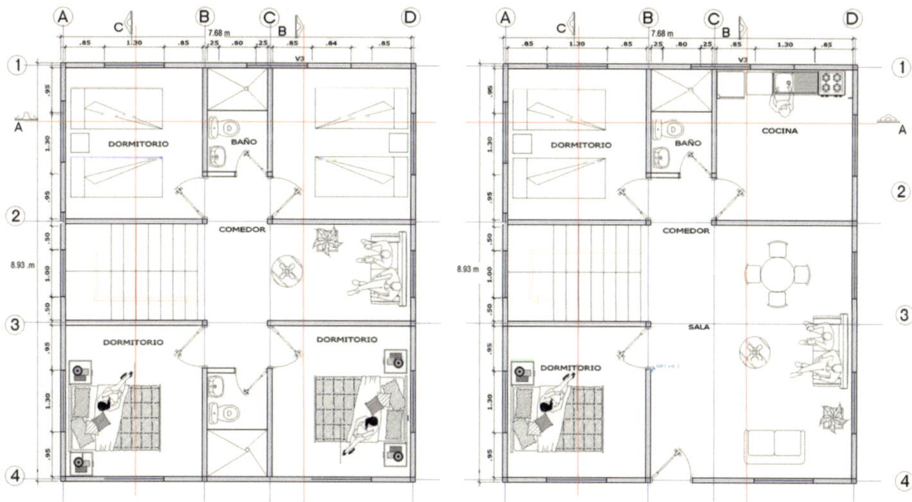

Figure 5.
VUF-04, first and second levels, 137.16 m².

electricity required have been identified. These needs have been calculated based on the number of inhabitants and the characteristics of each home.

5.2.1 Potable water consumption and wastewater

The World Health Organization (WHO) recommends a minimum consumption of 50–100 L of water per person per day to meet basic needs. However, in developed urban areas, this consumption can increase to 150–200 L per person per day [37].

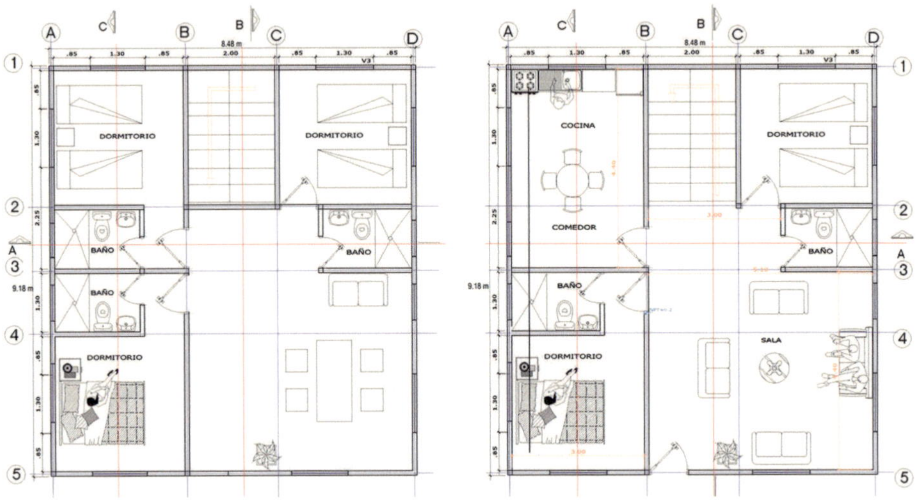

Figure 6.
VUF-05, first and second levels, 155.69 m^2.

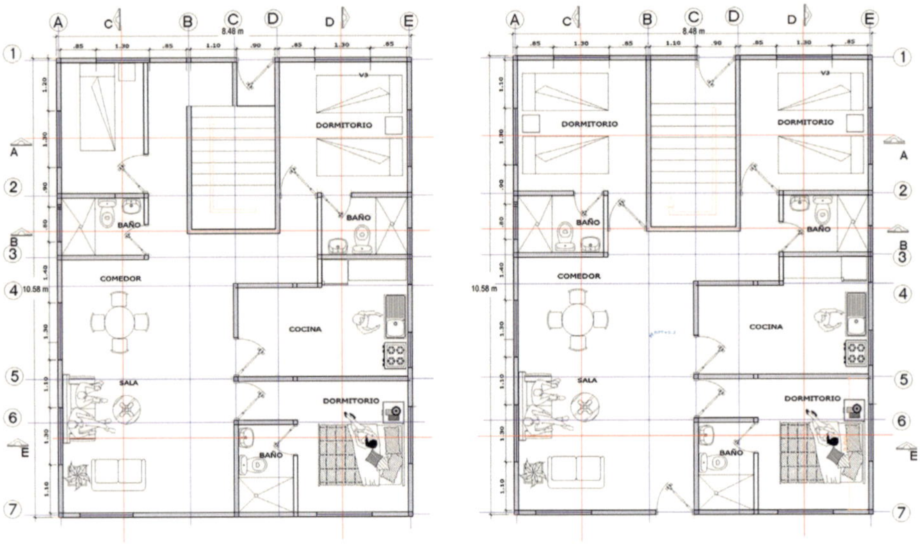

Figure 7.
VUF-06, first and second levels, 179.44 m^2.

For this study, a daily consumption of 150 L per person has been adopted. It is assumed that wastewater discharge is equivalent to potable water consumption, according to the National Building Regulations (RNE) of Peru, which considers an 80% efficiency in wastewater disposal [5].

5.2.2 Gas consumption and biogas production

According to the *2020 National Household Survey (ENAHO)* conducted by the *National Institute of Statistics and Informatics (INEI), LPG* is the main fuel used in Peruvian households for cooking, being used by approximately *80% of households in*

Peru [38]. The regions of *Arequipa (93.7%)*, *Tacna (92%)*, and *Tumbes (90.1%)* report the highest levels of usage [39].

The *Ibero-American Association of Liquefied Petroleum Gas (AIGLP)*, in its study on the *bottled LPG market in Latin America*, determined that the monthly LPG consumption per family in Peru is approximately *10 kg of bottled LPG*, which is consistent with the study conducted by *OSINERGMIN-GPAE* [39]. At the national level, the average monthly consumption of LPG is *1.05 cylinders of 10 kg per household*, with slightly higher consumption in urban areas (*1.06 cylinders*) [39]. Additionally, the average number of household members is *3.2* [38], resulting in an average LPG consumption of *3.31 per person per month*. However, for the calculations in this study, an average consumption of *5 kg of LPG per person per month* has been estimated.

In this study, it is considered that *1 kg of LPG* is approximately equivalent to *1.25 m^3 of biogas*. This equivalence is valid when the biogas has undergone *purification processes* that increase its methane concentration to levels comparable to *natural gas*. In its raw form, biogas typically contains between *50% and 70% methane*, resulting in a lower calorific value [40, 41]. However, through purification processes such as *pressure swing adsorption (PSA)* [42], the use of *semipermeable membranes* [40], and the removal of *carbon dioxide (CO_2)* and *hydrogen sulfide (H_2S)* [40, 42], biogas can reach methane concentrations of *85% or more* [41, 43]. In this context, purified biogas acquires a higher calorific value, similar to that of natural gas, allowing *1.25 m^3 of purified biogas* to provide energy equivalence close to *1 kg of LPG* [40]. In its natural form, unpurified biogas (*1 m^3*) has an energy equivalent of *2.10 kg of LPG* [44].

5.2.3 Electricity consumption and photovoltaic panels

Monthly electricity consumption per person is estimated at 100 kWh, according to IEA data. The average output of a standard 300 Wp (Watt peak) solar panel, with an average of 5 peak sun hours per day, determines the number of photovoltaic panels required to meet the energy needs of each home, ensuring a continuous and sustainable electricity supply.

The resource needs for each home, calculated based on the number of inhabitants and the specific characteristics of each home, are presented in **Table 2**.

This analysis allows for the precise estimation of the resources needed for the operation of the family homes studied. By identifying the specific needs of each home in terms of water, gas, biogas, and electricity, personalized and sustainable solutions can be implemented to improve resource efficiency and sustainability in urban areas.

5.3 Relationship between housing area and number of inhabitants

The relationship between the built area in m^2 and the number of inhabitants can be modeled using polynomial regression. The regression equation is $y = -0.0005x^2 + 0.1612x - 5.3476$, with a coefficient of determination $R^2 = 0.9274$. This R^2 value indicates a good fit of the model to the observed data, explaining approximately 92.74% of the variability in the number of inhabitants based on the housing area (**Figure 8**).

The regression curve is parabolic, with a downward concavity, suggesting that as the housing area increases, the number of inhabitants also increases until reaching a maximum point, after which it begins to decrease. This behavior reflects the existence

House	Total area (m^2)	Inhabitants (couple + children)	Water consumption (L/day)	Wastewater (L/day)	Gas (kg/month)	Biogas (m^3/month)	Electricity (kWh/month)	Photovoltaic panels required
VUF-01	77.34	4	600	600	20	25	400	9
VUF-02	108.2	7	1050	1050	35	43.75	700	16
VUF-03	119.94	7	1050	1050	35	43.75	700	16
VUF-04	137.16	7	1050	1050	35	43.75	700	16
VUF-05	155.69	8	1200	1200	40	50	800	18
VUF-06	179.44	8	1200	1200	40	50	800	18

Table 2.
Total area of homes in relation to number of inhabitants.

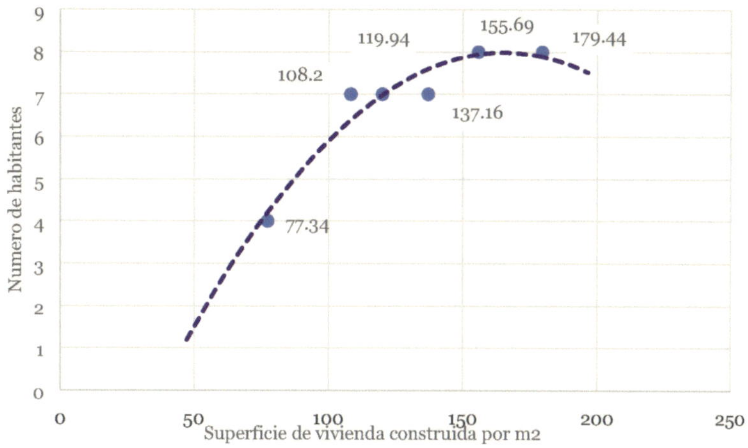

Figure 8.
Total housing area in relation to number of inhabitants.

of an optimal housing size in terms of built area that maximizes the capacity to accommodate inhabitants.

The maximum number of inhabitants, according to the regression equation, is found at a housing area of approximately 161.2 m^2. To determine this maximum point, the regression equation is derived and set to zero, confirming that the optimal housing size to maximize the number of inhabitants is around 161.2 m^2.

The graph shows how the observed data points (77.34, 108.20, 119.94, 137.16, 155.69, 179.44 m^2) align with the regression curve, suggesting a significant correlation between housing area and the capacity to accommodate inhabitants. The results indicate that:

- *Initial growth*: As the housing area increases, the number of inhabitants also rises. This initial growth can be attributed to the greater ability of larger homes to comfortably accommodate more people.

- *Maximum point*: The number of inhabitants reaches its highest value around an area of 161.2 m^2. This point indicates optimal efficiency in space usage, where the home maximizes its capacity to accommodate inhabitants without compromising comfort.

- *Subsequent decrease*: After the maximum point, the number of inhabitants begins to decrease. This phenomenon could be explained by factors such as a reduced need for additional space for the same family units or a preference for larger homes with fewer inhabitants for greater comfort.

The trend observed in the graph is consistent with findings from studies on housing design and space usage efficiency. Vidales et al. [11], in their study on bioclimatic and sustainable housing design models, concluded that there is an optimal housing size that maximizes both resource efficiency and occupant comfort. Exceeding this size can result in underutilized space and lower efficiency. Rojas García [12] also discusses how housing design and proper orientation can significantly influence thermal comfort and energy efficiency. Excessively large homes may not be as efficient in terms of energy consumption and space usage.

5.3.1 Optimization of space and sustainability

The graph suggests that designing homes around the optimal size of 161.2 m^2 maximizes the capacity to accommodate inhabitants efficiently and sustainably. Beyond this size, efficiency and sustainability tend to decrease, potentially leading to less effective use of space and energy resources.

This analysis highlights the importance of considering an optimal housing size to maximize efficiency in space and resource usage, contributing not only to improved capacity but also to the sustainability and comfort of the homes.

5.4 Determination of biogas needs

To calculate the biogas needs of each home, the energy equivalence between liquefied petroleum gas (LPG) and biogas is used, where 1 kg of LPG is equivalent to 1.25 m^3 of biogas. From the previously calculated LPG needs, these are converted into biogas requirements to meet the energy demands of the homes (**Table 3**).

The graph below illustrates the relationship between the constructed housing area (in m^2) and two key variables: LPG consumption (kg/month) and biogas production (m^3/month). Both relationships are represented by polynomial regressions, whose equations and determination coefficients (R^2) indicate the accuracy of the model fit to the observed data (**Figure 9**).

Analysis of the relationship between housing area, LPG consumption, and biogas production:

- *LPG consumption (kg/month):* The polynomial regression curve for LPG consumption is represented in blue and is described by the equation: $y = -0.0024x^2 + 0.806x - 26.738$, with a determination coefficient $R^2 = 0.9274$. This curve indicates that LPG consumption increases as the housing area grows, reaching a maximum around 168 m^2, after which it begins to decrease. This point of maximum LPG efficiency is calculated by deriving the regression equation and setting it to zero: $dy/dx = -0.0048x + 0.806 = 0$, where $\Longrightarrow x \approx 168$.

- *Biogas production (m^3/month):* The polynomial regression curve for biogas production is represented in green and is described by the equation: $y = -0.003x^2 + 1.0075x - 33.422$, with a determination coefficient $R^2 = 0.9274$. Similar to the LPG curve, this curve shows an increase in biogas production as the

House	Total area (m^2)	Inhabitants (couple + children)	LPG (kg/month)	Biogas (m^3/month)
VUF-01	77.34	4	20	25
VUF-02	108.2	7	35	43.75
VUF-03	119.94	7	35	43.75
VUF-04	137.16	7	35	43.75
VUF-05	155.69	8	40	50
VUF-06	179.44	8	40	50

Table 3.
Total area of homes in relation to LPG and biogas needs.

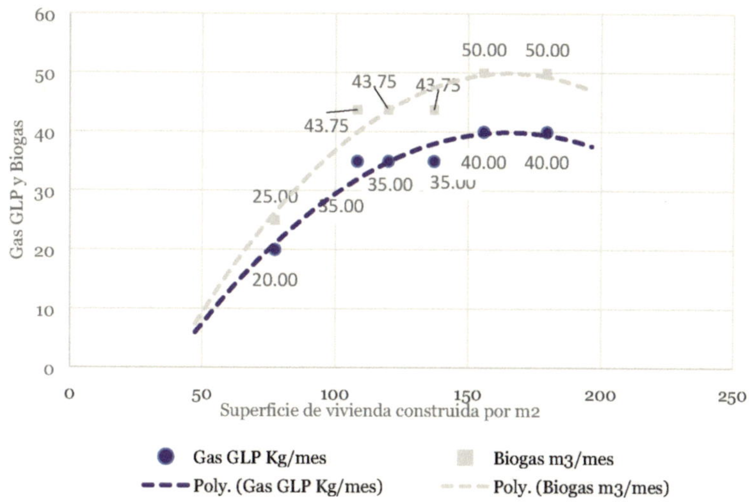

Figure 9.
Total housing area in relation to LPG and biogas.

housing area increases, peaking around 168 m^2. After this point, biogas production begins to decrease. The maximum point is determined similarly: $dy/dx = -0.006x + 1.0075 = 0$, where $\Longrightarrow x \approx 168$.

Both curves indicate that there is an optimal housing area of approximately 168 m^2, where both LPG consumption and biogas production reach their maximum values. Beyond this size, energy efficiency and biogas production decrease, due to resource distribution efficiency factors and available technology.

Interpretation of the curves:

- *Initial growth:* Both curves show an initial increase in LPG consumption and biogas production as the housing area grows. This increase can be explained by the greater capacity of larger homes to integrate more energy generation and consumption technologies efficiently.

- *Maximum point:* Both LPG consumption and biogas production reach their peak around a housing area of 168 m^2. This point represents maximum efficiency in terms of space and energy, where energy generation and consumption capacity are balanced most effectively.

- *Subsequent decrease:* After the maximum point, both curves show a decline in LPG consumption and biogas production. This suggests that beyond a certain housing size, efficiency in space usage and energy production does not increase proportionally with the area.

The observed relationship between housing area and energy consumption/production is consistent with previous studies on energy efficiency in sustainable housing. The FAO "Biogas Handbook" [30] emphasizes the importance of scale and design of biodigesters for efficient biogas production. Vidales et al. [11] found an optimal housing size that maximizes both resource efficiency and comfort. Exceeding this size can lead to decreased efficiency due to disproportionate use of space and energy resources. Rojas García [12] discusses how housing design and size can significantly influence energy efficiency.

Implications for sustainable housing design:

Designing homes around the identified optimal size of 168 m^2 can maximize energy efficiency, both in terms of LPG consumption and biogas production. Staying within this optimal range not only improves energy efficiency but also enhances the sustainability of the homes, reducing dependence on external energy sources and optimizing the use of renewable resources.

The analysis of the graph and its comparison with previous studies emphasize the importance of considering an optimal housing size to maximize energy efficiency and resource usage, contributing to the sustainability and comfort of homes.

5.5 Determination of electricity needs

To determine the electricity needs of prefabricated family homes (VUF) using photovoltaic panels, the number of required panels and the type of system needed to meet the energy demands of each home must first be calculated. Below is the methodology for calculating the number of solar panels required, compared with the previously analyzed variables (**Table 4**).

Reference values for solar panels:

- *Average monthly electricity consumption per person:* 100 kWh.
- *Average production of a standard solar panel:* 300 Wp (Watt peak) per panel.
- *Average peak sun hours per day:* 5 hours.

The trend of the polynomial regression curve in the graph shows the relationship between the constructed housing area (in m^2) and electricity consumption in kWh/month (**Figure 10**).

The graph illustrates the relationship between the constructed housing area (in m^2) and electricity consumption in kWh per month. The polynomial regression equation is: $y = -0.0487x^2 + 16.12x - 534.76$, with a determination coefficient $R^2 = 0.9274$. This equation suggests a non-linear relationship where electricity consumption

House	Total area (m^2)	Inhabitants (couple + children)	Bedrooms	Bathrooms	Electricity (kWh/month)	Photovoltaic panels required
VUF-01	38.48	4	2	1	400	9
VUF-02	76.96	7	4	2	700	16
VUF-03	78.57	7	4	2	700	16
VUF-04	74.35	7	4	2	700	16
VUF-05	74.7	8	4	2	800	18
VUF-06	89.4	8	4	2	800	18

Table 4.
Total housing area in relation to electricity and photovoltaic panels.

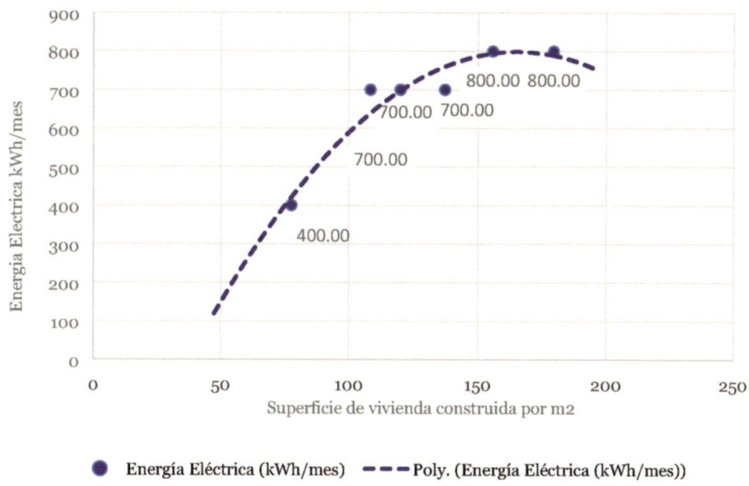

Figure 10.
Total housing area in relation to electricity.

increases with housing area until reaching a maximum point, after which it begins to decrease.

Electricity consumption reaches its maximum at some point within the data range. This maximum point can be calculated by deriving the regression equation and setting it to zero: $dy/dx = -0.0974x + 16.12 = 0$, where $\Longrightarrow x = 165.5$.

Therefore, the housing area of approximately 165.5 m^2 is where maximum electricity consumption is reached according to this regression.

The determination coefficient R^2 indicates that approximately 92.74% of the variability in electricity consumption can be explained by the housing area, suggesting a good fit of the polynomial model to the observed data. The graph shows observed data points and their correspondence with the regression curve. The specific points (400, 700, 700, 700, 800, 800) indicate constructed areas and the associated electricity consumption in each case.

The trend shows that there is an optimal housing size in terms of constructed area that maximizes electricity consumption. This size is approximately 165.5 m^2.

Interpretation of the trend curve:

- *Initial growth:* As the housing area increases, electricity consumption also rises. This initial growth can be attributed to the greater capacity of larger homes to accommodate appliances and lighting systems, resulting in higher energy consumption.

- *Maximum point:* Electricity consumption reaches its peak around a housing area of 165.5 m^2. This point represents optimal efficiency in terms of space and energy consumption, where the home makes the most of its installed capacity for electricity consumption.

- *Subsequent decrease:* After the maximum point, electricity consumption begins to decrease. This may be due to greater efficiency in space usage in larger homes, where the additional increase in area does not correspond to a proportional increase in energy consumption.

The observed relationship in the graph between housing area and electricity consumption is consistent with studies on energy efficiency in sustainable housing. According to Vidales et al. [11], in their study on bioclimatic and sustainable housing design models, they found that there is an optimal housing size that maximizes resource efficiency and occupant comfort. Exceeding this size may result in decreased efficiency due to disproportionate use of space and energy resources. Rojas García [12] also discusses how housing design, including orientation and size, can significantly influence thermal comfort and energy efficiency. Large homes may not be as efficient in terms of energy consumption, which aligns with the observation that electricity consumption decreases after a certain housing size.

The International Energy Agency (IEA) [45], in its studies on residential energy consumption, suggests that energy efficiency improves with housing size up to an optimal point, after which additional consumption does not result in proportional benefits in terms of comfort or functionality.

Implications for sustainable housing design:

Designing homes around the identified optimal size (165.5 m^2) can maximize energy efficiency, both in terms of electricity consumption and occupant comfort. Staying within this optimal range not only improves energy efficiency but also enhances the sustainability of the homes. This is crucial for reducing dependence on external energy sources and improving the use of renewable resources.

The analysis of the graph and its comparison with previous studies emphasize the importance of considering an optimal housing size to maximize energy efficiency and resource usage. This approach not only contributes to the sustainability of the homes but also improves comfort and reduces dependence on non-renewable sources.

5.6 Determination of photovoltaic panel needs

The attached graph shows the relationship between the constructed housing area (in m^2) and the number of photovoltaic panels required. The polynomial regression equation modeling this relationship is: $y = -0.0012x^2 + 0.3857x - 13.225$, with a determination coefficient $R^2 = 0.9271$. This equation suggests a non-linear relationship, where the number of photovoltaic panels increases with housing area until it reaches a maximum point, after which it begins to decrease (**Figure 11**).

The curve presents a parabolic shape with a downward concavity, indicating that as the housing area increases, the number of photovoltaic panels also increases to a certain maximum point, after which it starts to decrease.

The maximum number of photovoltaic panels required is reached at a point within the data range. This maximum point can be calculated by deriving the regression equation and setting it to zero: $dy/dx = -0.0024x + 0.3857 = 0$, where $\Longrightarrow x = 160.7$. Therefore, a housing area of approximately 160.7 m^2 is where the maximum number of photovoltaic panels is reached according to this regression.

The determination coefficient R^2 of 0.9271 indicates that approximately 92.71% of the variability in the number of photovoltaic panels can be explained by the housing area, suggesting a good fit of the polynomial model to the observed data. The graph shows observed data points and their correspondence with the regression curve. The specific points (9, 16, 16, 16, 18, 18) indicate constructed areas and the associated number of photovoltaic panels for each case.

The trend indicates that there is an optimal housing size in terms of constructed area that maximizes the number of photovoltaic panels required. This size is approximately 160.7 m^2. Beyond this size, the number of photovoltaic panels tends

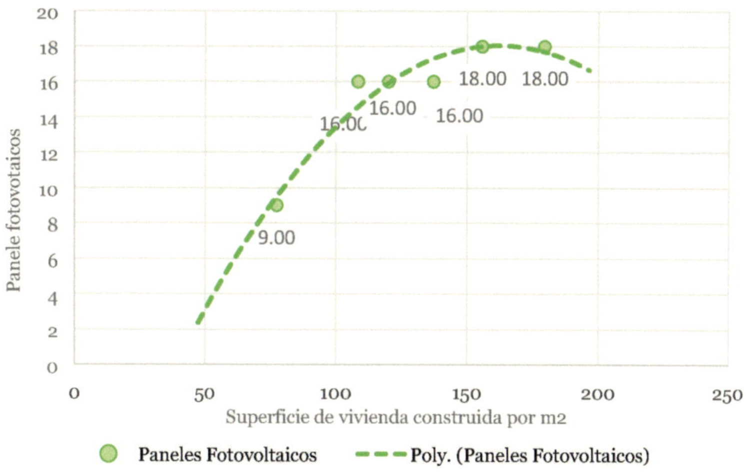

Figure 11.
Total housing area in relation to photovoltaic panels.

to decrease, which may be due to a range of factors such as space limitations for panel installation or energy efficiency considerations.

Interpretation of the curve:

- *Initial growth:* As the housing area increases, the number of photovoltaic panels also rises. This initial growth can be attributed to the greater capacity of larger homes to install more solar panels and, therefore, generate more renewable energy.

- *Maximum point:* The number of photovoltaic panels reaches its peak around a housing area of 160.7 m^2. This point represents optimal efficiency in terms of space and solar energy generation capacity, where the home maximizes its capacity to install photovoltaic panels.

- *Subsequent decrease:* After the maximum point, the number of photovoltaic panels begins to decrease. This could be due to a reduced need for additional space for solar panels in larger homes, where the increase in area does not correspond to a greater capacity for energy generation.

The observed relationship between housing area and the number of photovoltaic panels is consistent with studies on the efficiency of solar energy generation in sustainable housing. Lamigueiro [10], in his study on photovoltaic solar energy, highlights the importance of housing size and orientation to maximize efficiency in solar energy generation. This study suggests that there is an optimal size for solar panel installation, beyond which the generation capacity does not increase proportionally. Nandwani [26] also establishes how housing design, including orientation and size, can significantly influence energy efficiency and solar generation capacity. Large homes may not be efficient in terms of space for solar panel installation, which aligns with the observation that the number of photovoltaic panels decreases after a certain housing size.

Implications for sustainable housing design:

To optimize space and energy, designing homes around the identified optimal size (160.7 m^2) can maximize efficiency in photovoltaic panel installation and solar energy generation. Staying within this optimal range not only improves solar energy generation efficiency but also enhances the sustainability of the homes. This is crucial for reducing dependence on external energy sources and improving the use of renewable resources.

The analysis of the graph and its comparison with previous studies emphasize the importance of considering an optimal housing size to maximize efficiency in photovoltaic panel installation and solar energy generation. This approach not only contributes to the sustainability of the homes but also improves efficiency and reduces dependence on non-renewable sources.

5.7 Relationship between constructed housing area and the volume of wastewater, greywater, and blackwater

To determine the volume of potable water required for each home and the volume of wastewater discharged, water needs are broken down into three categories: potable water, greywater, and blackwater. It is assumed that 60–70% of the water consumed in a home becomes greywater, and the remaining 30–40% becomes blackwater.

For the wastewater analysis, the higher value is considered to ensure greater safety and support for the proposed system. Therefore, potable water is considered equivalent to wastewater. This equivalence reflects a situation that can occur in the daily life of a household (**Table 5**).

Reference values for percentages:

- *Greywater*: represents approximately 65% of total wastewater.
- *Blackwater*: represents approximately 35% of total wastewater.

This table provides a comprehensive comparison of potable water needs and the volume of wastewater, broken down into greywater and blackwater, along with other variables such as gas, biogas, and electricity, including the number of photovoltaic panels required for each of the VUF homes (**Figure 12**).

The trends of the polynomial regression curves in the graph show the relationship between the constructed housing area (in m^2) and three variables: wastewater (L/day), greywater (L/day), and blackwater (L/day).

Wastewater (L/day):

The polynomial regression curve for wastewater is represented in green, with the equation: $y = -0.0731x^2 + 24.18x - 802.14$, with a determination coefficient

House	Wastewater (L/day)	Greywater (L/day)	Blackwater (L/day)
VUF-01	600	390	210
VUF-02	1050	682.5	367.5
VUF-03	1050	682.5	367.5
VUF-04	1050	682.5	367.5
VUF-05	1200	780	420
VUF-06	1200	780	420

Table 5.
Total housing area in relation to wastewater, greywater, and blackwater.

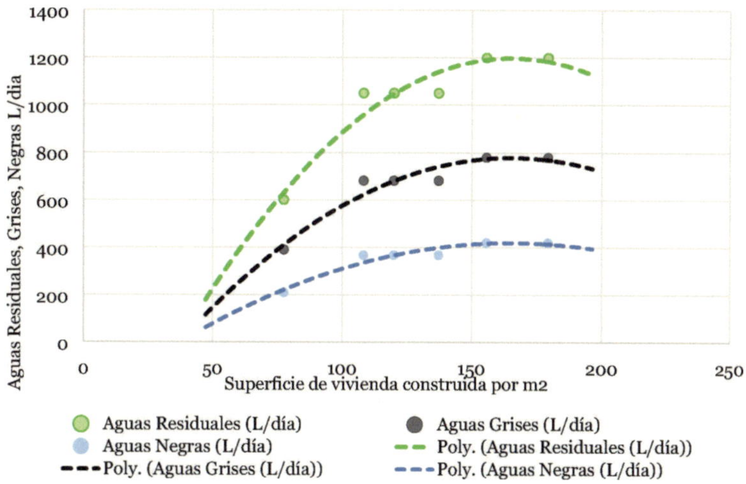

Figure 12.
Total housing area in relation to wastewater, greywater, and blackwater.

$R^2 = 0.9274$. This curve shows a trend of increasing wastewater as the housing area increases, reaching a maximum around 165.5 m^2, after which wastewater begins to decrease. The maximum point can be calculated by deriving the equation: $dy/dx = -0.1462x + 24.18 = 0$, where $\Longrightarrow x = 165.5$.

Greywater (L/day):

The polynomial regression curve for greywater is represented in black, with the equation: $y = -0.0475x^2 + 15.717x - 521.39$, with a determination coefficient $R^7 = 0.9274$. The curve shows a trend of increasing greywater as the housing area increases, reaching a maximum around 165.5 m^2, after which greywater begins to decrease. The maximum point can be calculated by deriving the equation: $dy/dx = -0.095x + 15.717 = 0$, where $\Longrightarrow x = 165.5$.

Blackwater (L/day):

The polynomial regression curve for blackwater is represented in purple, with the equation: $y = -0.0256x^2 + 8.463x - 280.75$, with a determination coefficient $R^2 = 0.9274$. The curve shows a trend of increasing blackwater as the housing area increases, reaching a maximum around 165.5 m^2, after which blackwater begins to decrease. The maximum point can be calculated by deriving the equation: $dy/dx = -0.0512x + 8.463 = 0$, where $\Longrightarrow x = 165.5$.

All the curves for wastewater (L/day), greywater (L/day), and blackwater (L/day) show that there is an optimal housing size in terms of constructed area that maximizes the generation of wastewater, greywater, and blackwater. This size is around 165.5 m^2. Beyond this size, the generation of wastewater, greywater, and blackwater tends to decrease, which may be related to water usage efficiency and other factors not specified in the graph. The determination coefficient $R^2 = 0.9274$ suggests that the polynomial equations provide a good fit to the observed data, explaining approximately 92.74% of the variability in each variable.

Interpretation of the curves:

- *Initial growth:* All curves show an initial increase in the volume of wastewater, greywater, and blackwater as the housing area increases. This initial growth can be attributed to the greater capacity of larger homes to generate higher volumes of wastewater due to an increase in the number of occupants and service usage.

- *Maximum point:* The volume of wastewater reaches its peak around a housing area of 165.5 m^2. This point represents optimal efficiency in terms of space and wastewater generation. Similarly, the volume of greywater and blackwater also reaches its maximum within the same area range.

- *Subsequent decrease:* After the maximum point, the volume of wastewater, greywater, and blackwater begins to decrease. This could be due to greater efficiency in space usage in larger homes, where the additional increase in area does not correspond to a proportional increase in wastewater generation.

The observed relationship in the graph between the housing area and the generation of wastewater, greywater, and blackwater is consistent with studies on water management efficiency in sustainable housing. Crites and Tchobanoglous [17], in their work on wastewater treatment in small communities, emphasize the importance of housing design and size to maximize efficiency in wastewater management. The graph's results align with the idea that there is an optimal size to maximize efficiency in wastewater generation and treatment. López et al. [46] discuss how greywater recycling can complement sustainable water management strategies, suggesting that there is an optimal housing size that maximizes efficiency in greywater management. García and Barbosa [20], in their study on the use of constructed wetlands for wastewater post-treatment, highlight the importance of the scale and design of the treatment system to maximize efficiency, which is consistent with the observation that blackwater volume decreases after a certain housing size.

Implications for sustainable housing design:

To optimize space and water resources, designing homes around the identified optimal size (165.5 m^2) can maximize efficiency in the generation and treatment of wastewater, greywater, and blackwater. Staying within this optimal range not only improves wastewater management efficiency but also enhances housing sustainability. This is crucial to reduce dependence on external water sources and improve water resource management.

The analysis of the graph and its comparison with previous studies emphasize the importance of considering an optimal housing size to maximize efficiency in the generation and treatment of wastewater, greywater, and blackwater. This approach not only contributes to the sustainability of homes but also improves water resource efficiency and reduces dependence on non-renewable sources.

5.8 Resource needs and renewable energy generation capacity

Table 6 summarizes the resource needs and renewable energy generation capacity for diverse types of family homes (VUF), categorized by their total area:

Data interpretation:

- *Wastewater (L/day):* The homes generate a wastewater volume proportional to their size and the number of inhabitants. Larger homes (VUF-05 and VUF-06) generate 1200 L/day, while smaller homes (VUF-01) generate 600 L/day.

House	Wastewater (L/day)	Greywater (L/day)	Blackwater (L/day)	LPG (kg/month)	Biogas (m^3/month)	Electricity (kWh/month)	Photovoltaic panels required
VUF-01	600	390	210	20	25	400	9
VUF-02	1050	682.5	367.5	35	43.75	700	16
VUF-03	1050	682.5	367.5	35	43.75	700	16
VUF-04	1050	682.5	367.5	35	43.75	700	16
VUF-05	1200	780	420	40	50	800	18
VUF-06	1200	780	420	40	50	800	18

Table 6.
Resource needs and renewable energy generation capacity based on housing area.

This trend aligns with studies indicating that water consumption and wastewater generation increase with the size of the home and the number of occupants.

- *Greywater and blackwater (L/day):* Greywater accounts for about 65% of total wastewater, while blackwater constitutes 35%. This breakdown is essential for the differentiated treatment of these waters, as indicated in studies on constructed wetlands and greywater recycling systems. Crites and Tchobanoglous [17] highlight the importance of housing design and size to maximize wastewater management efficiency. Additionally, García and Barbosa [20] emphasize the effectiveness of constructed wetlands in wastewater treatment, validating the relevance of the breakdown presented.

- *LPG (kg/month) and biogas (m^3/month):* LPG consumption and biogas production are directly proportional to the size of the home and the number of inhabitants. Larger homes require more LPG and generate more biogas. Biogas production, calculated based on LPG consumption, demonstrates how homes can achieve energy self-sufficiency through efficient biodigesters. According to the FAO "Biogas Manual" [30], biogas production aligns with these data, showing the viability of energy self-sufficiency in homes using biodigesters.

- *Electricity (kWh/month):* Electricity consumption increases with the size of the home. Larger homes (VUF-05 and VUF-06) consume 800 kWh/month, while smaller homes (VUF-01) consume 400 kWh/month. This pattern is consistent with studies on energy efficiency in homes, which highlight the need for efficient design to reduce energy consumption.

- *Photovoltaic panels required:* The number of photovoltaic panels required also increases with the size of the home, reaching a maximum of 18 panels for the larger homes. This reflects a direct relationship between housing area and the capacity to install solar panels, thus optimizing renewable energy generation. Nandwani [26] argues that housing design can significantly influence energy efficiency and solar generation capacity, as evidenced by the photovoltaic panel data presented. It also aligns with studies by the International Energy Agency (IEA) [45], which suggest that energy efficiency improves with optimal housing design, as reflected in the electricity consumption presented in the table.

Implications for sustainable housing design:

To optimize space and resources, designing homes around an optimal size can maximize efficiency in both wastewater generation and treatment, as well as renewable energy production. Staying within an optimal size range not only improves efficiency in water and energy resource management but also increases housing sustainability. This approach is crucial for reducing dependence on external energy sources and improving the management of renewable resources.

The table effectively summarizes the needs and capacities of diverse types of family homes in terms of water and energy resources. The analysis and comparison with previous studies highlight the importance of considering an optimal housing size to maximize efficiency in wastewater generation and treatment, as well as renewable energy production. This approach not only contributes to the sustainability of homes but also improves resource efficiency and reduces dependence on non-renewable sources.

5.9 Reference indices used

The indices used to calculate the needs for potable water, wastewater, gas, and electricity in this study come from a combination of standard bibliographic sources and common experiences in residential housing planning and design. Below are the sources and foundations for each of the indices used:

Potable water and wastewater:

- *Per person potable water consumption:*

  Source: World Health Organization (WHO). The WHO recommends a minimum water consumption of 50–100 L per person per day to meet basic needs. In developed urban areas, this consumption can reach between 150 and 200 L per person per day.
  Reference: WHO (2003). "Domestic Water Quantity, Service Level and Health" [37].

- *Distribution of wastewater into greywater and blackwater:*

  Source: Typical data used in the design of domestic wastewater treatment systems. The standard distribution of 65% for greywater and 35% for blackwater is based on studies and common practices in sanitary and environmental engineering.
  Reference: Metcalf and Eddy, Inc. (2003). "Wastewater Engineering: Treatment and Reuse" [47].
  Gas consumption:

- *Per person LPG consumption:*

  Source: Common practices and studies on the consumption of liquefied petroleum gas (LPG) in residential homes. A standard value of 5 kg of LPG per person per month is commonly used.
  Reference: Various data according to item 5.2.2.

- *Equivalence between LPG and biogas:*

  Source: Studies on the energy content in different fuels. One kilogram of LPG contains approximately 46 megajoules (MJ) of energy, while 1 m^3 of biogas contains approximately 20 MJ.
  Reference: "Biogas Handbook" 2013 [40, 44] and data from the International Renewable Energy Agency (IRENA) [42].
  Electricity consumption:

- *Per person electricity consumption:*

  Source: Average residential consumption data in developed countries. A typical value of 100 kWh per person per month is used in the design of electrical supply systems for homes.
  Reference: Energy consumption data from organizations such as the International Energy Agency (IEA).

Photovoltaic panels:

- *Solar panel production:*

  Source: Typical data on the production of standard 300 Wp (Watt peak) solar panels, considering an average of peak sunlight hours per day.
  Reference: Technical data provided by solar panel manufacturers and solar energy feasibility studies [48–50].

These values reflect common experience and best practices in the design of housing and public service systems, backed by technical literature and the recommendations of recognized organizations. This methodology ensures that the resource needs and energy generation capacity in homes are based on solid foundations and applicable to both urban and rural contexts.

6. Conclusion

6.1 Sustainable technologies and the optimal housing size

The technologies implemented in urban homes not only significantly optimize energy resources but also notably reduce carbon emissions and inefficient waste management. This analysis highlights those technologies such as biodigesters for biogas production, photovoltaic panels to generate electricity, and wastewater and greywater management systems synergistically contribute to the energy self-sufficiency of homes. In urban contexts, especially in high-growth areas, these technologies are not only viable but represent a solution to reduce the ecological footprint of cities.

The optimal size of homes, between 160 and 168 m^2, revealed by this study, maximizes resource efficiency and renewable energy generation. Beyond this range, efficiency begins to decrease. This strategic approach to housing design not only reduces the consumption of non-renewable resources but also positions homes as solutions to mitigate climate change in cities around the world.

6.2 Comparative analysis with previous studies

This study aligns with previous research that also underscores the importance of sustainable technologies in urban housing. For example, López et al. [46] highlight the importance of solar energy capture and efficient water management as essential elements to reduce environmental impact. However, our approach adds the quantitative dimension, showing the relationship between housing size and energy efficiency, a consideration that López et al. do not address in depth. These authors emphasize the need to adopt technologies that adapt to local conditions, an approach also reflected in this analysis. However, not only is the adaptation of technologies to local conditions crucial, but this study also demonstrates a more specific approach to the optimal size of urban homes to maximize energy efficiency and sustainability. This aspect is less discussed in López et al.'s work, but in this study, it is notably relevant.

Similarly, García Serrano and Corzo Hernández [7] highlight the importance of renewable energies and wastewater treatment to improve urban quality of life. Our research reinforces this approach but also offers a detailed analysis of how housing size directly impacts resource consumption and renewable energy production, providing a precise metric to maximize efficiency. While these authors emphasize the importance of renewable energy and wastewater treatment, this study adds the quantitative dimension of the relationship between housing size and resource consumption. Thus, not only is their focus on energy self-sufficiency confirmed, but a precise metric is provided that optimizes efficiency in the urban context.

Acurio et al. [24], on the other hand, emphasize the importance of proper solid waste disposal and its transformation into biogas, which coincides with the conclusions on biogas production in optimized homes. This analysis corroborates the assertion that the management of organic waste through compact biodigesters not only reduces dependence on LPG but also contributes to a circular economy in homes. While Acurio et al. focus on the disposal of solid waste and its transformation into biogas, this study reinforces this concept by applying it to a specific range of housing size. Additionally, by comparing the impact of compact biodigesters with biogas production, it is confirmed that the integration of sustainable technologies in urban homes is viable and efficient to generate a circular economy.

6.3 Corroboration with authors

This analysis confirms what authors like López et al. [46] proposed, who identify the integration of renewable energy as a key factor in reducing the environmental impact of urban homes. Moreover, the biogas production and use of photovoltaic panels observed in this study corroborate what Acurio et al. [24] mentioned, who saw in energy self-sufficiency a path to sustainability in cities. Our research goes one step further, showing how the implementation of compact biodigesters in optimally sized homes can reduce LPG dependence and promote a circular economy in urban contexts. This corroboration approach strengthens the validity of the study by aligning with and extending the findings of previous studies.

With López et al. [46]: In addition to the integration of renewable energy, this study not only confirms the relevance of these systems for urban sustainability but also shows the importance of combining them with waste management technologies, such as biogas systems. This integrated approach expands on what López et al. suggest, showing a more comprehensive path to energy self-sufficiency.

Additionally, the idea that there is an optimal housing size that maximizes resource efficiency is consistent with findings from previous studies. For example, García Serrano and Corzo Hernández [7] indicated that proper housing design is key to optimizing comfort and the use of natural resources. This study delves into that premise, showing how size directly influences energy consumption and waste production.

With Acurio et al. [24]: The analysis of biogas in optimized homes not only supports Acurio et al.'s ideas but goes beyond by demonstrating that proper implementation of compact biodigesters can transform urban homes into energy self-sufficient units, reducing LPG consumption and favoring the transition to a more circular economy.

6.4 Relationship between housing size and the analyzed variables

The study reveals a nonlinear relationship between the built area of homes and key variables such as LPG gas consumption, biogas production, electricity consumption, the number of photovoltaic panels required, as well as the generation of wastewater, greywater, and blackwater. Polynomial regression curves indicate that these variables increase with the housing area until reaching a peak, between 160 and 168 m^2, after which they begin to decrease.

6.5 Optimization of housing size

The results suggest that there is an optimal housing size, around 160–168 m^2, in which resource consumption and renewable energy generation are maximized. This optimal size corresponds to the point where variables of interest, such as LPG and electricity consumption, biogas production, and the amount of wastewater, greywater, and blackwater, reach their maximum value. Beyond this size, both resource consumption and renewable energy generation and waste production decrease, which may be related to a reduction in the efficiency of resource use.

6.6 Efficiency and sustainability

The determination coefficient (R^2) of approximately 0.9274 in all polynomial regression curves indicates a high degree of model fit to the observed data. This suggests that the polynomial equations are adequate to explain the variability of the variables concerning the housing area. The existence of an optimal size highlights a balance between the housing area and resource efficiency and waste generation. Therefore, designing homes around this optimal size can maximize sustainability and efficiency, reducing both non-renewable resource consumption and waste production.

6.7 Implications for housing design and infrastructure planning

To maximize energy and resource efficiency, as well as optimize waste management, it is advisable to design homes close to the identified optimal built area size. This recommendation not only improves housing sustainability and efficiency but also reduces dependence on external resources and minimizes environmental impact. Additionally, the observed data and trends are crucial for infrastructure and service planning, such as wastewater and greywater management systems, and the integration of renewable energy installations and gas consumption.

6.8 Contribution to urban sustainability

The implementation of sustainable techniques and procedures in urban family homes is both viable and beneficial. These practices offer significant improvements in terms of energy efficiency and environmental impact reduction. The combination of wastewater treatment systems, solid waste management, and solar energy capture allows for the transformation of traditional homes into self-sufficient and environmentally responsible units. This approach is supported by previous studies, such as those by López et al. [46], García Serrano and Corzo Hernández [7], and Acurio et al. [24], which highlight the effectiveness and efficiency of these technologies in various contexts.

6.9 Recommendations for future studies and developments

It is essential that future studies continue to explore and refine these technologies to ensure their adaptation to different geographical and socio-economic contexts. This will maximize their positive impact on the environment and people's quality of life. Furthermore, designing sustainable homes involves creating spaces that are not only efficient in the use of resources but also provide an elevated level of comfort, reducing dependence on external energy and improving the efficiency of natural resource use. Future studies could explore how the implementation of these technologies could be adapted to different climates or geographical contexts. Additionally, the social impact of the widespread adoption of these solutions in broader urban contexts could be investigated.

6.10 Global impact

The adoption of integrated systems for wastewater treatment, solid waste management, and solar thermal and photovoltaic energy capture is not only technically feasible but also economically advantageous. These systems can significantly reduce the carbon footprint of homes and contribute to more sustainable resource management. The evidence from this study underscores the need to continue promoting the design and implementation of sustainable homes that maximize resource efficiency and minimize environmental impacts.

Acknowledgements

The authors acknowledge the use of Meta AI to polish the language, and for the transduction of the manuscript.

Conflict of interest

The authors declare that they have no conflict of interest.

Author details

Guillermo Yorel Noriega Aquise* and Fredy Nicolás Molina Rodríguez
Catholic University of Santa María, Peru

*Address all correspondence to: gmonoriegae@gmail.com

References

[1] Arela Bobadilla R, Riesco Lind G, Chávez Contreras G. Una Mirada a la Expansion de la Ciudad de Arequipa en los Ultimos 40 Años. Arequipa Peru: Estudio Programatico Publicacion Universitaria; 2021

[2] OEFA Organismo de Evaluación y Fiscalización Ambiental. Informe Técnico, MINAM Ministerio del Medio Ambiente, Gobierno del Perú, Equipo de subdirección de supervisión a entidades de la Dirección de supervisión. Hecho en el Depósito Legal en la Biblioteca Nacional del Perú N° 2014-05991, febrero del 2014, Lima Perú. Available from: https://www.minam.gob.pe/el-ministerio/organismos-adscritos/oefa/

[3] OEFA Fiscalización ambiental en aguas residuales, Informe Técnico, MINAM Ministerio del Medio Ambiente, Gobierno del Perú. "La fiscalización ambiental en aguas residuales" Equipo de subdirección de supervisión a entidades de la Dirección de supervisión. Equipo de la Oficina de Comunicaciones y Atención al Ciudadano., Hecho en el Depósito Legal en la Biblioteca Nacional del Perú N° 2014-05991, abril del 2014, Lima Perú. Available from: https://www.minam.gob.pe/el-ministerio/organismos-adscritos/oefa/

[4] Orihuela Paredes JC. Un Análisis de la Eficiencia de la Gestión Municipal de Residuos Sólidos en el Perú y sus Determinantes. Lima Peru: Estudios Socioeconómicos y Estadísticos. Investigación Seleccionada en el Concurso Nacional de Investigaciones; 2018

[5] RNE – MVSC Reglamento Nacional de edificaciones, Ministerio de vivienda Saneamiento y Construcción, Gobierno del Perú DS N° 011-2006 - Vivienda. Lima Peru: Editora Oficial El peruano; 2006. Available from: https://www.gob.pe/institucion/vivienda/informes-publicaciones/2309793-reglamento-nacional-de-edificaciones-rne

[6] Romero Rojas JA. Tratamiento de aguas residuales. Teoría y principios de diseño. In: Rojas JAR, editor. Tratamiento de aguas residuales, teoría y principios de diseño. Bogotá Colombia: Editorial Escuela Colombiana de Ingeniería, Tercera reimpresión; 2010. p. 1207. Available from: https://www.escuelaing.edu.co/es/

[7] García Serrano J, Corzo Hernández A. Depuración con Humedales Construidos. Guía Práctica de Diseño, Construcción y Explotación de Sistemas de Humedales de Flujo Subsuperficial. 2008

[8] Rondón Toro E, Szantó Narea M, Pacheco JF, Contreras E, Gálvez A. Guía general para la gestión de residuos sólidos domiciliarios, Manuales de la Cepal. Santiago, Chile: Publicaciones de las naciones Unidas; 2016. ISSN 2518-3923. Avilable from: https://www.cepal.org/es/publicaciones/40407-guia-general-la-gestion-residuos-solidos-domiciliarios

[9] Varnero Moreno MT. Manual de biogás. In: Mini energía – PNUD – FAO – GEF, Ministerio de Energía, Programa de las Naciones Unidas para el Desarrollo, Organización de las Naciones Unidas para la Alimentación y la Agricultura, Global Environment Facility Editado por Proyecto CHI/00/G32, "Chile: Remoción de Barreras para la Electrificación Rural con Energías Renovables". Santiago de Chile: FAO; 2011. ISBN 978-95-306892-0. Available from: https://www.fao.org/4/as400s/as400s.pdf

[10] Lamigueiro OP. Energía Solar Fotovoltaica. España. 192 p: Creative Commons ebook; 2013

[11] Vidales ACV, Herrera LER, Cromeyer GFV. Diseño de un modelo de vivienda bioclimática y sostenible. Entorno. 2011;**49**:7-20

[12] Rojas García MA. Aprovechamiento de la Energía Solar, en Casa Habitación. Instituto Politecnico Nacional, Escuela Superior de Ingenieria y Arquitectura, Unidad Tecamachalco, Tesis para obtener el Título de Arquitecto Ingeniero Mexico; 2018. Available from: https://tesis.ipn.mx/bitstream/handle/123456789/25442/Aprovechamiento%20de%20la%20energía%20solar%20en%20casa%20habitación.pdf

[13] Givoni B. Climate Considerations in Building and Urban Design. John Wiley & Sons; 1998

[14] Duffie JA, Beckman WA. Solar Engineering of Thermal Processes. New York: Wiley; 1980

[15] Asdrubali F, D'Alessandro F, Schiavoni S. A review of unconventional sustainable building insulation materials. Sustainable Materials and Technologies. 2015;**4**:1-17

[16] González J, Fiorito F. Daylight design of office buildings: Optimisation of external solar shadings by using combined simulation methods. Buildings. 2015;**5**:560-580

[17] Crites R, Tchobanoglous G. Tratamiento de aguas Residuales en Pequeñas Poblaciones. Santa FE de Bogotá: McGraw-Hill Interamericana, S.A; 2000. ISBN: 0-07-289087-8, ISBN: 958-41-0042-4

[18] España A. Guía Técnica de Recomendaciones para el Reciclaje de Aguas Grises en Edificios. Asociación Española de Empresarios de Tratamiento y Control de Aguas; 2016

[19] Vymazal J. Constructed wetlands for wastewater treatment: Five decades of experience. Environmental Science & Technology. 2011;**45**(1):61-69

[20] García LEJ, Barbosa RG. Uso de humedales artificiales de flujo subsuperficial y verticales para postratamiento de aguas residuales en unidad escolar de la DICIVA. Jóvenes en la Ciencia. 2017;**3**(2):2443-2447

[21] Eawag W. Compendium of Sanitation Systems and Technologies. Dübendorf: Swiss Federal Institute of Aquatic Science and Technology (EAWAG); 2008

[22] Kadlec R, Wallace S. Treatment Wetlands. Boca Raton, London, New York: CRC Press Taylor & Francis Group; 2009. pp. 267-347

[23] Schluep M, Hagelueken C, Kuerh R, et al. Sustainable Innovation and Technology Transfer Industrial Sector Studies. Recycling From E-Waste to Resources and P. d. N. U. p. l'environnement, Recycling, from E-Waste to Resources. Berlin, Germany: United Nations Environment Programme, UNEP – STEP; Supervision and Technical Editing; 2009

[24] Acurio G, Rossin A, Teixeira PF, Zepeda F. Diagnóstico de la Situación del Manejo de Residuos sólidos Municipales en América Latina y el Caribe. Washington – EEUU: Publicación conjunta del Banco Interamericano de Desarrollo y la Organización Panamericana, Washington, D.C.; 1997. No.ENV.97-1071997

[25] Kossmann W et al. Biogas Digest Volume 1: Biogas Basics. Germany: Information and Advisory Service on Appropriate Technology; 1999

[26] Nandwani SS. Energía Solar. Conceptos Básicos y su Utilización. Heredia (Costa Rica): Universidad Nacional; 2005. pp. 1-26

[27] Kalogirou SA. Solar thermal collectors and applications. Progress in Energy and Combustion Science. 2004;**30**(3):231-295

[28] Luque A, Hegedus S. Handbook of Photovoltaic Science and Engineering. John Wiley & Sons; 2011

[29] López J, Burgos A, Cambeses H, Sánchez D, Rodríguez P. El reciclaje de aguas grises como complemento a las estrategias de gestión sostenible del agua en el medio rural. In: Río Mandeo, Cuenca Fluvial y Desarrollo Sostenible. 2012. Recuperado de: www.geama.org/sanitaria/index.php

[30] FAO. Manual de biogás. Editado por: Proyecto CHI/00/G32 "Chile: Remoción de Barreras para la Electrificación Rural con Energías Renovables". Ministerio de Energía, Programa de las Naciones Unidas para el Desarrollo, Organización de las Naciones Unidas para la Alimentación y la Agricultura, Global Environment Facility, ed: MINENERGIA/PNUD/FAO/GEF. Chile: Editado por Proyecto CHI/00/G32 Chile. 2011

[31] Skaalvik J, Sevaldsen P, Barnes R. Towards a Green Economy-Pathways to Sustainable Development and Poverty Eradication. 2011

[32] Weiland P. Biogas production: Current state and perspectives. Applied Microbiology and Biotechnology. 2010;**85**:849-860

[33] Nohlen D. Ciencia Política Comparada: El Enfoque Histórico-Empírico. Granada: Editorial Universidad de Granada; 2013

[34] Gómez Díaz de León C, León de la Garza EA. Método Comparativo. 2014

[35] Piovani JI, Krawczyk N. Los Estudios Comparativos: Algunas notas históricas, epistemológicas y metodológicas. Educação & Realidade. 2017;**42**:821-840

[36] Nohlen D. El método comparativo. Antologías para el Estudio y la Enseñanza de la Ciencia Política. 2020;**3**:41-57

[37] Howard G, Bartram J, Water S, World Health Organization. Domestic Water Quantity, Service Level and Health. 2d ed. Genova Switzerland: World Health Organization; 2020. ISBN 978-92-4-001525-8 print version

[38] ENAHO - INEI. Encuesta naciojnal de hogares – Instituto Nacional de Estadistica e informatica. In: Diccionario de la Encuesta Nacional de Hogares 2020. Lima Peru; 2021. Available from: https://www.datosabiertos.gob.pe/dataset/encuesta-nacional-de-hogares-enaho-2020-instituto-nacional-de-estad%C3%ADstica-e-inform%C3%A1tica-inei

[39] OSINERGMIN – GPAE. Organismo Supervisor de kla inversion en energia y Mineria, Gerencia de Politicas y Analisis Economico, Informe de Resultados, Consumo y Usos de los Hidrocarburos Líquidos y GLP, Encuesta Residencial de Consumo y Usos de Energía – ERCUE 2019-2020, O. O. S. d. l. I. e. E. y. M.-G. d. P. y. A. E. GPAE, ed., Lima Peru; 2022. [Online]. Available from: https://cdn.www.gob.pe/uploads/document/file/2560069/ERCUE%20Hidrocarburos%20%202019-2020.pdf

[40] Wellinger A, Murphy JD, Baxter D. The Biogas Handbook: Science, Production and Applications. Cambridge, United Kingdom: Elsevier, ed: Woodhead Publishing serie

in Energy. IEA Bioenergy; 2013. ISBN 978-0-85709-498-8 (print), ISBN 978-0-85709-741-5

[41] Deublein D, Steinhauser A. Biogas from Waste and Renewable Resources. Weinheim Alemania: Wiley-VCH Verlag GmbH & Co.; 2008. ISBN 978-3-527-31841-4

[42] IRENA. International Renewable Energy Agency, Biogas for Energy Production, Abu Dhabi, Emiratos Árabes Unidos; 2019. p. 40

[43] Encyclopedia of Energy, Natural Resource, and Environmental Economics. Laramie, WY, USA: Elsevier, University of Wyoming; 2013. ISBN 978-0-08-096452-2

[44] IRENA. International Renewable Energy Agency, Measuring Small-Scale Biogas Capacity and Production, Abu Dhabi Emirates Árabes Unidos. IRENA Headquarters Masdar City, Abu Dhabi, United Arab Emirates; 2016. p. 36. ISBN 978-92-95111-12-7

[45] F. Bank. World Energy Outlook 2018. Paris: IEA; 2018

[46] López J, Burgos A, Cambeses H d, Sánchez D, Rodríguez P. El Reciclaje de Aguas Grises Como Complemento a las Estrategias de Gestión Sostenible del Agua en el Medio Rural. Río Mandeo: Cuenca Fluvial y Desarrollo Sostenible; 2012. Recuperado de: www.geama.org/sanitaria/index.php

[47] Metcalf E, Eddy J, INC. Ingeniería de Aguas Residuales Tratamiento, Vertido y Reutilización Tercera Edición. Madrid, Buenos Aires y otros lugares: McGraw – Hill. ISBN: 0-07-041690-7, ISBN: 84-481-1727-1 (Vol. I), ISBN: 84-481-1612-7 (Obra completa), Dep6sito legal: M. 24.848-1995. Año 1995 Impreso en Madrid - España

[48] Torres JPN, Nashih SK, Fernandes CA, Leite JC. The effect of shading on photovoltaic solar panels. Energy Systems. 2018;**9**:195-208

[49] Plante RH. Solar Energy, Photovoltaics, and Domestic Hot Water: A Technical and Economic Guide for Project Planners, Builders, and Property Owners. Academic Press; 2014

[50] La Camera F. Renewable Power Generation Costs in 2020–IRENA. Masdar City, Abu Dhabi: International Renewable Energy Agency; 2020